MARINES IN CRISIS

MARINES IN CRISIS

The Cold War Transformation of the U.S. Marine Corps
1947–1995

Charles P. Neimeyer, PhD

MCUP
MARINE CORPS UNIVERSITY PRESS
Quantico, VA
2024

LIBRARY OF CONGRESS CATALOGING-IN-PUBLICATION DATA

Names: Neimeyer, Charles Patrick, 1954– author. | Marine Corps University (U.S.). Press, issuing body.

Title: Marines in crisis : the Cold War transformation of the U.S. Marine Corps, 1947–1995 / Charles P. Neimeyer.

Other titles: Cold War transformation of the U.S. Marine Corps, 1947–1995

Description: Quantico, Virginia : Marine Corps University Press, 2023. | In scope of the U.S. Government Publishing Office Cataloging and Indexing Program (C&I); Federal Depository Library Program (FDLP) distribution status to be determined upon publication. | Includes bibliographical references and index. | Summary: "Throughout the Cold War and into the 1990s, the Marine Corps faced multiple strategic inflection points. Some of these moments were fights for institutional survival, some were based on emerging technology or internal upheaval, and others were more concerned with developing operational doctrine. When compared to the development of its amphibious warfare doctrine between World War I and World War II, these Cold War decisions related to the Marine Corps came about at an astonishing pace. Many of these post-World War II moments came only after painful experiences in increasingly complex and multidimensional Joint combat operations or humanitarian interventions, where international politics, rapidly changing technology, new societal norms, and even culture played an ever-larger role on the battlefield. For the Marine Corps, the Cold War and beyond seemingly required its senior leaders to predict the rapid-fire changes that impacted the new way of war and evolving politics of conflict. In response, those leaders continually transformed the Marine Corps to ensure it played a significant role in U.S. military matters"— Provided by publisher.

Identifiers: LCCN 2023007493 (print) | LCCN 2023007494 (ebook) | ISBN 9798986259468 (paperback) | ISBN 9798986259475 (epub)

Subjects: LCSH: United States. Marine Corps—History—20th century. | Cold War.

Classification: LCC VE23 (print) | LCC VE23 (ebook) | DDC 359.960973—dc23/eng/20230308 | SUDOC D 214.513:C 67

LC record available at https://lccn.loc.gov/2023007493
LC ebook record available at https://lccn.loc.gov/2023007494

Published by
Marine Corps University Press
2044 Broadway Street
Quantico, VA 22134

1st Printing, 2024
ISBN: 979-8-9862594-68
DOI: 10.56686/9798986259468

This book is dedicated to the Marine Corps Historical Foundation, whose support made this study possible, and to Janet Louise Neimeyer, my beloved spouse.

CONTENTS

Illustrations ix

Foreword xiii

Preface xvii

Acknowledgments xxiii

Select Terms, Abbreviations, and Acronyms xxxi

Introduction 3

Chapter One 31
The Cold War Begins

Chapter Two 84
No More Vietnams

Chapter Three 136
Trials of the 1970s

Chapter Four 179
The Cushman-Wilson-Barrow Era

Chapter Five 246
Beirut, Grenada, and the Reagan Era

Chapter Six 302
General Alfred Gray and Maneuver Warfare

Chapter Seven 367
To the Crucible and Beyond

Conclusion 431

Select Bibliography 445

Index 465

About the Author 471

1. MajGen Omar Bundy and Col Albertus W. Catlin — 6
2. British troops blinded by poison gas — 8
3. Marine Corps Medal of Honor recipients — 12
4. Marines on Tarawa beach — 15
5. U.S. military personnel at a nuclear weapons test, 1951 — 20
6. U.S. Army M-28/M-29 Davy Crockett Weapons System — 23
7. President Harry S. Truman signing the National Security Act — 33
8. Gen Alexander A. Vandegrift before the Senate Committee on Armed Services — 37
9. Col Robert D. Heinl Jr. — 39
10. LtCol Victor H. Krulak — 47
11. President Harry S. Truman presents the Medal of Honor to Maj Louis H. Wilson Jr. — 51
12. Col Lewis B. "Chesty" Puller and BGen Edward A. Craig at the command post of the 1st Marines — 52
13. Lt Frank E. Petersen Jr. climbs out of the cockpit of a Vought F4U Corsair — 54
14. Marines in the Chosin Reservoir campaign — 61
15. USS *Valley Forge* (LPH 8) — 64
16. USS *Iwo Jima* (LPH 2) — 65
17. Two Marine Corps helicopters from Marine Corps Air Station, Tustin, CA — 67
18. BGen Frederick J. Karch led through a crowd near Da Nang, Republic of Vietnam — 76
19. Map of Corps boundaries in South Vietnam — 79
20. 1stLt George S. Dorgatt teaches a class in Vietnam — 81
21. The Marine Corps' six four-star generals on active duty — 83
22. Capt John W. Ripley — 89

23. Refugees evacuate from Saigon, April 1975 — 111
24. SS *Mayaguez* — 115
25. The wreckage of two U.S. Air Force helicopters at Koh Tang Island — 125
26. President Gerald R. Ford and his national security team in a midnight meeting — 129
27. MajGen Charles F. Bolden Jr. — 153
28. Astronaut Charles F. Bolden Jr. — 155
29. Marine Corps recruiting poster from the 1980s — 177
30. Gen Robert E. Cushman Jr. — 181
31. Soviet Cold War–era naval vessels — 183
32. Marine Corps M–50 Ontos in Vietnam — 185
33. LtGen Robert E. Hogaboom — 186
34. Hawker Siddeley AV–8A Harrier attack aircraft — 189
35. Tomcat aviation patch for Marine Fighter Squadron 531 — 190
36. Gen Louis H. Wilson Jr. — 194
37. Gen Robert H. Barrow — 195
38. Official logo for the Marine Corps Mountain Warfare Training Center — 205
39. USNS *Sisler* (T–AKR 311) — 215
40. Desert One crash site — 217
41. McDonnell Douglas F/A-18 Hornet fighter aircraft — 228
42. Landing Craft, Air Cushion (LCAC) — 229
43. USS *New Jersey* (BB 62) — 235
44. Comdt Paul X. Kelley at a news briefing, August 1983 — 244
45. Distribution of religious groups in Lebanon — 248
46. The zones of the mulitnational force in Beirut — 249
47. The band MEGA plays a USO show on 22 October 1983 — 264
48. A cloud of smoke rises from the Marine barracks shortly after the bombing on the morning of 23 October 1983 — 265
49. Sketch of route taken by the Marine barracks suicide bomber — 267
50. D-day map of Operation Urgent Fury — 276

51. A combat artist's rendition of Marine Corps helicopters launching from USS *Guam* (LPH 9) — 278

52. Gen Alfred M. Gray Jr. — 303

53. The fall of the Berlin Wall — 305

54. U.S. Air Force Col John R. Boyd — 317

55. Lockheed P-80 Shooting Star jet fighter — 318

56. U.S. Marshals Service mugshot of Panamanian dictator Manuel A. Noriega — 328

57. Marine Corps LAV-25 on patrol near the Panama Canal — 337

58. U.S. Army map of Operation Just Cause — 339

59. Navy officials inspect mine damage to USS *Tripoli* (LPH 10) — 361

60. U.S. Air Force chart shows the various levels of mission-oriented protective posture equipment — 363

61. Kuwaiti oil field burns during Operation Desert Storm — 365

62. Map of Somalia for Operation Restore Hope — 387

63. Haitian president Jean-Bertrand Aristide — 398

64. U.S. military and political leadership meet with Col Thomas S. Jones after violence breaks out between U.S. forces and Haitian police during Operation Uphold Democracy — 405

65. Gen Charles C. Krulak — 419

66. Marine recruits at Parris Island, SC, conquer "the Crucible" — 425

FOREWORD

In the conclusion of *First to Fight: An Inside View of the Marine Corps*, Lieutenant General Victor H. "Brute" Krulak observes:

> The Corps must always be mindful of what it stands for and realize that its willingness to meet any challenge and accept any hardship underlies both its attraction to the best of young recruits and its value to the country. Unpopular and difficult issues will continue to present themselves, but they must be met head-on and without compromise. This is what the Corps did in its unrelenting campaign against drugs. In that matter, the Marines proved once again that they are justified in holding themselves as something more than just a cross-section of society. This same dynamic attitude must prevail wherever societal conditions threaten the standards of the Corps.[1]

The U.S. Marine Corps exists to fight and win our nation's battles. Throughout its long and storied history, the Corps has undergone many transformations to become the naval expeditionary force-in-readiness we know today. Not one of these transformations occurred by accident. They were foreseen and carefully shaped by senior leaders throughout the Corps, with welcomed input coming from every echelon of the Service. Innovation and adaptation have always been part of the Marine Corps' DNA. Nearly every instance of significant institutional change was precipitated by a crisis. As Andrew Grove, the former CEO of Intel, has observed, "Bad companies are destroyed by crisis. Good companies survive them. Great companies are improved by them."[2] Dr. Charles P. Niemey-

---

[1] Victor H. Krulak, *First to Fight: An Inside View of the Marine Corps* (Annapolis, MD: Naval Institute Press, 1984), 224–25.
[2] As quoted in Ryan Holiday, *The Obstacle Is the Way: The Timeless Art of Turning Trials into Triumph* (New York: Portfolio, 2014), 3.

er's important and timely book, *Marines in Crisis: The Cold War Transformation of the U.S. Marine Corps, 1947–1995*, reveals how the Marine Corps overcame crises in the past, how it continues to be strengthened by crisis, and why it remains such a national treasure to the American people today.

As the United States enters a challenging new era in its history, characterized by rapid technological change, conflicts in Europe and the Middle East, and immeasurably consequential great power competition in the Western Pacific, it is worth asking: How will the Marine Corps remain ready, capable, and relevant for this dynamic environment? How will future Marines continue to emerge from adversary to create an even stronger Corps with a deeper commitment to victory? *Marines in Crisis* challenges readers to think about the ingredients that fuel institutional innovation and adaptation, how successful transformations are shepherded, and how organizations can seize opportunity from crisis.

Cultivating this spirt of innovation starts with the Commandant of the Marine Corps. The Commandants featured in this book responded to crises not by focusing on the exigencies of the moment but rather by deliberately looking after the long-term health of the organization—what Marines call "the soul of the Corps." When faced with a crisis, these Commandants doubled- and tripled-down on what made their Marines truly unique in the hearts, minds, and souls of the American people. By doing so, they strengthened and nourished the soul of the Corps. And as they did this during the Cold War and post-Cold War inflection points described herein, they created the modern Marine Corps, and their legacy looms large.

Today, the United States is facing a new crisis—the most severe recruiting shortfall since the advent of the all-volunteer force. Against this backdrop, the Marine Corps has been the only large Service to consistently meet its assigned accessions "mission" while exceeding the U.S. Department of Defense's quality standards. Marine recruiters can accomplish this because they employ a proven system of processes and techniques within a command architecture that was carved in the granite of combat-hardened wisdom by the 26th and 27th Commandants,

Generals Louis H. Wilson Jr. and Robert H. Barrow. Their insistence on and reinvigoration of inspiring and engaged leadership at every level in the making of Marines and on personal discipline fueled the manpower renaissance that continues to pay dividends today.

In the wake of the failed Operation Eagle Crisis in 1980; the bombing of the Marine barracks in Beirut, Lebanon, in 1983; and other crises at that time, the 29th Commandant, General Alfred M. Gray Jr., left an indelible mark on the Corps by changing how Marines think about their most fundamental task—warfighting. General Gray's unique personification of operational excellence was informed by passionate lifelong education. His far-ranging impacts included publishing *Warfighting*, Fleet Marine Force Manual 1, with its timeless lessons about the nature, character, and conduct of war, and establishing Marine Corps University, where servicemembers from across the U.S. Joint Force continue to exchange professional ideas and build critical relationships with civilians from across the U.S. government and servicemembers from 32 partnered nations.

Every week at the Marine Corps recruit depots at Parris Island, South Carolina, and San Diego, California, new Marines earn their coveted Eagle, Globe, and Anchor from their drill instructor after completing a grueling 54-hour culminating event known as "The Crucible." These newly minted Marines then become trained in their occupational fields, join operational units in the Fleet Marine Force, and are deployed around the world as today's "strategic corporals." This transformation of young men and women of character into strategic corporals who proudly and selflessly serve their nation was the vision of the 31st Commandant, Charles C. Krulak.

Although Dr. Neimeyer's study concludes with the reforms implemented by General Krulak, he looks forward to the era of hybrid war and the future crises that Marines will be required to respond to. Every successive Commandant since General Krulak has nourished the Marine Corps' spirit of innovation and adaptation while also sustaining the Service's role as the nation's premier crisis response force. It is worth noting that the Corps accomplished this while supporting two prolonged campaigns in Iraq and Afghanistan. Yet, as long as the United States remains a maritime nation with global interests, there will always be a need for forward-deployed, forward-engaged Marines who are ready to respond to any crisis at a moment's notice. As the Marine Corps navigates another strategic inflection point with the rise of China as the pacing threat to the United States, Marines at every level are working furiously to ensure that the Corps continues to not only remain ready for any crisis but emerge from it stronger.

Marines in Crisis describes how all of these brilliant innovations were engendered by crises while offering a relevant, timely, and historically informed journey into the soul of the Corps. It could not have come at a better time.

Semper Fidelis,
William J. Bowers
Major General, U.S. Marine Corps
Commanding General, Marine Corps Recruiting Command

PREFACE

Throughout the Cold War era, the U.S. Marine Corps, the smallest of the nation's four military Services, was forced to remake itself several times over. These transformations emerged from several external and internal factors. In every case, however, the primary driver behind these transformations was the need for the Marines to prove their warfighting relevance in the modern era. Consequently, the Marine Corps found itself in the most intense fight of its organizational life immediately following the conclusion of World War II. With the advent of nuclear bombs, most strategists believed that amphibious warfare, a critical skill that the Marines had honed during numerous and bloody World War II campaigns against the Empire of Japan, had become obsolete. Furthermore, some key political leaders, including President Harry S. Truman, felt that the time had finally come to end the intense inter-Service rivalry that had plagued the U.S. war effort. Truman's solution was to unify all the Services into a single administrative entity, the newly created U.S. Department of Defense, in 1947. Had Congress accepted Truman's idea, the Marine Corps would have had no future.

To be fair, the Marine Corps was an easy target for those who supported Service unification in the late 1940s. Starting in 1946, the Marine Corps had steadily reduced from a high of six full combat divisions and five aircraft wings during World War II to just a single division and aircraft wing on the eve of the Korean conflict in 1950. To make matters worse, the Marines were saddled with outdated equipment and the allegedly obsolete amphibious warfare operational doctrine. As a result, the Marine Corps had to prove to leaders in both the executive and legislative branches of the U.S. government that it could still help win future conflicts. Ironically, on 5 August 1945, the day prior to the atomic bombing of Hiroshima, Japan, General Alexander A. Vandegrift, a Medal of Honor recipient from the bloody Guadalcanal campaign in 1942–43 and the 18th Commandant of the Marine Corps, wrote in *The New York Times* that the long-anticipated amphibious invasion of Japan would be tough and costly. Yet, he wrote that thanks to lessons learned from earlier campaigns, the United States was ready to pull this operation off and would prevail in the end. After a Boeing B-29 Superfortress heavy bomber dropped the nuclear bomb "Little Boy" on Hiroshima the following day and then another Superfortress dropped the nuclear bomb "Fat Man" on Nagasaki on 9 August 1945, the nuclear age arrived like a thunderclap, hitting with a suddenness that surprised everyone, including Vandegrift. From that moment forward, all the Services needed to reassess their roles in future wars, none more so than the Marine Corps.[1]

Nevertheless, many senior Marine Corps leaders suspected that amphibious warfare as a warfighting concept was not entirely dead. With the U.S. Navy now the most dominant seapower in the world, total control of the seas was nearly guaranteed for decades to come. Consequently, amphibious warfare remained a potential game changer in situations in which the likelihood of the use of nuclear weapons remained remote. The hugely successful 1st Marine Division landing at Inchon, South Korea, on 15 September 1950 seemed to quiet any amphibious warfare naysayers for a while, but just to a small degree. They reemerged after the United States' principal antagonist in the Cold War, the Soviet Union, successfully developed its own nuclear weapons program in 1949, meaning that the United States no longer possessed a monopoly over this type of destructive weaponry.

If Vandegrift is known for leading the way toward a tentative post-World War II renaissance in amphibious warfare, then Commandants Alfred M. Gray Jr. and Carl E. Mundy Jr. in the late Cold War era and especially Commandant Charles C. Krulak in the post-Cold War era should be credited with creating a path for the Marine Corps to incorporate the tenets of operational maneuver from the sea into what was needed for a twenty-first-century fighting force. For instance, Gray's championing of the maneuver warfare concept throughout the 1980s prepared the Ma-

---

[1] Gen Alexander A. Vandegrift, "From Guadalcanal to the Shores of Japan," *New York Times*, 5 August 1945, 7, 28–29.

rine Corps to effectively operate during Operation Desert Storm in 1991 and potentially deploy to frozen northern Norway. Gray's *Warfighting*, Fleet Marine Force Manual 1 (FMFM 1), published in 1989, clearly belongs in the pantheon of other prestigious Marine Corps doctrinal publications such as the *Tentative Manual for Landing Operations*, published in 1934, and the *Small Wars Manual*, published in 1940.[2] None of this could have been achieved without the personnel renaissance that Commandants Louis H. Wilson Jr. and Robert H. Barrow created during the years immediately following the Vietnam War. Wilson and Barrow, who served as Commandant between 1975 and 1983, have long been credited with creating a successful personnel campaign that enabled the Corps to maintain a much larger peacetime establishment than had ever been previously attempted. Moreover, they did this when many young Americans had little interest in voluntary military service. This renaissance in personnel and training was a long, hard slog that eventually paid dividends throughout the 1980s and 1990s, when the Corps was tasked with short-notice combat deployments to locations such as Lebanon, Grenada, Panama, and Kuwait.

When General Krulak became Commandant in 1995, he was quick to capitalize on the success of *Warfighting*. He wrote that the manual "changed the way Marines think about warfare. It has caused energetic debate and has been translated into several foreign languages, issued by foreign militaries, and published commercially. It has strongly influenced the development of doctrine by our sister Services." In keeping with his philosophy that doctrine should never become stagnant, Krulak published an updated version of *Warfighting*, which was designated as Marine Corps Doctrinal Publication 1 (MCDP 1). The now-retired Gray heartily concurred with Krulak's assessment and even authored the preface. Gray wrote that "like war itself, our approach to warfighting must evolve. If we cease to refine, expand, and improve our profession,

---

[2] *Warfighting*, Fleet Marine Force Manual 1 (Washington, DC: Headquarters Marine Corps, 1989); *Tentative Manual for Landing Operations, 1934* (Washington, DC: Headquarters Marine Corps, 1935); and *Small Wars Manual*, Fleet Marine Force Reference Publication 12-15 (Washington, DC: Headquarters Marine Corps, 1940).

we risk becoming outdated, stagnant, and defeated. Marine Corps Doctrinal Publication 1 refines and expands our philosophy on warfighting, taking into account new thinking about the nature of war and the understanding gained through participation in extensive operations over the past decade."[3] With *Warfighting*, Krulak desired to define a Service-specific way of fighting wars. He strongly felt that this philosophy needed to become inculcated into the marrow of every active-duty Marine, both officers and enlisted. Krulak later expanded on the nature of twenty-first-century warfare when he published his ideas about the "Strategic Corporal" and "Three-Block War," concepts more fully explored in the epilogue of this study.

Throughout its modern history, at least since World War II, the Marine Corps seemed to have had an uncanny knack of usually finding the right leader for the right job at the right moment. For example, during the Vietnam War, U.S. Marine Corps advisors to the South Vietnamese Marine Corps known as Co-Vans gained experience that produced an extraordinary crop of postwar officers for the U.S. Marine Corps. Many of these former Co-Vans were later found in crucial leadership positions long after the last Marines left Saigon in 1975. Former Co-Van Brigadier General James R. Joy commanded the 22d Marine Amphibious Unit (22d MAU) that replaced the bomb-shattered 24th MAU in Beirut, Lebanon, in 1983. Former Co-Van Lieutenant Colonel Ray L. Smith was Joy's battalion landing team commander during Operation Urgent Fury in Grenada that same year. Another former Co-Van, General Walter E. Boomer, became the commander of U.S. Marine Corps Forces Central Command (USMARCENT) and eventually the Assistant Commandant of the Marine Corps. Finally, former Co-Van General Anthony C. Zinni played a crucial role in the success of Operation Restore Hope in Somalia in 1992–93. By the late 1990s, Zinni was the commanding general of U.S. Central Command (USCENTCOM).

After 1975, other combat-tested Marine Corps officers and enlisted veterans of the Vietnam War led the way toward rebuilding the Service

---

[3] *Warfighting*, Marine Corps Doctrinal Publication 1 (Washington DC: Headquarter Marine Corps, 1997).

and other enterprises. For example, Captain George R. Christmas, a veteran of the Battle of Hue City and a Navy Cross recipient, later rose to the rank of lieutenant general. Following his command of I Marine Expeditionary Force (I MEF), he served as the deputy chief of staff for manpower and Reserve affairs at Headquarters Marine Corps in Washington, DC. After retirement, Christmas was later appointed to the positions of president and chief executive officer of the Marine Corps Historical Foundation and was instrumental in leading a multimillion-dollar effort toward the establishment of the now-renowned National Museum of the Marine Corps in Quantico, Virginia, which opened in 2006. Christmas was just one of the many Vietnam veterans who made substantial contributions toward the betterment of the Marine Corps in the post-Vietnam War era. As had happened following World War II, those Marines who returned home from Vietnam decided to ensure that their fellow Marines killed on the battlefields of Southeast Asia did not die in vain. Although it took time, Vietnam veterans, such as former Marine captain, Navy Cross recipient, future Secretary of the Navy, and later U.S. Senator James H. Webb Jr. (D-VA) and General Gray, the 29th Commandant of the Marine Corps, joined forces to remake the Marine Corps and ensure that the institution remained relevant and ready for the future.

The following chapters will describe for readers how past transformational Marine Corps leaders, using an amazing amount of foresight and understanding about the essential nature of warfare, made sure that the Marine Corps never again fell into the trap of fighting the previous conflict. They will also examine the numerous strategic inflection points that all the U.S. military Services reached throughout the Cold War and into the immediate post-Cold War years. The primary focus of this study is to illustrate how the Marine Corps institutionally adapted to the challenges of the Cold War while remaining relevant as a fighting Service. Marine Corps leaders correctly anticipated many of these inflection points but missed others, causing the Marines to pay the price in blood for a lack of Service foresight. As a result, while drawing on leadership lessons learned from World War II and the major Cold War conflicts in Korea and Vietnam, this study will more appropriately focus on

what these leaders did when they had responsibility for the future direction of the Corps. The Cold War presented the Marine Corps with its
greatest organizational challenge since it transitioned from a disparate
collection of individually numbered rifle companies that recognized the
advantages the sea offers to a maneuver force in an extended campaign
ashore during World War I. This study also traces the operational and administrative peacetime challenges faced by critical Marine Corps leaders
who tried to keep the Service from lapsing into irrelevance and, at the
same time, maintain a combat reputation with quality personnel trained
and equipped for worldwide expeditionary service.

ACKNOWLEDGMENTS

First and foremost, this book would not have been possible without the strong and consistent support of certain key people and organizations. I got to know many of the people in the chapters that follow during my time on active duty, when I was on the faculty of the Naval War College, or when I served as the director and chief of the Marine Corps History Division (HD) between 2006 and 2017. At the top of this list must be the Marine Corps Heritage Foundation (MCHF) and the National Museum of the Marine Corps (NMMC). Few people are aware of this today, but Navy Cross recipient Lieutenant General George R. Christmas and Lieutenant General Robert R. Blackman, who succeeded Christmas as the foundation's president, raised record amounts of money in support of the NMMC and even provided scholarship funding for students studying and writing on Marine Corps history around the country. Christmas and Blackman were two dynamic leaders who proved that fundraising, while certainly not a core competency for any active-duty general officer, is a valuable skill that can be developed in retirement. I first got to know Blackman years ago when I was assigned to Headquarters Marine Corps as a brand-new captain. He was one of those people who seemed destined for flag rank—he was that impressive—and of course my prediction turned out to be true. The MCHF has been made especially better due to the tireless efforts of its former vice president for operations, Susan Hodges. Hodges has been a tremendous asset for the foundation over the years, and while quietly working behind the scenes, she made sure the "trains were running on time." Now in retirement, I believe her successor, Jennifer M. Vanderveld, will do the same outstanding job.

Thanks to the efforts of all the presidents of the MCHF, especially those who have held the position since the museum opened its doors in 2006, the NMMC is now the world-class institution we know and love today. However, the quality level of the museum—something that is immediately evident as soon as one enters through its front doors—would not have been possible without the strong and sustained performance of the first director of the NMMC, Lin Ezell. I got to know Ezell very well when I was hired as HD director about a year after she was successfully lured away from the Smithsonian Institution's National Air and Space Museum Steven F. Udvar-Hazy Center to manage the new museum by retired Major General Donald R. Gardner, the president of Marine Corps University (MCU), in 2005. Ezell used to sit next to me at the president's bimonthly directors meeting, and it was rare if she did not have something exciting to report about the museum or Marine Corps history in general. She, along with her deputy, retired Captain Charles G. Grow, worked tirelessly on the production of phase II of the museum, which included coverage of many of the years focused on in this book. Few are aware that Grow, as a Marine Corps reservist, worked for HD as an accomplished combat artist. His work was featured on the cover of several HD bulletins known as *Fortitudine*. I now consider both very good friends, and their help and assistance to me in the field of Marine Corps history over the years has been uniformly superb. I wish them both well in retirement.

It is also important to mention the role that MCU played in the production of this manuscript. Major General Gardner led MCU as its president for five years between 2004 and 2009. Many consider his tenure as the golden years for the university. He oversaw the creation of the NMMC; the movement of HD from the Washington Navy Yard in Washington, DC, where it had been located since 1971, to the MCU campus in Quantico, Virginia; and the establishment of Marine Corps University Press (MCUP) in 2008, all the while running rigorous educational efforts that resulted in the professional military education (PME) and training of thousands of students in his time. Gardner also must have seen something in me when he selected me to direct HD in late 2006. It was no secret at the time that most of HD was not happy with suddenly relocating to Quantico after having been comfortably ensconced at the Navy Yard for 35 years. The move was understandably traumatic, but Gardner had faith that the new museum and HD would work better being colocated at MCU rather than being separated by 35 miles of Interstate 95. He was right. I was also fortunate to attend Gardner's second retirement from federal service in 2009 and witnessed two former Commandants of the Marine Corps make the statement to the assembled guests that Gardner was the best leader either one of them had ever seen. I fully concurred with their estimation of the qualities of this fine Marine Corps general officer.

Several key faculty members and administrators at MCU have also been especially helpful. First is Dr. Nathan Packard, a Marine Corps Reserve officer who also has the sterling reputation of being one of the best faculty members in the entire university. I consider him to be yet another friend I made while directing HD. I was honored to have served on his dissertation committee at Georgetown University (also my alma mater), where he knocked his dissertation defense out of the park. Packard's advice, both written and oral, proved extremely beneficial to me in the production of this manuscript. I believe he remains the national expert on numerous aspects of Marine Corps Cold War history, and his knowledge of the race, discipline, and morale problems extant in the Marine Corps during the 1970s and into the 1980s is simply superb. Packard also pointed me to very important publications and documents located in the Gray Research Center (GRC) at MCU that I might have missed. Next is retired Colonel Keil Gentry. He is another friend who I first got to know when he was assigned to the Office of Legislative Affairs on Capitol Hill. Later, he became the director of the Marine Corps War College, the most senior PME educational institution in the entire Service. On retirement, Gentry was immediately hired as the vice president for business affairs at MCU. His selection was a no-brainer. He has an amazing interest in Marine Corps history, and I consider him an unparalleled peer in that regard. He served with me on two of the three boards appointed by the Commandant of the Marine Corps to confirm the actual membership of the now-famous flag raising parties during the historic Battle of Iwo Jima in World War II. All the board members noted his attention to detail and keen eye. I appreciated his deliberate and careful methodology throughout the identification process.

I also must acknowledge the sage advice and wisdom of General Alfred M. Gray, the 29th Commandant of the Marine Corps. In my last year as HD director, thanks to the urging of the then-president of MCU, Brigadier General William J. Bowers, we put together a 100-year doctrinal study from the Battle of Belleau Wood in 1918 to the present day. One of the lectures featured the wit and wisdom of Gray among others who led the Marine Corps in the post-World War II era and especially those who ascended to leadership roles after the end of the Vietnam War. MCU was fortunate to host such key persons as General Gray; General Charles C. Krulak, the 31st Commandant of the Marine Corps; and General John M. Paxton Jr., the 33d Assistant Commandant of the Marine Corps, in a series of lectures spaced out during the course of an academic year. Bowers's foresight and intuition as to what students at MCU needed to hear at that moment was extraordinary. To this day, Gray remains at the head of my personal pantheon of Marine Corps heroes who I got to know over the years.

I am especially indebted to the director of MCUP, Angela Anderson, and all her incredible staff. I had the foresight to hire Anderson as the second senior editor for MCUP after the first senior editor, Ken Williams, moved on to work for the U.S. Air Force. At the time, the senior editor worked for the HD director. Let me say that once in the position, Anderson hit the ground running. Her work ethic and production capacity were truly amazing to behold. Her recent editorial review with this manuscript has been very helpful. She remains consistently diligent toward ensuring that every manuscript she reviews for potential publication—and there are dozens of them during any given year—remains a quality product. My dissertation mentor at Georgetown University, Dr. Marcus Rediker, once sagely informed me that the quality of any written effort is not necessarily found in the original draft produced but is eventually achieved in the rewrites. Anderson and her staff kindly and tactfully guided me through this effort. This manuscript is a much better product for it.

I need to also mention the help and support of all the HD membership, past and present. Few realize today that HD, established in 1919, is one of the longest continuously standing organizations in the entire Marine Corps. For example, Annette Amerman, a longtime reference historian at HD who has since moved on to other work for the U.S. Navy History and Heritage Command, was often and directly consulted by numerous Commandants of the Marine Corps for critical historical information. Amerman provided me with the same service and sent me key background documents related to the 1970s and 1980s eras of Marine Corps history. Her robust files went back to World War I, and the material they contained were usually not found anywhere else. I also believe that HD possesses no better historian than Paul Westermeyer, whose superb knowledge of the historiography of the Marine Corps is simply unsurpassed by anyone that I can name. Westermeyer, a former enlisted Marine, is also the national expert on Marine Corps participation in Operation Desert Shield and Operation Desert Storm. Retired U.S. Army colonel Douglas Nash, now retired from federal service, helped me with a lot of technical details related to Cold War equipment that the North Atlantic Treaty Organization (NATO) used over the years. For the past three years, thanks to the diligent efforts of HD, much of the material I needed to write this manuscript was often found in their extensive files and publications, and they pointed out some of the most important troves of information directly to me. In sum, HD, as well as the librarians and staff at the GRC, were crucial in assisting me in the research of primary and secondary sources.

Finally, I would like to extend my thanks to all Marines, past and present, who may have served on active duty or in the Reserve during the 50-year-long Cold War. Even with the publication of this book, this history remains a work in progress. One of the amazing things about Marine Corps history is that, at least since the early 1970s, operational units down to the battalion level have been required to submit a unit command chronology report, usually on an annual or semiannual basis. As readers will see in the chapters that follow, defining the actual moment when a unit was in a "combat-like" situation became increasingly hard to tell, especially after the Vietnam War ended. My old unit, 2d Battalion, 11th Marines, saw significant combat during the Vietnam

War. However, its command chronology rarely listed anything of note happening in its official record from the end of Vietnam until its next large-scale combat deployment during Operation Desert Storm in 1991. Consequently, if the command chronologies are taken at face value, many units simply appear to have large blank spaces between 1973 and 1990. We now know this to not be true. These units were all quite busy, just in different ways. Much more work and credit need to be given to Marine Corps veterans of the Cold War for discovering the critical strategic inflection points that emerged during hundreds of peacetime deployments and even hybrid operations such as those in Lebanon, Somalia, and Haiti, where differentiating between kinetic and benign situations was often difficult or even impossible to accurately do. In other cases, an entire decade of debate and work on maneuver warfare resulted in the extraordinary success of Marine forces during Operation Desert Storm. Can anyone imagine the Marine Corps seamlessly getting through those thick Iraqi minefields and simultaneously incorporating a U.S. Army armored brigade on short notice without the maneuver warfare debate of the previous decade having taken place? Training and education matter, and the need to "get it right" before the next rapidly evolving crisis has been never more important than it is today.

I remain a proud Cold Warrior today. Although I do not have the campaign ribbons to show for it, my time as a Marine during this era was still important to me for learning important operational and doctrinal imperatives that later proved their worth in future contingency situations. During 1984, for example, I was privileged to attend the advanced artillery officers course at Fort Sill, Oklahoma. As students, we received the latest and greatest on the Army's then-emerging AirLand Battle concept, which was very much like the Marine Corps' maneuver warfare theory. I was later amazed to see that what I had just learned at Fort Sill was being practically applied when I reported for duty as a member of the 11th Marines in the fall of 1984. Although the 2d Marine Division, thanks to Gray, is renowned for its early embrace of maneuver warfare, I found that the concept was alive and mostly well even in the 1st Marine Division in the mid-1980s. During my subsequent three-year tour of duty at

Camp Pendleton, California, I was deployed for 314 days, or nearly one-third of my entire time in the 1st Marine Division. Much of it occurred at Marine Corps Air Ground Combat Center in Twentynine Palms, California, or during a unit deployment to Okinawa, Japan, and the Philippines—a first for an artillery battery at that time. I learned more in those 314 days than any other time in my 20-year-long career. Consequently, the Cold War needs to be seen in a different light. Rather than seeing a blank spot in a unit's command chronology, it might be best to study what these units were doing during this important moment in Marine Corps history. It is my sincere hope that other (better) historians will take up this challenge in the future.

SELECT TERMS, ABBREVIATIONS, AND ACRONYMS

AAV	amphibious assault vehicle
AFQT	Armed Forces Qualification Test
ARVN	Army of the Republic of Vietnam
AVF	all-volunteer force
AWS	Amphibious Warfare School
BLT	battalion landing team
CAP	Combined Action Program
CNO	Chief of Naval Operations
Co-Van	U.S. Marine Corps advisor to the Vietnamese Marine Corps
DMZ	demilitarized zone
DOD	U.S. Department of Defense
DON	U.S. Department of the Navy
FMF	Fleet Marine Force
FSSG	Force Service Support Group
GIUK	Greenland-Iceland-United Kingdom
JFACC	Joint force air component commander
JSOTF	Joint Special Operations Task Force
JTF	Joint task force
LAV	light armored vehicle
LZ	landing zone
MAB	Marine amphibious brigade
MAF	Marine amphibious force
MAGTF	Marine air-ground task force
MAU	Marine amphibious unit
MCU	Marine Corps University
MEB	Marine expeditionary brigade
MEF	Marine expeditionary force
MEU	Marine expeditionary unit

MPF	Maritime Prepositioning Force
MPS	Maritime Prepositioned Shipping
MWTC	Mountain Warfare Training Center
NATO	North Atlantic Treaty Organization
NCA	National Command Authority
NLF	National Liberation Front
NSC	National Security Council
OPLAN	operation plan
PAVN	People's Army of Vietnam
PGM	precision-guided munition
PLO	Palestine Liberation Organization
PME	professional military education
POMCUS	Prepositioning of Materiel Configured to Unit Sets
RDJTF	Rapid Deployment Joint Task Force
SAC	Strategic Air Command
SOC	special operations capable
UN	United Nations
USCENTCOM	U.S. Central Command
USEUCOM	U.S. European Command
USLANTCOM	U.S. Atlantic Command
USPACOM	U.S. Pacific Command
USSOCOM	U.S. Special Operations Command
USSOUTHCOM	U.S. Southern Command
VNMC	Vietnamese Marine Corps

MARINES IN CRISIS

INTRODUCTION

In June 2018, the U.S. Marine Corps celebrated the 100th anniversary of what is arguably the most important battle in its institutional history—the Battle of Belleau Wood. In an All Marines Message (ALMAR) commemorating the event, General Robert B. Neller, the 37th Commandant of the Marine Corps, remarked, "For our Corps, Belleau Wood has become a symbol of Marine courage and tenacity. Many consider this battle the birth of the modern day Marine Corps."[1] General Neller was not the only Commandant to think this way. The 31st Commandant, General Charles C. Krulak, certainly did. So did the 13th Commandant, Major General John A. Lejeune, who had previously witnessed the extreme carnage of World War I, although arriving in France just after Belleau Wood. Appointed as the commanding general of the U.S. Army's 2d Division, a hybrid Army-Marine Corps unit, in July 1918, he became convinced, following the horrific Battle of Blanc Mont Ridge, that future twentieth-century warfare was going to verge toward unprecedentedly large-scale, casualty-intense conflicts. Lejeune believed that those conflicts could potentially be even worse than what he was then experiencing, and he thought the Marine Corps needed to be ready for it. As things turned out, Lejeune was proven right.

So why is Belleau Wood, and World War I in general, so widely recognized—then and today—as a watershed moment in the development of the modern Marine Corps? While military historians have focused on the extraordinary valor and courage of the Marines during the intense three-week campaign to take Belleau Wood, this emphasis has caused them to miss the true significance of the battle. Rather than solely exemplifying their bravery, Belleau Wood shocked the Marines and anyone who unfortunately experienced the fight. In fact, it was a bloodletting

---

[1] *Belleau Wood 100th Anniversary*, All Marines Message 019/18 (Washington, DC: Headquarters Marine Corps, 23 May 2018).

that went well beyond what the Marine Corps had ever experienced in conflicts dating back to 1775. In truth, even with more than 10 months of intensive training after their arrival in France in July 1917, the Marines of 1918 were still institutionally unprepared for what they were to face on the battlefield, and the servicemen ultimately paid a heavy price for it.

Despite the leatherneck courage on display at Belleau Wood, it was the sheer volume of casualties that got everyone's attention. In heavy fighting on 6 June 1918 near the town of Bouresches, elements of the 5th and 6th Marine Infantry Regiments, as part of the 4th Marine Brigade, initially attacked German machine gun emplacements and rifle pits across an open field of wheat in a loose line, "five yards apart, in four ranks, twenty yards between each rank."[2] The 4th Marine Brigade "suffered 5,711 casualties [more than 1,000 killed in action], and lost half of its officers" in just 19 days.[3] One German officer who observed the attack noted that "the Americans were obliged to come down from the heights they were occupying before the eyes of the Germans. They did this in thick lines of skirmishers, supported by columns following immediately behind. The Germans could not have desired better targets; such a spectacle was entirely unfamiliar to them."[4] The U.S. Army of the Potomac under Lieutenant General Ulysses S. Grant had attacked the entrenched lines of the Confederate Army of Northern Virginia under General Robert E. Lee at the Battle of Spotsylvania Court House on 12 May 1864 in much the same fashion. The rebels, however, did not possess machine guns or rapid-firing artillery.[5]

Conversely, Colonel Albertus W. Catlin, the commanding officer of the 6th Marine Regiment, thought the attack had gone remarkably well.

---

[2] Barrie Pitt, *1918: The Last Act* (New York: W. W. Norton, 1962), 156.

[3] Pitt, *1918*, 159. Maj Edwin N. McClellan, the officer in charge of the Marine Corps Historical Section and a World War I veteran, in a short 15-page pamphlet he wrote on the battle, listed approximately 4,700 casualties for the 4th Marine Brigade, which included 902 Marines who had been gassed. See Maj Edwin N. McClellan, *The Battle of Belleau Wood* (Washington, DC: Marine Corps History Division, ca. 1930), 13, 15.

[4] LtCol Ernst Otto (German Army), quoted in David Bonk, *Chateau Thierry & Belleau Wood 1918: America's Baptism of Fire on the Marne* (Oxford, UK: Osprey, 2007), 62.

[5] For more on the fighting at Spotsylvania Court House on 12 May 1864, see Gordon C. Rhea, *The Battles for Spotsylvania Court House and the Road to Yellow Tavern, May 7–12, 1864* (Baton Rouge: Louisiana State University Press, 1997), 232–307.

He noted that the Marines "went in as if on parade, and that is literally true. There was no yell and wild rush, but a deliberate forward march, with lines at dress right. They walked a regulation pace, because a man is little use in a hand-to-hand bayonet struggle after a hundred-yard dash."[6] Before the first day at Belleau Wood closed, "the Marines had suffered 1,087 dead or wounded, more casualties than the Corps had taken thus far in its 143-year history."[7] While leading the assault with a trench cane in hand, Colonel Catlin was shot through the chest by a German sniper and evacuated back to the United States, taking no further part in the war. French observers of the attack thought that the Americans were incredibly brave but also naïve about the realities of combat on the western front.[8]

At Belleau Wood, the Marines also faced poison gas for the first time. On 11 June 1918, Corporal Don V. Paradis of the 80th Company, 2d Battalion, 6th Marines, commanded by Major Thomas Holcomb, described just such an attack. "The Germans plastered the entire area with artillery of all sizes," he recorded. "Added to the horror of mustard gas was the inclusion in their high explosives of a vomiting gas that made it almost impossible to keep the mask on and made eyes water to obstruct vision."[9] Although the Marines had practiced being subjected to a gas attack prior to going into the line, their chemical warfare gear was barely adequate. A few Marines, to their eternal regret, had even decided to lighten their battle burden by discarding their gas masks. Further, while a gas mask protected a Marine's face, eyes, and airway from mustard

---

[6] Bonk, *Chateau Thierry & Belleau Wood 1918*, 62.

[7] David T. Zabecki, "The U.S. Marines' Mythic Fight at Belleau Wood," *Marine Corps Times*, 26 June 2021.

[8] Catlin had a remarkable career in the Marine Corps. He graduated from the U.S. Naval Academy in 1890 and commanded the Marine detachment aboard USS *Maine* (1889) when it exploded while moored in Havana Harbor in 1898. He was later awarded the Medal of Honor for bravery while leading the 3d Marine Regiment during the intervention at Veracruz, Mexico, in 1914. His Belleau Wood wound affected him for the rest of his life and forced him into early retirement in 1919. For more on this episode at Belleau Wood, see Richard Suskind, *The Battle of Belleau Wood: The Marines Stand Fast* (Toronto: Macmillan, 1969), 41; and Alan Axelrod, *Miracle at Belleau Wood: The Birth of the Modern Marine Corps* (Guilford, CT: Lyons Press, 2007), 153.

[9] Cpl Don V. Paradis, quoted in Michael A. Eggleston, *The 5th Marine Regiment Devil Dogs in World War I: A History and Roster* (Jefferson, NC: MacFarland, 2016), 69.

Figure 1. MajGen Omar Bundy and Col Albertus W. Catlin

Source: U.S. Army War College Historical Section.

gas, many still became casualties if enough exposed skin or especially their eyes encountered the chemical agent, which could persist on the surface of things long after a gas attack had ended and a Marine had removed their mask. As a result of the prolonged vicious fighting at Belleau Wood, including facing assaults from poison gas in various forms, Holcomb's 2d Battalion "suffered a shocking 764 casualties out of a paper strength of 900 marines."[10]

Things did not get any better for the surviving Marines after Belleau Wood. Near the war's conclusion, as U.S. forces prepared for the final Meuse-Argonne offensive, Major General Lejeune's 2d Division, as part of France's 4th Army, received orders to take an initial objective of Blanc Mont, a heavily fortified position that the Germans had held since September 1914.[11] Lejeune decided that the best way to approach these heights was to conduct a "converging attack by both [2d Division] brigades."[12] Despite a heavy preliminary bombardment on the German defensive network, the Marines and soldiers sustained an extraordinary number of casualties taking Blanc Mont during the week of intense combat in early October 1918. The 4th Marine Brigade alone reported "448 killed or dead from wounds, 1,902 wounded or gassed, and 310 missing. The total—2,660 men—representing 30 percent of the Brigade's strength on 1 October. . . . The 5th [Marine] Regiment . . . suffered 1,097 casualties on [just a single] terrible day [4 October]. It marked the 4th Brigade's costliest day of the war."[13]

In a later article about the experiences of the Marines in France during World War I, General Krulak wrote that "the Corps [in World War I] found itself on a futuristic battlefield that it had not prepared for, one that it did not anticipate, and the Marines who fought there paid the

---

[10] David J. Ulbrich, "The Importance of Belleau Wood," *War on the Rocks*, 4 June 2018.

[11] *Blanc Mont (Meuse-Argonne-Champagne)* (Washington, DC: War Department, 1922), 5; and Lt Peter F. Owen, USMC (Ret) and LtCol John Swift, USMC (Ret), *A Hideous Price: The 4th Brigade at Blanc Mont, 2–10 October 1918* (Quantico, VA: Marine Corps History Division, 2019), 6. Owen is also the author of the award-winning book *To the Limit of Endurance: A Battalion of Marines in the Great War* (College Station: Texas A&M University Press, 2014). He remains the leading expert on the Marine Corps in World War I.

[12] Owen and Swift, *A Hideous Price*, 6.

[13] Owen and Swift, *A Hideous Price*, 54–55.

Figure 2. British troops blinded by poison gas

Source: Imperial War Museum.

price in blood."[14] Krulak concluded that one of the most important lessons from the costly Battle of Belleau Wood was the creation of an overarching institutional mindset that embraced change. He noted:

> Belleau Wood, in many ways, constituted a strategic inflection
> point for the Marine Corps. In the business world, a strategic
> inflection point occurs when your competition develops a new
> product or your market changes so that what you produced
> in the past is no longer desired. At Belleau Wood, the Marine
> Corps discovered that warfare had changed, and we had failed
> to adapt to these changes. . . . Those who survived never for-
> got that lesson, and they vowed that the Corps would never
> again be caught unprepared. They became the innovators, risk
> takers, and visionaries who championed amphibious assault in

[14] Gen Charles C. Krulak, "Through the Wheat to the Beaches Beyond: The Lasting Impact of
the Battle for Belleau Wood," *Marine Corps Gazette* 82, no. 7 (July 1998): 13.

the 1920s, close air support in the 1930s, and vertical envelopment in the 1950s. They were the architects that built the force-in-readiness that we are the proud stewards of today.[15]

This study will primarily focus on later rapid-fire strategic inflection points that the Marine Corps reached during the Cold War and into the early 1990s. Some of these inflection points were fights for institutional survival, some were based on emerging technology or even internal upheaval, and others were more concerned with developing future operational doctrine. All were dominated by the domestic political process extant in the United States at the time. Nevertheless, Marine leadership made these Cold War decisions at an astonishing pace compared to when the Marine Corps famously developed its amphibious warfare doctrine between the World Wars. Many of these post-World War II moments came only after painful experiences in increasingly complex and multidimensional Joint combat operations or humanitarian interventions, where international politics, rapidly changing technology, new societal norms, and culture played an ever-larger role on the modern-day battlefield.

When looking back on the interwar period of Marine Corps institutional history (1919–41), this era has been conceived as a time of relaxed yet fiscally constrained experimentation and professionalization that simultaneously included frequent kinetic activity against insurgencies in the Caribbean and Central America. Lejeune's rise to the office of the Commandant in 1920, as well as his trust in the prophetic talents of Lieutenant Colonel Earl H. "Pete" Ellis, also made a huge difference. A graduate of the U.S. Naval Academy in Annapolis, Maryland, and the U.S. Army War College in Carlisle, Pennsylvania, Lejeune used his experience on the battlefields of France in World War I as well as his participation in Joint U.S. Navy–Marine Corps fleet exercises off the Caribbean Island of Culebra prior to World War I to envision a significantly larger and more amphibiously oriented Service. Ellis used "a chance assignment to the Naval War College" to evolve into a self-made military

---

[15] Krulak, "Through the Wheat to the Beaches Beyond," 17.

strategist, as it "exposed him to the intellectual currents sweeping the Navy." During the interwar period, "naval theorists recognized the requirement for advance bases in support of the fleet and argued that the Marine Corps defend them."[16]

Ellis immediately sensed that amphibious warfare was the future warfighting niche that the rapidly modernizing Marine Corps had been long seeking. Developing into a gifted naval strategist, he authored a series of cutting-edge papers on advanced base operations while on the faculty of the Naval War College, which he joined at the direct request of its president. In 1921, Ellis's papers were combined into a course of study at Headquarters Marine Corps, and Lejeune made the decision to fully embrace Ellis's scholarship. Throughout his work, Ellis "accurately predict[ed] not only war with Japan but also that the United States will need to take the strategic offensive and [that] Japan will first try 'to reduce the naval superiority of the United States'."[17]

Lejeune and Ellis, however, took things a step further, and worked to create Marine Corps forces that could defend advanced bases as well as seize them by amphibious assault if needed. This realization meant that the Marine Corps had to create both mobile and defensively oriented forces for the Navy's use in the furtherance of any future naval campaign. Consequently, Ellis famously authored *Advanced Base Operations in Micronesia*, Operation Plan 712, in 1921.[18] *Advanced Base Operations in Micronesia* was nothing short of revolutionary because most military strategists at the time believed that the British debacle in the Gallipoli campaign (1915–16) had proven that an amphibious assault against a well-equipped opposition force would most likely fail. Fortunately, the strategic shift for the Corps under Lejeune and Ellis won out in the end. The Marines

---

[16] Dirk Anthony Ballendorf and Merrill L. Bartlett, *Pete Ellis: An Amphibious Warfare Prophet, 1880–1923* (Annapolis, MD: Naval Institute Press, 1997), x.
[17] Brent A. Friedman, ed., *21st Century Ellis: Operational Art and Strategic Prophecy for the Modern Era* (Annapolis, MD: Naval Institute Press, 2015), 46–47.
[18] *Advanced Base Operations in Micronesia*, Operation Plan 712 (Washington, DC: Headquarters Marine Corps, 1921).

continued to refine their amphibious warfare doctrine through successive Joint fleet experiments with the Navy throughout the interwar era.[19]

Nevertheless, this interwar doctrinal shift faced criticism from within the Marine Corps, most significantly from double Congressional Medal of Honor recipient Brigadier General Smedley D. Butler. In truth, Butler was not a well-educated Marine, making up for this deficiency by vigorous participation in far-flung expeditionary operations in the Philippines, China, the Caribbean, and Central America. In many ways, Butler and Lejeune were the two most famous Marines alive between 1914 and 1929. Coming from the school of hard knocks vice the classrooms at Newport, Rhode Island, Butler was awarded his first Medal of Honor for actions during the brief but violent U.S. military incursion at Veracruz, Mexico, in 1914. He was awarded his second Medal of Honor just a year later during counterinsurgency operations at Fort Rivière, Haiti. At this time, Butler's view that the Marine Corps needed to focus more on counterinsurgency operations seemed to make eminent sense due to the Service taking part in such engagements in Panama, Haiti, Santo Domingo, and especially Nicaragua. In fact, the Marine Corps seemed so involved in Caribbean and Central American affairs that it tried to capture all its lessons from those incursions in an amazing publication titled *Small Wars Manual* (1940).[20]

Yet, it is important to understand that the *Small Wars Manual* did not represent a strategic or operational shift for the Marines in the interwar years. Rather, as noted by contemporary defense analyst Bradley L. Rees:

> It is not doctrine; it is not an operational analysis of expeditionary operations, nor is it necessarily a strategy. Its uniqueness, however, lies in how it conveys a philosophy—an underlying theory—that addresses complexity, the necessity for adaptability, and the criticality given to understanding the social, psy-

---

[19] James W. Hammond Jr., *The Treaty Navy: The Story of the U.S. Naval Service between the World Wars* (Victoria, BC: Trafford, 2001), 156–61; and Hans Schmidt, *Maverick Marine: General Smedley D. Butler and the Contradictions of American Military History* (Lexington: University Press of Kentucky, 1987), 125–27.
[20] Schmidt, *Maverick Marine*, 58–95; and *Small Wars Manual*, Fleet Marine Force Reference Publication 12-15 (Washington, DC: Headquarters Marine Corps, 1940).

Figure 3. Marine Corps Medal of Honor recipients

Source: official U.S. Department of Defense photo.

chological, and informational factors that affect conflict. The [*Small Wars Manual*] reflects how ill-defined areas of operations, open-ended operational timelines, and shifting allegiances are just as relevant today, if not more so than relative combat power analyses and other more materially oriented planning factors have been in most of two century's [*sic*] worth of war planning. More so, the [*Small Wars Manual*] places significant weight on how behavior, emotions, and perceptions management are central in shaping decision-making processes.[21]

While many at the time believed that Butler was the heir apparent to the office of the Commandant when Lejeune retired from active service in 1929, the position instead went to Lejeune protégé Major General Wendell C. Neville, yet another in a line of Naval Academy graduates who would serve as Commandant until the appointment of Major General Thomas Holcomb in 1936. Neville died just one year after taking of-

---

[21] Bradley L. Rees, "An Assessment of the *Small Wars Manual* as an Implementation Model for Strategic Influence in Contemporary and Future Warfare," *Small Wars Journal*, 29 April 2019.

fice and, much to the increased ire of Butler, was replaced by Brigadier General Ben H. Fuller, who was then promoted to major general. Like Lejeune, Fuller had attended the Army War College, and like Ellis, he also attended the Naval War College, further mirroring the latter's career when he returned there as a member of the faculty. Fuller later continued within the field of professional military education (PME) when he was placed in charge of Marine Corps Schools at Marine Corps Base Quantico, Virginia (1922–23). He retired as Commandant on 1 March 1934 and was replaced by his own hand-picked successor, Major General John H. Russell Jr., who carried on Fuller's amphibious warfare doctrinal legacy. From the tenure of Lejeune through that of Holcomb, every Marine Commandant of the interwar era strongly supported the amphibious warfare concept rather than counterinsurgency as the primary operational role and mission for the Corps going forward.

Leading up to the U.S. entry into World War II, the Marine Corps continued to focus on amphibious operations in support of a larger naval campaign. As early as 1931 at the Marine Corps Schools, "a committee started work on *Marine Corps Landing Operations*." However, this work was not yet ready for widespread distribution because the status of the East and West Coast-based Marine Corps expeditionary forces remained "undefined." Consequently, just before retiring, Fuller ordered classes at the schools to be suspended and divided the staff and students into committees "to study various aspects of landing operations. . . . By January 1934, they had produced [a] *Tentative Manual for Landing Operations*." Additionally, Fuller convinced the Navy to include designated Marine expeditionary forces as a part of future fleet operations. The chief of naval operations, Admiral William H. Standley, "suggested that [these forces] be called the Fleet Marine Force (FMF). The FMF replaced the East and West Coast Expeditionary Forces. . . The FMF was [also] to be available for operations and exercises afloat or ashore as part of annual Fleet Problems."[22]

---

[22] Hammond, *The Treaty Navy*, 158–59.

Between 1934 and 1941, the Navy and Marine Corps conducted six major Fleet Landing Exercises. Fortunately, World War II-era amphibious warfare expert Lieutenant General Holland M. Smith recorded a detailed analysis of all of them. He concluded that "the six Fleet Landing exercises resulted in a widespread interest in amphibious tactics and a general recognition of their complexity in both services." Smith also noted, however, that the Navy-Marine Corps team lacked the forces and specialized equipment—especially suitable landing craft—necessary to successfully carry off an amphibious assault at that time. Smith stated that "although the major deficiencies and needs in personnel and material, apparent during the conduct of the first six fleet landing exercises, were recognized and reported, the exercises were carried out year after year on an improvised and skeletonized basis. Urgency came only with the [Second World] War."[23]

During its successful island-hopping campaign in the Pacific throughout World War II, the Navy-Marine Corps team proved the efficacy of amphibious assault. It could and did work, although at an extraordinarily high cost. In fact, both Services rather consistently ran into a myriad of operational problems with nearly every major amphibious assault conducted in the Pacific between 1942 and 1945. Once again, the specter of high casualties caused many strategic planners to reconsider whether the cost in lives was worth the taking of an objective that would likely soon fall under Admiral Chester W. Nimitz's great Central Pacific drive toward the home islands of Japan.

No single battle during the entire course of the Pacific War was more controversial than Operation Galvanic, the 20–24 November 1943 struggle for Tarawa in the Gilbert Islands by the 2d Marine Division. During the fight against fanatical Japanese defenders, the Marines and Navy suffered 3,407 total casualties. Having covered the fighting on the small atoll, U.S. war correspondents alarmed both Congress and the American public. For the first time, to emphasize the true cost of amphibious assaults, wartime censors allowed U.S. newspapers to publish photographs

---

[23] LtGen Holland M. Smith, *The Development of Amphibious Tactics in the U.S. Navy* (Washington, DC: Marine Corps History Division, 1992), 29–31.

Figure 4. Marines on Tarawa beach

Source: National World War II Museum.

of Marines killed in action at Tarawa. In fact, the Marine Corps suffered more casualties at Tarawa in just 76 hours than it had during the entire 6 months of fighting for Guadalcanal the previous year. "Moreover," Colonel Joseph H. Alexander wrote in a brief history of the campaign, "the ratio of killed to wounded at Tarawa was significantly high, reflecting the savagery of the fighting."[24]

The outcry over the casualties at Tarawa was the opposite of what the Marine Corps received for Belleau Wood during World War I. Instead of stories about leatherneck courage, which also abounded at Tarawa, a major U.S. newspaper ran the headline: "Grim Tarawa Defense a

---

[24] Col Joseph H. Alexander, USMC (Ret), *Across the Reef: The Marine Assault of Tarawa* (Washington, DC: Marine Corps History Division, 1993), 50.

15

Surprise, Eyewitness of Battle Reveals; Marines Went in Chuckling, To Find Swift Death Instead of Easy Conquest."[25] U.S Army general Douglas MacArthur complained that "these frontal attacks by the Navy, as at Tarawa, are a tragic and unnecessary massacre of American lives."[26] Congress called for a special investigation and found plenty of issues, from not having enough amphibious assault vehicles (AAVs) to ineffective and poorly planned naval and aerial bombardment fires. "The bloody cost of Tarawa," historian James P. McGrath III argues, "caused military leaders, politicians, and the American public to question the wisdom of opposed amphibious assault."[27]

The 1st Marine Division's capture of Peleliu in the Palau Islands during Operation Stalemate II (15 September–27 November 1944) was another noteworthy example. Despite receiving significant assistance from the U.S. Army's 81st Division later in the operation, "U.S. casualties on Peleliu numbered 1,544 killed in action and 6,843 wounded. Notably, the 1st Marine Regiment suffered 70 percent casualties—1,749 men—in six days of fighting while the 7th Marine Regiment suffered 46 percent casualties."[28]

While the 1st Marine Division's commander, Major General William H. Rupertus, originally estimated that the island would be taken in approximately 4 days after the initial landing on 15 September 1944, it took the Marines and follow-on Army forces nearly 10 weeks to declare the island fully secured on 27 November. Nevertheless, Japanese diehards remained on the island well into 1947, a full two years after the war ended. What was even more astounding about Peleliu was that the battle probably did not need to have happened. Peleliu's location primarily drove the operation, as its capture was necessary to secure General MacArthur's flank as he prepared to invade the Philippine Island of Leyte

---

[25] Alexander, *Across the Reef*, 50.

[26] Gen Douglas MacArthur, USA, quoted in Alexander, *Across the Reef*, 50.

[27] James P. McGrath III, "Missing the Mark: Lessons in Naval Gunfire Support at Tarawa," in *On Contested Shores: The Evolving Role of Amphibious Operations in the History of Warfare*, ed. Timothy Heck and B. A. Friedman (Quantico, VA: Marine Corps University Press, 2020), 234, https://doi.org/10.56686/9781732003149.

[28] Carsten Fries, "Operation Stalemate II: The Battle of Peleliu, 15 September–27 November 1944," Naval History and Heritage Command, 10 January 2020.

500 miles away. During a mid–August pass by Peleliu, carrier aviation forces commanded by Navy admiral William F. Halsey Jr. noticed exceptionally light resistance from Japanese aircraft. Halsey consequently recommended that the amphibious assault on 15 September be canceled and that MacArthur move up the Leyte operation. MacArthur agreed to launch his campaign early but still desired that Stalemate II occur to fully negate the possibility of a Japanese air attack on his flank. The discussion between Halsey, MacArthur, and Nimitz dragged on so long that Nimitz finally admitted that it was too late to stop operational preparations already well underway and, "assessing that the operation remained a prerequisite of the planned Leyte Gulf landings, did not countermand it."[29]

Moving on from the horrific bloodlettings at Tarawa and Peleliu, the Navy and Marine Corps created tremendous improvements in naval gunfire support, preinvasion bombardment techniques, armored amphibious assault platforms, the use of dedicated ships for command and control, improved fire support coordination ashore, and even formed "a number of amphibious warfare training centers throughout Hawaii, including a naval gunfire training center on Kahoolawe Island, Hawaii." Still, amphibious assaults remained tremendously difficult to carry out.[30] The Navy–Marine Corps team incurred even higher casualty levels during the multidivisional invasions of Iwo Jima and Okinawa in early 1945. In fact, the titanic struggle for Iwo Jima (19 February–26 March 1945), ended up being the worst fight of the entire war for the Marines. The U.S. government awarded an astounding 27 Medals of Honor for actions during this single battle, 22 of which went to Marines. At Iwo Jima, the fighting "cost U.S. forces 6,871 killed and 19,217 wounded."[31] The battle was the only Pacific island assault in which the total Navy–Marine Corps casualties in killed and wounded exceeded that of the enemy. Likely due to its longer duration, the total campaign casualties for Operation Iceberg on Okinawa (1 April–22 June 1945) were even worse. Even with U.S.

---

[29] Fries, "Operation Stalemate II."
[30] McGrath, "Missing the Mark," 236.
[31] Carsten Fries, "Battle of Iwo Jima, 19 February–26 March 1945," Naval History and Heritage Command, 16 March 2022.

Army units making up more than half the forces ashore, the III Marine Amphibious Corps (III MAC), which included the 1st and 6th Marine Divisions, and supporting Marine aircraft wings once again suffered egregious casualties, with III MAC losing 2,779 killed and 13,609 wounded. In total, "Victory at Okinawa cost more than 49,000 American casualties, including about 12,000 deaths."[32] Perhaps the most prominent casualty was the Tenth U.S. Army commander, Lieutenant General Simon Bolivar Buckner Jr., who was "killed on June 18 by enemy artillery fire during the final offensive. He was the highest-ranking American general killed in action during World War II."[33]

The impact of the sudden arrival of the nuclear age on the entire U.S. military establishment beginning in 1945 cannot be underestimated. The situation was further compounded when the United States' primary Cold War adversary—the Soviet Union—joined the nuclear club in 1949 and rapidly expanded its supply of these special weapons throughout the 1950s and 1960s. Moreover, while the United States largely demobilized its conventional forces in the years immediately following World War II, the Soviet Union did not. Its Red Army likely maintained approximately 175 divisions on active duty in the late 1940s, enabling the Soviets to further strengthen their iron grip over most of the still-occupied eastern Europe, especially East Germany. Comparatively, from 1945 to 1950, with much of its former force structure in reserve, the now diminutive active-duty Marine Corps shrank from a wartime high of six combat divisions and five aircraft wings totaling 485,000 wartime personnel down to two reduced divisions and aircraft wings by 25 June 1950—the date North Korea invaded South Korea.[34]

---

[32] Adam Givens, "Okinawa: The Costs of Victory in the Last Battle," National WWII Museum, 7 July 2022.

[33] "Iwo Jima and Okinawa: Death at Japan's Doorstep," National WWII Museum, accessed 8 December 2021.

[34] Allan R. Millett, *Semper Fidelis: The History of the United States Marine Corps*, rev. ed. (New York: Free Press, 1991), 445, 447, 480–81; and Richard A. Bitzinger, *Assessing a Conventional Balance in Europe, 1945–1975* (Santa Monica, CA: Rand, 1989). Bitzinger noted that the West most likely had faulty data on Soviet force strengths due to various reasons, but many experts accepted the number of 175 divisions at that time and for years after.

Starting in the late 1940s and continuing into the administration of President John F. Kennedy (1961–63), U.S. war planners intended to address this potential imbalance in future conflicts by blending conventional forces and the use of nuclear weapons of various size and lethality. For example, as early as 1953, the chairman of the Joint Chiefs of Staff, Navy admiral Arthur W. Radford, commented that "today atomic weapons have virtually achieved a conventional status within our armed forces." The only problem with this way of thinking was that the military leadership soon discovered that "nuclear weapons could not be used just as if they were conventional weapons. Their radius of destruction was too large and their aftereffects too pervasive to employ them in such a precise and discriminating fashion." American military strategist Bernard Brodie, an original architect of escalatory response to potential Soviet aggression against Western Europe, summarized this idea, noting that "a people saved by us through our free use of nuclear weapons over their territories would probably be the last that would ever ask us to help them."[35]

The fact that nuclear weapons could imperil large-scale amphibious operations, with their hundreds of tightly packed support vessels and slow-moving armored amphibians, convinced many strategists that the era of opposed amphibious assault was over. In late 1949, Army general Omar N. Bradley, then chairman of the Joint Chiefs of Staff, testified before the House Armed Services Committee and predicted that "large-scale amphibious operations will never occur again."[36] After Bradley was proven wrong when the rushed and partially reconstituted 1st Marine Division captured the port of Inchon, South Korea, less than a year later, Marine Corps leadership was convinced that amphibious warfare still had a role to play in the post–World War II era. Of course, the North Ko-

---

[35] Lawrence Freedman, "The First Two Generations of Nuclear Strategists," in *Makers of Modern Strategy: From Machiavelli to the Nuclear Age*, ed. Peter Paret (Princeton, NJ: Princeton University Press, 1986), 747–48; and Bernard Brodie, "More About Limited War," *World Politics* 10, no. 1 (October 1957): 117.

[36] Gen Omar N. Bradley, USA, quoted in Col Robert D. Heinl, USMC (Ret), "The Inchon Landing: A Case Study in Amphibious Planning (May 1967)," *Naval War College Review* 51, no. 2 (1998): 118. This article was a reprint of a lecture by Heinl that was published in the *Naval War College Review* in May 1967.

Figure 5. U.S. military personnel at nuclear weapons test, 1951

Source: official U.S. Department of Defense photo.

reans possessed no nuclear weapons of their own at that time, nor did their primary benefactor, the People's Republic of China (PRC).[37]

However, due to a host of difficulties experienced by U.S. armed forces in the Korean War (1950–53), the new administration of President Dwight D. Eisenhower reviewed its national security force structure and ultimately made the decision to refocus on the primary threat that the conventionally powerful and now–nuclear–capable Soviet Union posed.

---

[37] Col Robert D. Heinl Jr., *Victory at High Tide: The Inchon–Seoul Campaign* (Baltimore, MD: Nautical and Aviation Publishing, 1979), 3, 7–10, 14–20; BGen Edwin H. Simmons, *Over the Seawall: U.S. Marines at Inchon* (Washington, DC: Marine Corps History and Museums Division, 2000), 2–3; and "Brief History of the United States Marine Corps," Marine Corps University, accessed 27 February 2023.

Consequently, the Eisenhower administration and Congress heavily invested in the recently independent U.S. Air Force, focusing funds more specifically on its nuclear weapons-centric Strategic Air Command (SAC), then led by legendary World War II-era bomber commander General Curtis E. Lemay. The Eisenhower administration was going to call this shift in national security strategy the "New Look Defense."[38]

The New Look Defense envisioned the creation of 143 Air Force air wings to offset cuts to American ground combat power, and SAC was going to lead the way. By the close of 1953, SAC "had fully equipped 11 of the 17 wings in the atomic strike force. . . . Strategic Air Command personnel numbered almost 160,000 at 29 Stateside and 10 overseas bases."[39] National security planners in the United States hoped that the New Look emphasis from 1953 onward would keep the nation properly focused on its principal enemy, the Soviet Union, and avoid allowing the Communists in the Soviet Union or the PRC to determine the time and place for yet another debilitating regional conflagration, such as the recently concluded Korean War.

In fact, John Foster Dulles, secretary of state in the Eisenhower administration, was convinced that the implication of the United States possibly using nuclear weapons against the PRC caused it and North Korea to sign the Korean Armistice Agreement in July 1953. Dulles argued that "no local defense, could, by itself, contain Communist land forces. Consequently, the Administration would 'depend primarily upon a great capacity to retaliate instantly, by means and at places of our choosing. . . . Instead of having to try to be ready to meet the enemy's many choices, . . . it is now possible to get, and share, more basic security at less cost'."[40]

In 1962, the U.S. Army, in an effort to prepare for operations on a future nuclear battlefield, conducted a series of Nevada desert wargames called Operation Ivy Flats. These wargames led the Service to emphasize smaller nuclear warheads that could be fired from artillery tubes or even

---

[38] Herman S. Wolk, "The 'New Look'," *Air Force Magazine*, 1 August 2003.
[39] Wolk, "The 'New Look'."
[40] Wolk, "The 'New Look'."

a highly unusual weapons system dubbed "the Davy Crockett" that fired a small nuclear projectile from a recoilless rifle mounted on a tripod or the back of a jeep. Designed to literally fry Soviet tank crews from within via gamma rays emitted by the small nuclear blast while limiting collateral damage to the surrounding community, the Army decided to purchase more than 2,000 of these Davy Crockett weapons systems. It was only live tested once during the Ivy Flats games. Attorney General Robert F. Kennedy witnessed this trial of the weapon, "dubbed the Little Feller II shot," which "detonated on-target less than 40 feet above the ground and had a yield equivalent to 18 tons of TNT."[41] Needless to say, during the 1950s and 1960s, thousands of U.S. Army soldiers and Marines were directly exposed to nuclear fallout from these atomic wargames.

Although the Marine Corps agreed to require some of its larger-caliber artillery batteries, such as the self-propelled M110 eight-inch howitzers, as well as some of its tactical aviation squadrons to be nuclear capable by the 1960s, after the 1950s the Marine Corps largely avoided getting too heavily involved in plans to fight in a nuclear environment, continuing to focus on the twin concepts of vertical and amphibious assault instead. The advent of nuclear weapons, however, was likely one of the reasons that the Marine Corps became enamored with the new helicopter technology of the 1950s. Rather than solely storming ashore in slow-moving armored amphibians, helicopters provided the possibility for the Navy-Marine Corps amphibious warfare team to distribute its assault forces over a much wider area of the ocean, limiting its vulnerability to nuclear weapons.

In addition to its already proven armored amphibious vehicles, the Marines needed to consider the helicopter as a reliable assault platform, something that the technology was going to famously fail to deliver during the Vietnam War (1959–73). For instance, helicopters played a predominant role in the Battle of Ia Drang early in the conflict (14–18 November 1965). In that fight, the 1st Battalion, 7th Cavalry Regiment, 1st Air Cavalry Division, commanded by U.S. Army lieutenant colonel Har-

---

[41] Paul Huard, "This Nuke Proved that Size Doesn't Matter," *Medium*, 14 February 2015.

Figure 6. U.S. Army M-28/M-29 Davy Crockett Weapons System

Source: official U.S. Department of Defense photo.

old G. Moore, were attacked by three People's Army of Vietnam (PAVN) regiments intent on destroying the Americans before reinforcements could arrive at Moore's principal landing zones X–Ray and Albany. Consequently, at just these two locations alone, "234 [soldiers] were killed and more than 250 were wounded in a period of four days. In the 43-day Ia Drang campaign, 545 Americans were killed."[42] The Battle of Ia

---

[42] "Ia Drang Valley Incident," Defense POW/MIA Accounting Agency, accessed 8 December 2021. The battles for the landing zones in the Ia Drang Valley was later the subject of LtGen Harold G. Moore, USA (Ret), and Joseph L. Galloway, *We Were Soldiers Once . . . and Young: Ia Drang, the Battle that Changed the War in Vietnam* (New York: Random House, 1992); and later the film *We Were Soldiers*, directed by Randall Wallace (Hollywood, CA: Paramount Pictures, 2003), starring Mel Gibson as LtCol Moore and Sam Elliott as Moore's outstanding SgtMaj Basil L. Plumley. While Moore fought in the battle, Galloway was an eyewitness to the difficulty Moore's soldiers experienced in trying to land reinforcements via helicopters being subjected to heavy ground fire.

Drang demonstrated that opposed assaults against units that relied on helicopters for transport directly into battle could be just as costly as any of the amphibious variety.

The Cold War was replete with rapidly shifting political decisions that directly impacted the U.S. military, its organization, and its fighting doctrine. Early nineteenth-century Prussian strategist Carl von Clausewitz wrote that "war is not merely an act of policy but a true political instrument, a continuation of political intercourse, carried on with other means."[43] Clausewitz included this statement in the first of eight books that comprised his seminal work, On War, a text that is still studied in all the U.S. military war colleges. Yet, it is also important to note that in Book 8 of On War, Clausewitz stated that "no one starts a war—or rather, no one in his senses ought to do so—without first being clear in his mind what he intends to achieve by that war and how he intends to conduct it. The former is its political purpose; the latter its operational objective."[44] While Clausewitz was clear on the preeminence of politics in warfare, he was less definitive with regard to fighting limited wars with purposely constrained means, a hallmark of the type of conflict that took place throughout the Cold War and continues to this day.

As the wars and interventions during the Cold War and beyond illustrate, nearly every post-World War II operation involving the use of U.S. military force were conflicts of political choice. Because of its nuclear stockpile, the United States believed that a direct attack on its homeland was unlikely—at least until 11 September 2001. Additionally, U.S. leadership supposed that decisions could be made easily by the National Command Authority (NCA), sometimes with or without the consent of Congress, to use military force as a means of policy continuation.[45] This simple reality allowed U.S. politicians to consider fighting limited wars

---

[43] Carl von Clausewitz, On War, trans. and ed. Michael Howard and Peter Paret (Princeton, NJ: Princeton University Press, 1976), 87.

[44] Clausewitz, On War, 579.

[45] The term National Command Authority was used by the U.S. Department of Defense in the late twentieth century to refer to the president as commander in chief and the secretary of defense as the top sources of lawful military orders impacting the nation's armed forces and the use of nuclear weapons.

with limited means while reserving the bulk of their strategic and even conventional forces to keep an eye on the Soviet Union, the nearest peer competitor to the United States for the last half of the twentieth century. In many ways, the United States seems to be conducting a similar general strategy toward a rising China today.

For the Cold War–era Marine Corps, modernity was going to play an ever-larger role in future combat operations, especially following the end of the predominately conventional warfighting that took place in Korea. Emerging strategic inflection points, as General Krulak described, were now going to come at a rapid pace. Changes in technology and its application to the battlefield was nearly continuous throughout the Cold War and into the twenty-first century. Moreover, the politics of fighting limited wars had a direct and enduring impact on the organization and operational doctrine of all the Services. It literally constrained what forces could or could not do on the ground, in the air, or at sea. This new reality also required on-the-scene battlefield leaders to make instant decisions, similarly to fighter pilots, that could potentially upset the political situation of any intervention or military campaign. Starting with the Korean War, U.S. presidents would use the military in directly political ways, and all the Services needed to understand this new paradigm. Since the beginning of the Cold War, the Marine Corps has seemingly required Commandants and other senior leaders to have the ability to anticipate both the rapid-fire changes impacting the new way of war and the evolving politics of conflict. The exigencies of the Cold War demanded that the Marine Corps refocus a significant amount of time toward PME or risk the consequences on the battlefield.

The following chapters and the events described in them are generally arranged in chronological order during the Cold War and briefly beyond. The intent is not to focus in detail on the major Marine Corps combat operations, such as actions in Korea, Vietnam, or even during Operation Desert Storm. Indeed, other Marine Corps historians have covered these affairs in extensive detail. Rather, the thesis of the entire study is to point out those strategic inflection points, based on those that Krulak mentioned in his analysis of Belleau Wood, throughout the Cold War.

The first chapter primarily focuses on the unification fights of the late 1940s and the advent of the National Security Act of 1947, a time when the Marine Corps had the potential of going out of existence. Just two years after the raising of the iconic American flag on Mount Suribachi during the Battle of Iwo Jima, U.S. defense planners and even President Harry S. Truman were more than ready to end the institutional existence of the Marine Corps. The chapter also includes a discussion of the impact that new 1950s helicopter technology was going to have on all the Services, none more so than the Marine Corps and its amphibious warfare mission.

Chapter 2 focuses on the political decisions made toward the end of the Vietnam War that portended a new template for future combat operations within a limited war context. It exposes the failure of the American war effort in the last years of U.S. involvement in Southeast Asia but also reveals the incredible leadership bonus that the Marine Corps accrued from its stay-behind advisor corps, also known as Co-Vans. During these last years of involvement in Southeast Asia, American domestic politics and the national media fully inserted itself into future national security affairs. Soon after, the entire nation came to understand the negative effects that ambiguous political objectives could have on its armed forces as well as its allies. Throughout the end of the American experience in Southeast Asia in the early 1970s, phrases such as *safe havens*, *hearts and minds*, and *mission creep* became more widely used. In fact, the mantra of "no more Vietnams" that emerged from the conflict caused the Marine Corps to consider, for the first time since Belleau Wood, combat operations on the continent of Europe—this time against a more powerful and fully mechanized potential adversary, the Soviet Union.[46]

In chapter 3, the detritus of the Vietnam imbroglio and the social unrest inside the United States from 1968 to 1972, coupled with the end of the national draft in 1973 and the advent of the all-volunteer force (AVF), came full circle for the Marine Corps and for all the Services. The immediate post-Vietnam War era caused Marine Corps leadership to

---

[46] Richard M. Nixon, *No More Vietnams* (New York: Arbor House, 1985), 212.

search for answers to its worsening manning and disciplinary situation. During the early 1970s, racial tension affected nearly every single Marine Corps unit. At the same time, the Marine Corps officer corps was likely the least diverse of all the Armed Services. Throughout the 1970s, the Marine Corps spent a great deal of time on self-reflection, with leadership asking whether it was truly the elite fighting force that it had long claimed to be.

Chapter 4 covers the crucial commandancies of General Louis H. Wilson and General Robert H. Barrow in the late 1970s and early 1980s. As the 26th and 27th Commandants, respectively, these two officers moved to decisively shake the Marine Corps out of its post-Vietnam-era doldrums, embrace the AVF, and finally prepare to reset the Service for major contingency operations during the 1980s. While Barrow had the benefit of increased defense budgets during the administration of President Ronald W. Reagan, Wilson corrected the personnel situation throughout the Corps that seemed to hold everyone back, setting the stage for further Marine Corps success. Today, many credit these two Commandants with saving the Marine Corps from itself during this difficult time in its institutional history.

Chapter 5 confirms the tremendous importance that the Goldwater-Nichols Department of Defense Reorganization Act of 1986 had on all the Services. The law literally forced the Services to finally cooperate more effectively in future warfighting and contingency operation environments. Prior to Goldwater-Nichols, as had occurred throughout much of the Cold War, all the U.S. Services struggled with communications and a common warfighting doctrine, which caused needless casualties during overseas contingency operations, such as ones in Beirut, Lebanon, and in Operation Urgent Fury in Grenada. Further, the humiliating debacle in 1980 of Operation Eagle Claw—the failed attempt by the administration of President James E. "Jimmy" Carter to liberate American hostages seized during the 1979 Iranian Revolution—demonstrated the need to create an independent U.S. Special Operations Command (USSOCOM). Despite this clarity, the Marine Corps still resisted allowing its servicemembers to become part of the USSOCOM command structure until the turn of the twenty-first century. Finally, the chapter reveals that the largely successful and skillfully executed Operation Just Cause in Panama showed how well the Services had incorporated the tenants of the Goldwater-Nichols Act. While there remained areas of concern after Just Cause, the Services never looked back to the days before Goldwater-Nichols.

The sixth chapter focuses on the 29th Commandant of the Marine Corps, General Alfred M. Gray Jr., arguably the most important Marine Commandant since Lejeune. His strong advocacy of maneuver warfare doctrine and PME came along at just the right time and enabled the Marine Corps to participate with great skill in operations in support of the North Atlantic Treaty Organization (NATO) above the Arctic Circle as well as in actual combat in the deserts of Kuwait and Saudi Arabia. The intent of the chapter is not to overly focus on the details of Operation Desert Storm. Instead, it covers in greater part the struggle at the various schools of Marine Corps University in Quantico, Virginia, and schools in other locations to incorporate maneuver warfare as the Service's official warfighting doctrine. It was not always smooth sailing.

Finally, chapter 7 explores the initial period of General Krulak's commandancy. As Commandant after the Cold War had ended, he was prescient with his advocacy of operational maneuver from the sea and urban warfare. Most importantly, his demand that the Marine Corps extend professional military education even throughout the enlisted ranks paid great dividends for the Service. Krulak saw, far better than others, the necessity in the modern era to produce dynamic noncommissioned officers able to understand and take appropriate action, when required, in the increasingly complex battlefields of the coming twenty-first century.

The epilogue briefly returns to all the various strategic inflection points that emerged for the Marine Corps during the Cold War and its immediate aftermath. It also posits a challenge to other scholars to continue research into the history of the Cold War, a long-overlooked field of study. Further, the advent of hypersonic missiles, precision-guided munitions, drones, cyberwarfare, artificial intelligence, and advanced robotics will likely require the remaking of the organizational force structure and even doctrine for all the U.S. military Services, none more so than the still-amphibiously oriented Marine Corps. In many ways, the employment of Marines during the Global War on Terrorism was akin to the Vietnam period of the Cold War. As happened in Southeast Asia, U.S. servicemembers in Afghanistan largely fought from fixed forward operating bases and the nearly 20-year effort there ended remarkably similarly to that of Vietnam. To make matters worse, future trends in military affairs and technology will likely make large-scale amphibious assaults against dedicated opposition forces no longer feasible. Thanks to the current revolution in military affairs, the Marine Corps, along with the other Services, needs to discover and address today's latest strategic inflection point or risk a future battlefield shock that could possibly make Belleau Wood pale in comparison.

CHAPTER ONE

The Cold War Begins

As the smallest of the nation's Armed Services, the U.S. Marine Corps had always been concerned about its institutional existence. Furthermore, at times, the role and mission of the Marine Corps in the security affairs of the United States has lacked clarity. Throughout its long history, the failure to secure a sustained national mission has caused significant problems for the Marine Corps—so much so that at least two influential twentieth-century U.S. presidents, Theodore Roosevelt (1901–9) and Harry S. Truman (1945–53), vigorously questioned the need for maintaining a Marine Corps at all.

The years immediately after World War II were arguably the most difficult that the Marine Corps ever experienced. Despite having proven the utility of a large, robust Fleet Marine Force (FMF) in seizing military objectives in the furtherance of a naval campaign, many defense experts in the early dawn of the atomic age questioned the possibility of ever again conducting large-scale amphibious operations. To make matters even more complicated for the Service, the National Security Act of 1947, one of the most far-reaching and influential pieces of legislation relating to defense policy that Congress ever passed, created an overarching U.S. Department of Defense (DOD) to manage the nation's military affairs.[1] The motivation behind the department's creation was

---

[1] National Security Act of 1947, Pub. L. No. 117–103 (1947).

a good one. Throughout World War II, rancorous Service department rivalry often hindered mission success. In the immediate years after World War II, the senior leadership of the U.S. Army, including General of the Army Dwight D. Eisenhower, advocated for unifying the Services under a single department, similar to the unity of commands during the war, with both a single military leader and a single civilian leader to reduce this infighting and inefficiency.[2]

The National Security Act also furthered the idea that the United States might not need separate Services at all and spurred a strongly supported movement toward Service unification. The U.S. Army primarily championed this idea. By the end of World War II, the Army had five active-duty officers who held the rank of general of the Army, or a five-star general. The highest-ranking Army officer was General George C. Marshall Jr., who even outranked the vainglorious General Douglas MacArthur. Marshall possessed tremendous personal gravitas with both President Franklin D. Roosevelt and then Truman, so much so that Truman later appointed him secretary of state in 1947 and secretary of defense in 1950. The U.S. Navy had four admirals appointed to the five-star rank of fleet admiral. However, the only full general in the Marine Corps was Medal of Honor recipient and Commandant of the Marine Corps Alexander A. Vandegrift. Vandegrift had held the four-star rank since 1944 and was the first and only Marine Corps officer to achieve that distinction during World War II. Despite his elevated rank, Vandegrift was not considered a permanent standing member of the Joint Chiefs of Staff. Besides the Commandant, the Marine Corps did not have an additional active-duty full general until 1968, when General Lewis W. Walt was appointed Assistant Commandant of the Marine Corps.[3]

---

[2] Charles A. Stevenson, "Underlying Assumptions of the National Security Act of 1947," *Joint Forces Quarterly* 48 (1st Quarter, 2008): 129–33.

[3] *Fleet Admirals, U.S. Navy* (Washington, DC: Naval Historical Foundation, 1966); MajGen David T. Zabecki, USA, "Review of *Generals of the Army: Marshall, MacArthur, Eisenhower, Arnold, Bradley*, edited by James H. Willbanks," *Parameters* 44, no. 2 (Summer 2014): 116–18; Jon T. Hoffman, "Alexander A. Vandegrift," in *Commandants of the Marine Corps*, ed. Allan R. Millett and Jack Shulimson (Annapolis, MD: Naval Institute Press, 2004), 294, 301; and "Assistant Commandants of the Marine Corps," Marine Corps University, accessed 1 March 2023.

Figure 7. President Harry S. Truman signing the National Security Act

Source: National Archives and Records Administration.

Consequently, less than five years after raising the American flag on top of Mount Suribachi on Iwo Jima in 1945, the Marine Corps had been allowed to atrophy from a personnel level of approximately 485,000 officers and enlisted men and women to a single division of fewer than 75,000 active–duty Marines. Secretary of Defense Louis A. Johnson imposed significant cuts, reducing the FMF by 19 percent and the Marine aviation force by 48 percent between 1949 and 1950.[4] This steady and relentless devolution in size and mission was so disconcerting that, as early as 1946, Vandegrift addressed Congress about this trend. Known today as the "Bended Knee" speech, Vandegrift eloquently laid out the issue before a spellbound session of the Senate Committee on Naval Affairs. He stated:

---

[4] Paolo E. Coletta, *The United States Navy and Defense Unification, 1947–1953* (Newark: University of Delaware Press, 1981), 160.

Marines have played a significant and useful part in the military structure of this Nation since its birth. But despite that fact, passage of the unification legislation as now framed will in all probability spell extinction for the Marine Corps. . . . For some time I have been aware that the very existence of the Marine Corps stood as a continuing affront to the War Department General Staff, but had hoped that this attitude would end with the recent war as a result of its dramatic demonstration of the complementary and nonconflicting roles of land power, naval power, and air power. But following a careful study of circumstances as they have developed in the past 6 months I am convinced that my hopes were groundless, that the War Department's intentions regarding the Marines are quite unchanged, and that even in advance of this proposed legislation it is seeking to reduce the sphere of the Marine Corps to ceremonial functions and to the provision of small ineffective combat formations and labor troops for service on the landing beaches. Consequently I now feel increased concern regarding the merger measure, not only because of the ignominious fate which it holds for a valuable corps, but because of the tremendous loss to the Nation which it entails.[5]

Vandegrift went on to conclude that while the Marine Corps was proud of its past, "we do not rest our case on any presumed ground of gratitude owing us from the Nation. The bended knee is not a tradition of our Corps." If the Marine had not "made a case for himself after 170 years of service," Vandegrift argued, then the Service "must go." Even then, the Commandant imagined that they had "earned the right to depart with dignity and honor, not by subjugation to the status of uselessness and servility planned for him by the War Department."[6] Brigadier

---

[5] *Hearings on the Unification of the Armed Forces, before the Senate Committee on Naval Affairs,* 79th Cong. (6 May 1946) (statement of Gen Alexander A. Vandegrift, Commandant of the Marine Corps), hereafter Vandegrift statement.
[6] Vandegrift statement.

General Merrill B. Twining, a well-regarded member of Vandegrift's staff during the Guadalcanal campaign (August 1942–February 1943), was shocked at the sudden rise in the governmental and even public perception of the Marine Corps after World War II. He lamented that, in many people's minds, the Marine Corps seemed to have gone "from heroes to bastards overnight."[7]

Vandegrift's speech was a dramatic moment for the Marine Corps, but the effort to unify the Services in some fashion did not end with his plea for institutional survival. As a result, Marine Corps leadership believed that the key to future viability lay with having the Commandant become a permanent voting member of the Joint Chiefs of Staff, which the senior leadership of the other Services vehemently opposed. Furthermore, many Marine Corps leaders thought that Vandegrift needed to be much more proactive on this issue. Truman believed, to the dismay of nearly everyone inside the Marine Corps, that he had the authority to organize the Services through executive order rather than legislation.[8] The president's opinion, however, was later tempered in January 1947 when Secretary of the Navy James V. Forrestal and Secretary of War Robert P. Patterson sent him a joint letter that stated: "We are agreed that the proper method of setting forth the functions (so-called roles and missions) is by the issuance of an executive order concurrently with your approval of appropriate legislation." Attached to the letter was a "draft executive order [that] defined the service functions."[9] While the letter and its attachment, later known as the Forrestal-Patterson agreement, contained wording that retained the Marine Corps as part of the Navy's functions, Vandegrift had zero input into its contents.

---

[7] Alan Rems, "Semper Fidelis: Defending the Marine Corps," *Naval History Magazine* 31, no. 3 (June 2017): 36–41; and Benis M. Frank, interview with Gen Merrill B. Twining, 1 February 1967, transcript (Oral History Section, Marine Corps History Division, Quantico, VA), hereafter Twining interview.

[8] Gordon W. Keiser, *The U.S. Marine Corps and Defense Unification, 1944–1947: The Politics of Survival* (Baltimore, MD: Nautical and Aviation Publishing Company of America, 1996), 56–57; and Allan R. Millett, *Semper Fidelis: The History of the United States Marine Corps*, rev. ed. (New York: Free Press, 1991), 457–62.

[9] Keiser, *The U.S. Marine Corps and Defense Unification*, 71.

Consequently, Vandegrift commissioned several boards of officers assigned to both Headquarter Marine Corps and the Marine Corps Schools in Quantico, Virginia, to study the role and mission problem. While the Forrestal-Patterson agreement avoided any political controversy from abolishing the Marine Corps outright, it seemed to leave the size and usage of the Service entirely in the hands of the U.S. Navy, which meant that the Navy could relegate the greatly diminished Marine Corps to a strictly minor constabulary role in the future. In fact, Forrestal even said as much when he earlier proposed that the Marine Corps could be useful to "provide a balance of order in China" during the massive and bloody Chinese Civil War (1945–49) that resulted in the near-total victory of Mao Zedong's Communist revolutionaries.[10]

Two of the most important boards that Vandegrift convened was a panel of officers headed by Major General Lemuel C. Shepherd Jr., who eventually became the 20th Commandant of the Marine Corps, and a study group headed by Brigadier General Merritt A. Edson, a recipient of the Medal of Honor for actions during the Guadalcanal campaign, and Brigadier General Gerald C. Thomas, a veteran of both the Battle of Belleau Wood (1–26 June 1918) and Guadalcanal. While there was some overlap between the members of the two groups, the Edson-Thomas group was not really a formal board like the Shepherd-led group but more akin to an ad hoc investigatory body that eventually focused on preparing the Commandant for congressional testimony related to the ongoing unification and roles and mission fights. Much of the research effort fell on Lieutenant Colonel James D. Hittle, who had written a book on the history of the Joint staff, and Colonel Merrill B. Twining and Lieutenant Colonel Victor H. Krulak of Marine Corps Schools. Forming an internal working group called the "Chowder Society," Twining and Krulak aggressively pursued hard data to make a case for a future Marine Corps that was free from threats of extinction from any newly ensconced secretary of defense or antipathy from the other Armed Services.[11]

---

[10] Keiser, *The U.S. Marine Corps and Defense Unification*, 73.
[11] Keiser, *The U.S. Marine Corps and Defense Unification*, 73–76; and Rems, "Semper Fidelis."

Figure 8. Gen Alexander A. Vandegrift before the Senate Committee on Armed Services

Source: Harry S. Truman Presidential Library.

A series of unrelated events took place that had far-reaching consequences for the Corps and indirectly involved Lieutenant Colonel Hittle.
First, in 1946 Congress voted to reorganize its antiquated committee system and created powerful Armed Services Committees in the House and
Senate. The eminent Representative Carl Vinson (D-GA) already dominated the new House Armed Services Committee. A stalwart friend of
the U.S. Navy, Vinson was elected to the House of Representatives at the
age of 30, eventually serving 26 consecutive terms. Before Vinson passed
away in 1981 at the age of 97, the Navy took the extraordinary step of
naming one of their new nuclear-powered aircraft carriers after him
while he was still living, an honor rarely accorded anyone.[12]

---

[12] Millett, *Semper Fidelis*, 462–64; Aaron B. O'Connell, *Underdogs: The Making of the Modern
Marine Corps* (Cambridge, MA: Harvard University Press, 2012), 123–25; Roger H. Davidson,
"The Advent of the Modern Congress: The Legislative Reorganization Act of 1946," *Legislative
Studies Quarterly* 15, no. 3 (August 1990): 357–73, https://doi.org/10.2307/439768; George B.
Galloway, "The Operation of the Legislative Reorganization Act of 1946," *American Political
Science Review* 45, no. 1 (March 1951): 41–68; and Melvin B. Hill Jr., "Carl Vinson: A Legend in
His Own Time," Carl Vinson Institute of Government, accessed 4 March 2023.

However, the new Senate Armed Services Committee appeared to be more problematic for the Marines, as it was considered "a bastion of pro-Army sentiment."[13] At the height of the debate, U.S. Army Air Forces brigadier general Frank A. Armstrong Jr., a legendary figure in the strategic bombing campaign against Germany in World War II, stated publicly that the Marine Corps was "a small bitched-up Army talking Navy lingo. We are going to put those Marines in the Army and make efficient soldiers out of them."[14] Moreover, some members of the Chowder Society feared that Vandergrift had a "lukewarm" opposition to the pending national security legislation, arguing that he had not been "playing an aggressive hand in the Congressional poker game."[15]

No one seemed more worried about Vandegrift's political acumen than Lieutenant Colonel Victor Krulak. He believed that the Commandant was "ill-fitted for the gut fight that he faced in the unification controversy. This was to hurt the Marines before it was all over."[16] Edson and the Chowder Society's so-called "minister of propaganda," Lieutenant Colonel Robert D. Heinl Jr., were also concerned.[17] Consequently, a cadre of Marine officers and other allies, using means "both legal and illegal," consistently worked the halls of Congress on behalf of the Corps. As contemporary historian and Marine Corps Reserve officer Aaron B. O'Connell noted, "in a manner characteristic of the Marines, they treated the legislative arena as a theater of war, seeing the other services, partic-

---

[13] Millett, *Semper Fidelis*, 462.

[14] BGen Frank A. Armstrong, USAF, quoted in Demetrios Caraley, *The Politics of Military Unification: A Study of Conflict and the Policy Process* (New York: Columbia University Press, 1966), 151n; and Millett, *Semper Fidelis*, 461. Interestingly, the character of Col Frank Savage, the bomb group commander in the famous World War II-era action movie *Twelve O'clock High*, was said to have been based on BGen Armstrong. In the film, Savage, played by actor Gregory Peck, drives his bomber group pilots so relentlessly that he eventually suffers a nervous breakdown. Armstrong's hyperbolic speech in favor of the U.S. Air Force most likely got the attention of everyone in the naval Services.

[15] Keiser, *The U.S. Marine Corps and Defense Unification*, 105; and Richard Tregaskis, "The Marine Corps Fights for Its Life," *Saturday Evening Post*, 5 February 1949, 104–5.

[16] LtGen Victor H. Krulak, *First to Fight: An Inside View of the Marine Corps* (Annapolis, MD: Naval Institute Press, 1984), 26–27; and Rems, "Semper Fidelis."

[17] O'Connell, *Underdogs*, 98. O'Connell's entire chapter on this episode, titled "the Politicians and the Guerrillas," should be required reading for every student interested in civil-military relations and background to the activity of individual Marine Corps officers during the sensational Service unification fight of 1947.

Figure 9. Col Robert D. Heinl Jr.

Source: official U.S. Marine Corps photo.

ularly the Army, as the enemy."[18] Indeed, Heinl characterized the times as a fight for institutional survival, and they saw themselves as "guerrillas and almost fugitives."[19]

In 1950, during yet another effort to make the Commandant a permanent member of the Joint Chiefs of Staff, Truman disagreed and famously stated in a personal letter to Representative Gordon L. McDonough (R-CA) that "for your information the Marine Corps is the Navy's police force and as long as I am President that is what it will remain. They have a propaganda machine that is almost equal to Stalin's"—a comment the president almost immediately regretted.[20] The story broke on 5 September 1950 after McDonough added it into the congressional record. The outcry over comparing the Marines to Joseph Stalin while elements of the Marine Corps were engaged in heavy combat against Communist forces in Korea—and especially after the 1st Marine Division landed at Inchon, South Korea, 10 days later—caused Truman to make a rare public apology.[21]

The actual upshot of Truman's gaffe was later revealed in congressional legislation passed in 1952. Called the Douglas-Mansfield Act for Senator Paul H. Douglas (D-IL) and Representative Michael J. Mansfield (D-MT), both former Marines who cosponsored the bill, this legislation corrected some definitional deficiencies originally created by the Na-

---

[18] O'Connell, *Underdogs*, 101.
[19] O'Connell, *Underdogs*, 113.
[20] Robert D. Heinl Jr., *Soldiers of the Sea: The United States Marine Corps, 1775–1962* (Annapolis, MD: Naval Institute Press, 1991), 546.
[21] Krulak, *First to Fight*, 56–57.

tional Security Act of 1947 regarding the actual size of the Marine Corps and the role played by the Commandant within the Department of the Navy and on the Joint staff. The Douglas–Mansfield Act secured a seat on the Joint Chiefs of Staff for the Commandant when it discussed matters of concern to the Marine Corps. Further, the law specified that the Marine Corps would be comprised of a "minimum of three combat divisions and three aircraft wings and raised the ceiling on active-duty personnel strength to 400,000." The Marine Corps "became the only armed service whose principal mission, minimum size, and basic structure were detailed by public law."[22] Most importantly, the Douglas–Mansfield Act made the Commandant a true peer of the chief of naval operations within the Department of the Navy.

During the unification fight just a few years earlier, things were not always so clear. Brigadier General Edson took the extraordinary step of purloining a copy of classified papers known as Joint Chiefs of Staff Series 1478. These papers contained the opinions of the other Services on the unification process. What was worse was that "the Chowder Marines made additional copies and leaked their contents to key players in the unification fights—including journalists—which nearly caused the Commandant of the Marine Corps to be relieved of his duties." Rather incredibly, due to the Commandant not being a permanent member of the Joint Chiefs of Staff, the Marines had not been asked to contribute to the paper series. The series of papers especially highlighted the bitter antipathy U.S. Army leadership held toward the post–World War II Marine Corps. This activity also caused Truman to order the Commandant

---

[22] *Semper Fidelis: A Brief History of Onslow County, North Carolina, and Marine Corps Base, Camp Lejeune* (Camp Lejeune, NC: U.S. Marine Corps, 2006), 61–62. Both Paul H. Douglas and Michael J. Mansfield were tremendously proud of their service with the Marine Corps. Douglas volunteered for service as a private during World War II at the age of 50. He was later appointed an officer and participated in combat with the 1st Marine Division during the battles of Peleliu and Okinawa. Michael J. Mansfield served as a congressman, senator, and U.S. ambassador over the course of an exceptionally long public career. During World War I, he served in the U.S. Navy and immediately after the war enlisted in the Marine Corps as a private, serving for two years. Despite the impressive titles he earned over the course of his lifetime, Mansfield believed his two years as a Marine was the most important and had only his Marine Corps military rank emplaced on his tombstone when he passed away at the age of 98 on 5 October 2001.

to "get those lieutenant colonels [meaning Chowder Marines] off the Hill and keep them off." Nevertheless, Corps advocates not in uniform kept up the pressure to the point that Vandegrift threatened Edson with a court-martial if he did not cease with his activity on Capitol Hill and in the media. This warning possibly caused Edson to retire from active service, but he could now continue to safely lobby Congress as a civilian.[23]

The Marine Corps' search for a legislative solution to its dilemma was relentless. Due to the intense opposition of the Army and indifference of the Navy, the Marine Corps rightfully believed that it had few allies to assist it in its effort to survive the unification effort. The saving legislation that the Marine Corps had long been looking for turned out to be House bill 2319—the House version of the administration's unification bill. In a twist of fate, and "hoping to avoid the pro-Navy House Armed Services Committee," the Truman administration asked that the bill be transferred to the House Committee on Executive Expenditures, chaired by Representative Clare E. Hoffman (R-MI). Many believed that Hoffman lacked interest in any unification legislation and would turn the bill over to his subcommittee chairman, Representative James W. Wadsworth Jr. (R-NY), who was "a longtime pro-Army expert on defense matters."[24] Yet, as luck would have it, Hoffman was a close friend of Hittle's father. Suddenly, one of the most vociferous advocates for the Marine Corps now played a major role on its behalf. Hittle and his father showed Hoffman copies of the still classified Joint Chiefs of Staff papers and persuaded him to personally take an interest in the legislation.[25] Hoffman also demanded and received a number of sensitive internal Joint staff documents from the War Department, which enabled open testimony to reveal the anti-Marine bias of senior Army leadership. He even went so far as to personally grill the eminent General Dwight D. Eisenhower, then the Army chief of staff, about his alleged anti-

---

[23] O'Connell, *Underdogs*, 121–24. Well after the unification controversy ended, Edson, one of the most decorated Marines in the history of the Service, took his own life in 1955 for unknown reasons.

[24] Millett, *Semper Fidelis*, 462.

[25] Caraley, *Politics of Military Unification*, 171–72; O'Connell, *Underdogs*, 123–24; and Millett, *Semper Fidelis*, 462–63.

Marine Corps views. Nevertheless, Eisenhower vehemently denied any desire on his part to abolish the Marine Corps.[26]

From that point forward, opinion within the halls of Congress and with the American public trended favorably toward the Marines. Desperate to salvage a deal to get the National Security Act passed, the House drafted House bill 4214, which provided guarantees that the Marine Corps would be protected from diminishment by executive fiat or by any future secretary of defense. Most importantly, it also included language that Twining, Krulak, and Hittle drafted that "gave the Marine Corps primary responsibility for developing amphibious warfare doctrine and equipment, reaffirmed all the Corps's traditional duties, asserted the Corps's wartime utility and right to expand, and provided" that it remained "a separate service within the Department of the Navy."[27] The Chowder Society had clearly prepared the legislative battlefield very well. During early hearings on the matter, Secretary of War Robert P. Patterson became so exasperated about the sheer volume of questions he received about the fate of the Marine Corps that he exclaimed, "Marines, Marines! That's all I hear. They're not treated any differently than any of the other branches."[28] While the legislation received some further modifications, it went on to become public law.

The Marine Corps looked at the language contained in the National Security Act as a landmark victory in its constant struggle for institutional survival. The fact that the Marine Corps was even mentioned was considered progress. In previous iterations of the National Defense Act in 1903, 1916, and 1920, for example, the legislation did not include a single mention of the Marine Corps as a fighting Service.[29] Further,

---

[26] Caraley, *Politics of Military Unification*, 171–72.

[27] Millett, *Semper Fidelis*, 463–64; and Dwight Jon Zimmerman, "The Chowder Society," in *United States Marine Corps: Creating Stability in an Unstable World* (Tampa, FL: Faircount, 2007), 128–32.

[28] Maj Robert P. Patterson, USA, quoted in O'Connell, *Underdogs*, 124. Patterson served in the Army during World War I. He was awarded the Distinguished Service Cross and Silver Star for heroism in France. The office of the Secretary of War became that of the Secretary of the Army after the passage of the National Security Act of 1947 and the establishment of the Office of Secretary of Defense.

[29] LtCol James D. Hittle, "The Marine Corps and the National Security Act of 1947," *Marine Corps Gazette* 31, no. 10 (October 1947): 57–59.

the latest version of the act guaranteed that the Marine Corps would retain its organic aviation assets and maintain its relationship with the U.S. Navy by providing it with necessary FMF forces. While Truman was less than pleased with the outcome, especially as it related to the management of the newly formed DOD, his administration still got much of what it had originally sought.

Although the National Security Act of 1947 was indeed a positive step toward the permanent survival of the Marine Corps as a separate Service, it was still not clear what its role was going to be within the Navy. For example, in March 1948, Forrestal, now holding the newly established position of secretary of defense, held a major planning conference with all the Joint Chiefs at Key West, Florida. The new Commandant of the Marine Corps, General Clifton B. Cates, was pointedly not invited. Consequently, when questions were raised at the conference concerning the Marine Corps' role and mission in the early Cold War era, Forrestal deferred to Admiral Louis E. Denfeld, the chief of naval operations, as the Department of the Navy representative for the Marines.[30]

The Key West Conference had serious implications for the Marine Corps. It was clear from the start that Forrestal desired to solidify the future role and mission of the three larger Services—the Army, Air Force, and Navy—and bring them into conformity with the new National Security Act. Moreover, all the conference participants agreed that the most likely warfighting scenario was with the Soviet Union and the need to defend Western Europe. Consequently, this belief meant that the Army and especially the newly established Air Force were going to receive the majority of funds in future defense budgets. As for the Marines, their primary role as the nation's amphibious war experts was seen as obsolete. Further, at the heart of the conference debate was the desire of the Air Force to ultimately take control over all military aviation and the strong pushback delivered by both Forrestal and Denfeld for the Navy to retain control over all naval aviation platforms, indirectly including that of the

---

[30] Alan Rems, "A Propaganda Machine Like Stalin's," *Naval History Magazine* 33, no. 3 (June 2019): 36–41.

Marines.[31] In the end, the Navy won its point and saved naval aviation, including the Marine Corps' assets, keeping them from falling under control of the Air Force. Nevertheless, this issue is still debated today.

At the Key West Conference, it was not just the future of amphibious warfare being debated. Rather, the discussion also expanded to include the efficacy of the Navy's most sacred cow—carrier and land-based naval aviation. None other than Medal of Honor recipient Air Force lieutenant general James H. Doolittle, who had led a squadron of 16 North American B-25 Mitchell medium bombers over Japan during the famed "Doolittle Raid" in April 1942, believed as early as 1945 that the aircraft carrier had become as obsolete as the immediate postwar Army leadership believed amphibious warfare was for the Marines. Doolittle testified that "as soon as airplanes are developed with sufficient range so that they can go any place that we want them to go, or when we have bases that will permit us to go any place we want to go, there will be no further use for aircraft carriers."[32] One can only imagine how Cold War operations may have turned out for the Navy and Marine Corps if the Air Force had prevailed at Key West, since for most of the Cold War the Air Force was more focused on long-range bombers and the Strategic Air Command than aircraft launched from the deck of a carrier. A pro–Air Force decision at Key West "would have left the Navy [and the Marines] . . . unprepared to fight the 'limited wars' of the Cold War."[33]

Nevertheless, things got worse for the Marine Corps in 1949. Truman grew tired of Forrestal's resistance to his proposed reforms. The new secretary of defense job had placed Forrestal under a tremendous amount of mental strain. There were even sensational reports in the media that Forrestal had secretly met with Truman's rival in the 1948 presidential election, New York governor Thomas E. Dewey, who had been widely favored to win. In a narrow victory that surprised nearly every-

---

[31] Mark D. Vital, "The Key West Agreement of 1948: A Milestone for Naval Aviation" (master's thesis, Florida Atlantic University, Boca Raton, 1999), 110–11.

[32] *Hearings before the Committee on Military Affairs, United States Senate*, 79th Cong. (9 November 1945) (statement of LtGen James H. Doolittle, U.S. Army Air Forces), 308; and Vital, "The Key West Agreement of 1948," 18–19.

[33] Vital, "The Key West Agreement of 1948," 110.

one, Truman won another term and quickly asked for Forrestal's resignation. To replace Forrestal, Truman appointed a close political associate, Louis A. Johnson, as the new secretary of defense. Almost immediately, Johnson went after the Marine Corps with a vengeance. Colonel Twining later remarked that Johnson "made [Commandant] Cliff Cates' life miserable, treated him with contempt. . . . Cates hated him and he hated Cates and the Marine Corps."[34]

In 1984, looking back on the Service unification affair, then-retired Lieutenant General Krulak offered an "insider" point of view of the activity related to the unification struggle in his seminal book, *First to Fight: An Inside View of the U.S. Marine Corps.* Krulak pointed out that the book's genesis was in a series of letters between himself and the 21st Commandant of the Marine Corps, General Randolph M. Pate. Directly asked by Pate as to "why" the nation needed the Marine Corps, Krulak took a full five days before replying. When he finally did, he bluntly told Pate that the Service "exists today—we flourish today—not because of what *we* know we are, or what *we* know we can do, but because of what the grassroots of our country *believes* we are and *believes* we can do."[35] Years later, in thinking about his response to Pate's query, Krulak believed that he had "not adequately analyzed the rich and complex soil in which the durability of the Marines is rooted." So, he consequently added:

> While the mystique of the Corps transcends individuals, there were—in the early days, in my day, and still—people whose behavior exemplifies one or more of the qualities that characterize the Corps. Also, the Corps is in a sense like a primitive tribe where each generation has its own medicine men—keepers of

---

[34] Twining interview; and Rems, "A Propaganda Machine Like Stalin's." It should be noted that Twining was one of the leading operational planners for the Guadalcanal campaign in 1942. He provided a very informative memoir on his experience there. See Twining, *No Bended Knee: The Memoir of General Merrill B. Twining* (Novato, CA: Presidio Press, 1996). Twining is certainly one of the little-known intellectual giants of the Marine Corps, and his impact on it after World War II was significant. Unfortunately for Secretary of Defense James V. Forrestal, by 1949, he was overworked and severely depressed. Truman's request for his resignation was the final straw. He died by suicide on 22 May 1949 while undergoing psychiatric treatment at the Bethesda Naval Hospital.

[35] Victor H. Krulak, *First to Fight: An Inside View of the Marine Corps* (Annapolis, MD: Naval Institute, 1984), xvi.

the tribal mythology, protectors of tribal customs, and guardians of the tribal standards. Without them the tribe would wither, suffering from poverty of the soul.[36]

These transformational leaders in both the enlisted and officer ranks were occasionally referred to as the "Old Breed." They strive to imprint new generations of Marines with the values and mores that have transcended the Service during its institutional existence. Fortunately for the Marine Corps, throughout its long struggle for institutional survival and mission relevance during the twentieth century, it was blessed with many Marines whose transformational leadership made a direct difference in how it survived future challenges to its existence. These Marines were sometimes Commandants, future Commandants, or, like Krulak, were those who had been imbued with the spirit of the Marine Corps seemingly from birth. Nevertheless, they were all guardians of its standards and diligently ensured that the men and women of the Marine Corps always remained up to task.

Several of these senior Marine Corps leaders during the Cold War started their careers during World War II, immediately following the unification controversy, or during the Korean War. While they were not alone in their efforts, these officers led the Marine Corps through many critical strategic inflection moments during the Cold War. These future leaders also held one commonality: they had experienced the intense combat of a violent World War II amphibious assault; a major Cold War–era combat intervention such as Korea or Vietnam; or sometimes all three.

One of the most important leaders to emerge from the maelstrom of World War II was General Louis H. Wilson Jr., the 26th Commandant of the Marine Corps. Wilson later translated much of what he experienced in direct close combat with the Japanese on Guam in 1944 into leadership skills that worked quite well for him while holding the post of Commandant between 1975 and 1979. Born on 11 February 1920 in Brandon,

---

[36] Krulak, *First to Fight*, xvi.

Figure 10. LtCol Victor H. Krulak

Source: Marine Corps History Division.

Mississippi, Wilson grew up during the Great Depression, experiencing the economic catastrophe in one of the most impoverished regions in the country. Wilson's father passed away when he was young, so he later took on additional responsibilities for his immediate family members.[37]

As the war clouds gathered between the United States and Imperial Japan in the late 1930s and early 1940s, Wilson enlisted in the Marine Corps Reserve. Selected for officer's candidate school, he was commis-

---

[37] Gen Louis H. Wilson Jr., biographical file, Historical Reference Branch, Marine Corps History Division, Quantico, VA; and Katie Lange, "Medal of Honor Monday: Marine Corps Gen. Louis Wilson Jr.," Defense.gov, 26 July 2021.

sioned in November 1941. By 1942, he was assigned as an infantry officer in the new 9th Marine Regiment. Deploying in the Pacific theater as part of the 3d Marine Division, Wilson first saw action during the Bougainville campaign. By April 1943, he was promoted to captain and, one year later, was in command of Fox Company, 2d Battalion, 9th Marines, as the 3d Marine Division prepared an assault on the Japanese-occupied island of Guam, beginning on 21 July 1944.

The Battle of Guam (21 July–10 August 1944) became a defining moment in Wilson's life. As had happened with Vandegrift at Guadalcanal, Wilson, who would be awarded a Medal of Honor for actions during the fighting, learned the hard lessons of amphibious assault firsthand on Guam. As naval guns roared, the assault waves containing the 9th Marines as well as the 3d Division's two other infantry regiments, the 3d and 21st Marines, landed abreast on Asan beach. On Guam, the Japanese waited for U.S. supporting arms to lift before coming out to engage the infantry for the first time. The Japanese defenders were firmly dug in everywhere on the island. The topography of Guam greatly aided them, as the land rose from the beaches in a relentless series of ridgelines and slopes. To overcome this disadvantage, the Marines planned the landing of a separate Joint provisional brigade to assist with the reduction of the Japanese defenders on the Orote Peninsula.[38] The campaign to take Guam was an immense seaborne operational maneuver. Since U.S. forces controlled the sea around the entire island, the Japanese defenders could never know exactly where U.S. ground troops might appear on the battlefield. Even as a junior officer, Wilson recognized the difficulty of amphibious warfare. Yet, he also noted that it could still be successfully carried out and that the Marine Corps was especially suited for such a mission.

Nevertheless, the U.S. forces on Guam faced a stout defense from the Japanese forces. Mortar fire on Beach Red One was especially damaging. The first day's casualties in just this single location "exceeded the entire division casualties at Bougainville."[39] The 3d Marine Regiment had an es-

---

[38] Heinl, *Soldiers of the Sea*, 461–63.
[39] 1stLt Robert A. Aurthur, USMCR, and 1stLt Kenneth Cohlmia, USMCR, *The Third Marine Division* (Nashville, TN: Battery Press, 1988), 147.

pecially hard time taking the D-day objective known to Marine planners as "Chonito Cliff," an outcropping that dominated the Asan beachhead. Using tanks and flamethrowers to take out the Japanese troops defending from caves in the cliff face, the unit, while suffering the heaviest casualties of the 3d Division's three infantry regiments, eventually accomplished its mission.[40]

Due to casualties suffered by the 1st Battalion, 3d Marines, the 2d Battalion, 9th Marines, under Lieutenant Colonel Robert E. Cushman Jr., the future 25th Commandant of the Marine Corps, was taken out of reserve and fed into the front lines near Fonte Hill to support the 3d Marine Regiment. On the night of 25–26 July, Marine outposts reported heightened Japanese activity. The 9th Marines regimental scout platoon had been forced back into the main line, and combat outposts reported heavy infiltration taking place in front of the entire line. The Japanese planned a predawn assault against the overextended 21st Marines, who were "still groggy from five sleepless nights in which they had been counterattacked every night." In a controlled but highly violent attack, Japanese soldiers broke through the lines of understrength Baker Company, 1st Battalion, 21st Marines. Some of the assault force penetrated all the way to the area around regimental headquarters, just one ridge shy of the beach. Marine tankers, artillerists, pioneers, and other headquarters and service personnel formed a mobile reserve under the regimental executive officer, Lieutenant Colonel Ernest W. Fry Jr., to stop those who had broken through. At least 3,200 Japanese soldiers were killed on the division's front on this one single night, along with "at least 300 more in the rear areas near the beach."[41]

On this same night of 25 July, Wilson's Fox Company was about 200 yards short of the crest of Fonte Ridge, a key objective for the division. To make matters worse, Wilson's Marines were short on ammunition, and "nearly all hands were without water."[42] Throughout the night of 25–26 July, the 2d Battalion, 9th Marines, along with all the other bat-

---

[40] Aurthur and Cohlmia, *The Third Marine Division*, 147.
[41] Aurthur and Cohlmia, *The Third Marine Division*, 152–54.
[42] Aurthur and Cohlmia, *The Third Marine Division*, 155.

talions in the line, absorbed heavy Japanese counterattacks. The fighting for the Fonte Hill outcropping on the ridge cannot be described as anything less than surreal. Wilson's partially exposed Fox Company had to fight for their lives that night, and they mostly had to do it alone. Because of the terrain, George Company, positioned on Fox Company's left flank, was forced to pull back approximately 100 yards, leaving Wilson's company "in the center of the salient, holding tenaciously to a rocky mound well forward of the flank units."[43]

When the Japanese attack came that night, it was fought initially at extremely close range. While leading counterattacks to eject the Japanese who had entered his position, Wilson was wounded three times. In his citation for the Medal of Honor, it was noted that Wilson fought "fiercely in hand-to-hand encounters [and] furiously waged battle for approximately ten hours, tenaciously holding his line and repelling the fanatically renewed counterthrusts until he succeeded in crushing the last efforts of the hard pressed Japanese."[44] Even more incredibly, Wilson personally led a 17-Marine counterattack to take and hold the crest of the hill. At one point, this makeshift unit charged for 50 yards in the open under heavy fire to save one of his wounded Marines. Many of his Marines were killed or wounded, but Wilson and the survivors held the hill. Captain Fraser W. West, commanding officer of George Company, recalled the fighting that night as "bitter, close, and brisk."[45] The next morning, the remains of Japanese soldiers lay in front of or within the 2d Battalion's lines. They were especially thick in front of Fox Company. Two rifle companies of Cushman's 2d Battalion suffered 75 percent casualties, and the third company suffered 50 percent casualties. Japanese dead "ran near one thousand, including many officers of high rank."[46]

Like the survivors of Bloody Ridge on Guadalcanal, those who lived to tell the tale of Fonte Hill and the struggle for Guam passed along the

---

[43] Maj Orlan R. Lodge, *The Recapture of Guam* (Washington, DC: Marine Corps Historical Branch, 1954), 76.

[44] Cyril J. O'Brien, *Liberation: Marines in the Recapture of Guam*, Marines in World War II Commemorative Series (Washington, DC: Marine Corps History Division, 1994), 22–23.

[45] O'Brien, *Liberation*, 23.

[46] Authur and Cohlmia, *The Third Marine Division*, 156.

Figure 11. President Harry S. Truman presents the Medal of Honor to Maj Louis H. Wilson Jr.

Source: official U.S. Department of Defense photo.

lessons learned to a newer generation of Marines who followed in their footsteps during the Korean War. For example, Colonel Edward A. Craig commanded the 9th Marine Regiment during the recapture of Guam. In Korea, Brigadier General Craig's 1st Provisional Brigade, later called the "Fire Brigade," helped save the day during the desperate fighting in the Pusan Perimeter of South Korea between 4 August and 18 September 1950. There is no doubt that Craig's tenacity at Pusan, later during the Inchon landing (10–19 September 1950), and especially during the Chosin Reservoir campaign (26 November–13 December 1950) had its origins on Guam.

Figure 12. Col Lewis B. "Chesty" Puller and BGen Edward A. Craig at the command post of the 1st Marines

Source: Marine Corps History Division.

Another key early Marine Corps leader in the Cold War was Lieutenant General Frank E. Petersen Jr. In fact, Petersen's Marine Corps career would mirror nearly the entire length of the Cold War, missing just a few years on either end of it. Petersen had been commissioned in 1952 at the height of the Korean War, and it was not long before he was in the thick of things. He went on to stand out among what was only a handful of Black officers serving in Korea in general and in the Marine Corps as a whole. On 15 June 1953, during a critical combat mission, then-Second Lieutenant Petersen was flying in a four-plane division when his flight leader lost radio communications. Petersen, as the second section leader, without any hesitation took charge of the entire division and expertly led it on a bombing and strafing run against the enemy. All four planes returned safely to their base. This feat was so unique for a relatively new lieutenant that Petersen's commanding officer recommended him for the Distinguished Flying Cross. He later earned a total of six Air Medals in Korea as well. Petersen later candidly admitted that 15 June 1953 changed his life and his desire to remain on active duty as a Marine Corps aviator.[47]

Throughout his 38-year Marine Corps career, Petersen, a highly accomplished naval aviator, also focused on increasing the number of Black officers in the Marine Corps—a key area that the Service largely failed to improve on for several decades during the Cold War. The Marine Corps did not commission its first Black officer, Second Lieutenant Frederick C. Branch, until 1945. In fact, from 1942 to 1949, the Marine Corps operated a racially segregated enlisted training depot, then located at the newly established East Coast Marine Corps Base Camp Lejeune, North Carolina. During this timeframe, despite having trained "approximately 20,000 African-American men known as Montford Point Marines," the Marine Corps had few Black officers following the conclusion of World War II.[48] To make matters worse, soon after his commissioning, Branch was transferred into the Marine Corps Reserve. The following year, three other Black Marines received commissions. However, they too were all transferred into the Reserve. It took another executive order from President Truman to finally require Marine Corps leadership to accept Black officers into its regular establishment. Consequently, Petersen became

---

[47] LtGen Frank E. Peterson, *Into the Tiger's Jaw: America's First Black Marine Aviator* (Annapolis, MD: Naval Institute Press, 1998), 69–70, 73. A month after leading attack aircraft in Korea that resulted in Petersen receiving the Distinguished Flying Cross and having completed 50 combat missions, he received a letter from the then-Commandant of the Marine Corps, Gen Lemuel C. Shepherd Jr., notifying him that he was denied admission into the regular Marine Corps establishment. Petersen admitted that, at first, he believed his nonacceptance might have been due to the color of his skin, but he ultimately attributed it to the general military downsizing that followed the conclusion of the Korean War. Petersen was later augmented into the regular Marine Corps by a follow-on board.

[48] Bethanne Kelly Patrick, "The Montford Point Marines," Military.com, accessed 7 June 2021.

Figure 13. Lt Frank E. Petersen Jr. climbs out of the cockpit of a Vought F4U Corsair

Source: official U.S. Department of Defense photo.

one of the first Black officers who entered the active duty regular Marine Corps establishment, being commissioned as a second lieutenant. He had initially joined the U.S. Navy in 1950 but strongly desired to become a naval aviator with the Marine Corps. Despite the bigotry he faced during flight training, Petersen persevered through it all and became the Service's first Black pilot. Unfortunately, due to pervasive institutional racism, Petersen often found himself as the only Black officer wearing the Marine Corps uniform. For example, before shipping out for combat duty in Korea, in an incident at the Marine Corps Air Station El Toro officer's club, a White officer challenged Petersen's authenticity and threatened to have him arrested for "impersonating an officer."[49]

During the Vietnam War, Petersen went on to command Marine Fighter Squadron 314 (VMF-314), known as the "Black Knights," as a lieutenant colonel. While in command in 1968, Petersen's squadron won the Hanson Award, which was typically presented to the best aviation squadron in the Marine Corps. During the conflict, Petersen flew 280 more combat missions and was once shot down and rescued. He was awarded the Purple Heart and 17 more Air Medals. Incredibly, having been on active duty for nearly 20 years at that time, he was the senior Black officer in the entire Marine Corps.[50] Following his wartime command, Petersen was assigned as the special assistant for minority affairs to the Commandant. In this position, he implemented policy for increasing minority representation in the officer ranks in the 1970s. Petersen was also at the forefront of taking concrete steps toward addressing growing racial tension within the Marine Corps that manifested toward

---

[49] LtGen Frank E. Peterson, oral history, in *Pathbreakers: U.S. Marine African American Officers in Their Own Words*, comp. and ed. Fred H. Allison and Col Kurtis P. Wheeler, USMCR (Quantico, VA: Marine Corps History Division, 2013), 31–32. It should be noted that once the officer who challenged Petersen at the El Toro officer's club found out the real situation, he eventually rendered an apology and was sent overseas at the next opportunity. Nevertheless, it is clear from Peterson's oral history that he faced much difficulty and prejudice as a Marine officer in those early "pathbreaking" years. Amazingly, he never let it keep him down, and he made sure that his professional performance as a combat pilot and leader remained beyond reproach. For a fuller description of his travails as a junior officer, see Peterson, *Into the Tiger's Jaw*.

[50] Henry I. Shaw Jr. and Ralph W. Donnelly, *Blacks in the Marine Corps* (Washington, DC: Marine Corps History and Museums Division, 1988), 78.

the end of the Vietnam War. He also later taught himself how to fly the then-new but challenging Hawker Siddeley AV-8A Harrier jet aircraft and successfully commanded Marine Aircraft Group 32 (MAG-32). Petersen would become the Marine Corps' first Black officer to be promoted to the rank of brigadier general in 1979. In 1988, following his command of the Marine Corps Combat Development Command in Quantico, Virginia, he retired at the rank of lieutenant general.

Though other Black officers followed in Petersen's footsteps, it was often a struggle. Many Black officers noted that since the Vietnam War ended, recruiters' ability to meet the intended goal of increasing the presence of Black officers in the Marine Corps depended on the emphasis given by the Commandant or recruiting command at the time. Nevertheless, between the late 1960s and the early 1980s, some highly notable Black officers were recruited, many from historically Black colleges and universities or the U.S. Naval Academy, who later went on to achieve similar senior general officer rank, such as Lieutenant General Walter E. Gaskin Sr., Lieutenant General Ronald S. Coleman, Lieutenant General Ronald L. Bailey, and Lieutenant General Willie J. Williams. Other notable Black Marines, such as Major General Charles F. Bolden Jr., went beyond their military career as an aviator. Bolden became an acclaimed astronaut and was just one of three former astronauts to serve as administrator of the National Aeronautics and Space Administration (NASA) between 2009 and 2017. Other highly successful senior Black Marine Corps officers recruited during the 1970s and early 1980s became leaders as civilians in the federal government, education, or private industry. All can look to the pathbreaking role that Petersen played in the upward arc of their careers.[51]

Robert H. Barrow, a close contemporary of Wilson and the future 27th Commandant of the Marine Corps, was born on 5 February 1922 in the economically strapped Deep South in Baton Rouge, Louisiana, less than 250 kilometers from Wilson's birthplace. Barrow's family soon moved

---

[51] For an excellent description of the difficulties and progress of Black Marine Corps officers from the 1950s through the turn of the twenty-first century, see Allison and Wheeler, *Pathbreakers*, 29–226.

to the rural community of Rosalie, Louisiana. When he came of age, he decided to attend nearby Louisiana State University (LSU). However, he had to borrow "the 150 dollars he needed for the general school fee from his Episcopal minister and worked as a busboy waiting tables in the school dining hall and performed janitorial jobs on campus for his room and board fees."[52]

At the time, LSU required that all physically fit students be enrolled in the campus corps of cadets (the student body was then all-male). LSU had initially been a military school; even Marine Corps lieutenant general John A. Lejeune had enrolled at LSU as a student in preparation for his admission to the Naval Academy in 1884. Barrow, who had also enrolled in the Marine Corps Platoon Leaders Class program, was in his junior year when World War II broke out. Rather than wait to graduate, he dropped out of school and joined the Marine Corps as an enlisted man. According to his biography, Barrow was not overly impressed with the Marine Corps recruit training he received in San Diego, California. Due to wartime exigencies, the Service had been forced to shorten boot camp to six weeks. His drill instructors, both corporals, were not remarkable. Barrow said that the training was "not one that prepared someone to go off and be a fighting member of a fighting organization." Instead, "Probably the most important thing in boot training," Barrow recalled, "was learning something about how to shoot the rifle" as well as absorbing "the discipline and obedience to orders . . . although most of us were already adherents."[53] Barrow never forgot the importance of boot camp and the need to ensure that Marines being trained there arrived in the operating forces prepared for what they were to face.

On a personal level, Barrow must have done well in recruit training because he was asked to remain behind as a drill instructor after graduation. While performing in this capacity, he was sent to Officer Candidates

[52] BGen Edwin H. Simmons, USMC (Ret), "Robert Hilliard Barrow," in *Commandants of the Marine Corps*, ed. Allan R. Millett and Jack Shulimson (Annapolis, MD: Naval Institute Press, 2004), 437–39.
[53] Gen Robert H. Barrow, USMC (Ret), Oral History Transcript, 28 January 1986 session, BGen Edwin H. Simmons interviewer, Marine Corps History Division, Quantico, VA, 2015, 30–31; and Simmons, "Robert Hilliard Barrow," 439.

School, where he once again stood out. Initially assigned to a replacement battalion, Barrow was eventually sent to join the Sino-American Cooperation Organization located in China in World War II. This experience enabled Barrow to learn a lot about himself as a leader and, most importantly, about the capacity of the average Chinese soldier to endure severe privation.[54]

During the Korean War, Barrow learned the lessons that carried him through the rest of his storied Marine Corps career. In the fall of 1950, Barrow commanded Able Company, 1st Battalion, 1st Marines, a regiment then commanded by the legendary Colonel Lewis B. "Chesty" Puller. Barrow's battalion was in reserve during the famous Inchon landing but went ashore on the evening of 15 September 1950. With his company fighting in follow-on operations in the South Korean capital of Seoul, Barrow stood out among other company commanders of the division. Barrow's Able company, along with the rest of the 1st Marine Division, was soon transported to the east coast of Korea to begin what became known as the Chosin Reservoir campaign.[55]

The Chosin Reservoir campaign was a challenge to Marine Corps leadership, made worse by horribly cold weather and rugged mountainous terrain. Overextended and understrength after months of heavy fighting, the United Nations (UN) forces, commanded by General of the Army Douglas MacArthur, were surprised toward the end of November 1950 when Chinese Communist forces suddenly entered the war. In a precautionary move, near the last week of November, the commander of the 1st Marine Division, Major General Oliver P. Smith, ordered Puller to place one of "his battalions at Hagaru, another at Koto-ri, and the third at Chinhung-ni at the bottom of [Funchilin] Pass."[56] This movement was a fortunate and timely decision. On 27 November 1950, as elements of the 5th and 7th Marine Regiments attacked toward the village of Yudam-ni on the west side of

---

[54] Simmons, "Robert Hilliard Barrow," 440–41. The Sino-American Cooperation Organization was a covert intelligence group that operated behind Japanese lines in China.

[55] Simmons, "Robert Hillard Barrow," 443.

[56] Martin Russ, *Breakout: The Chosin Reservoir Campaign, Korea 1950* (New York: Penguin Books, 1999), 82.

the reservoir, Chinese Communist forces slammed into them, as they had against other UN forces all along the front. There was no longer any question of continuing the advance. Instead, it was about how to survive this massive onslaught. For approximately the next 30 days, the 1st Marine Division conducted a fighting retreat against extraordinary odds back to the port of Hungnam, along the peninsula's east coast, where the U.S. Navy evacuated them. During the retreat, Barrow's Able Company drew the assignment, along with the rest of the 1st Battalion, of guarding the critical chokepoint of Funchilin Pass along the main supply route. The terrain was steep and rocky, and the road was literally carved along the shoulder of the mountains in places and steeply dropped off into deep gorges on one side, with sheer rock walls on the other.

After some significantly hard fighting, especially by Fox Company, 2d Battalion, 7th Marines, at Toktong Pass, Smith had extricated the division from around the reservoir and made it south to the area around Koto-ri. However, Smith recognized that the 16-kilometer mountainous stretch of the main supply route that ran from Koto-ri to Chinhung-ni was the most vulnerable spot along the line of retreat. If the Chinese were going to make an all-out effort to cut the line, it would be there. To get through, Smith needed to airdrop bridge sections for damaged parts of the narrow road that snaked through Funchilin Pass. Marine combat engineers heroically worked around the clock to keep the road open while the infantry, helped by coordinated artillery and air strikes, held the Chinese back.

The Chinese concentrated several divisions to attack the supply route in the pass. A key terrain feature was Hill 1081. On 8 December 1950, as the 7th Marines moved southward out of Koto-Ri, the 1st Battalion attacked northward from Chinhung-ni to keep the pass open. To make matters more uncomfortable for the Marines, it began to snow. Yet, the blinding snowstorm turned out to be a hidden benefit for the Marines since it covered their approach toward the enemy, who had already dug in on Hill 1081. While "the temperature was five degrees below zero" and the 10-kilometer approach march was extraordinarily tricky, most of the Chinese defenders had no idea that the Marines were moving to evict them from this piece of crucial real estate.[57]

Barrow's Able Company, with support from the rest of his battalion on his flanks, was ordered to take the summit of Hill 1081. Thanks to the falling snow, Baker Company on Barrow's left surprised and overran a Chinese bunker complex on the hill's shoulder. Meanwhile, Able Company clawed its way to the summit with great difficulty. Spotting a Chinese defensive position, "Barrow decided to rely upon the element of surprise."[58] His plan worked like a charm. Losing 10 men in the assault, Barrow's Marines killed at least 60 Chinese soldiers and took control of the hill. Several hours later, they held off a Chinese counterattack and inflicted additional casualties on them.[59] With the main supply route secure, the retreat of the 1st Marine Division could continue. Barrow was later awarded the Navy Cross for his combat performance on Hill 1081.

Barrow's leadership of Able Company did not cease with its capture of the objective. In some ways, the weather and terrain proved a tougher enemy than the Chinese at that moment. The ground was so broken and rough that it took Barrow a full five hours to evacuate his seriously wounded Marines down the hill. Moreover, the temperature continued to drop and was close to an astounding 25 degrees below zero. Barrow noted that in such conditions "it is easy to say that a man should change his socks; but getting him to do so when the temperature is twenty-five degrees below is another matter. . . . I found it necessary to stay with the individual until he actually took off his boots and changed his socks and put his boots back on. Then I'd get him to walk about to restore circulation." Even so, Barrow's company eventually lost a number of Marines due to severe frostbite. Incredibly, Barrow did not freeze to death. During the long climb up the rocky hill, Barrow's radio operator, Corporal Daniel Fore, had lost his sleeping bag. Barrow later stated that he gave Fore his own bag, "not as a noble gesture," but because he thought

---

[57] Russ, Breakout, 407–8.

[58] Lynn Montross and Capt Nicholas A. Canzona, The Chosin Reservoir Campaign, vol. 3, U.S. Marine Operations in Korea, 1950–1953 (Washington, DC: Marine Corps Historical Branch, 1957), 314–16.

[59] Montross and Canzona, Chosin Reservoir Campaign, 316.

Figure 14. Marines in the Chosin Reservoir campaign

Source: official U.S. Department of Defense photo.

by doing so, he would remain alert to the details required of him in defense of the hill throughout the night, which was what he did. Barrow said that he "spent the remainder of the night moving from man to man and stamping [his] feet to keep the circulation going."[60]

Barrow survived the combat of the Chosin Reservoir in much the same way that his slightly older contemporary Wilson had done during the battle for Fonte Hill on Guam in 1944. Both men had been company commanders, and both had exhibited extraordinary leadership at critical moments in combat. Like Wilson, Barrow intended to carry these les-

---

[60] Russ, *Breakout,* 410.

sons forward. Both men would be selected to lead the Marine Corps in the critical immediate post-Vietnam War years.

During the Korean War, helicopter technology came into its own. In fact, this aircraft technology virtually transformed the Marine Corps in the early Cold War and continues to do so today. The issue, however, was that the technology was so leading-edge at that time that the Marine Corps made the bold decision to embrace helicopters while concurrently working on a doctrine to use them in combat. In November 1948, while assigned to Marine Corps Schools in Quantico, Virginia, Colonels Victor H. Krulak and Edward C. Dyer, a Marine aviator, wrote a pamphlet titled, "Amphibious Operations—Employment of Helicopters (Tentative)."[61] This work added to a series of manuals on amphibious operations and became known as *PHIB-31*. Krulak later noted in a letter to the director of the Marine Corps History and Museums Division that "we had so little to go on, no data; just conviction."[62] Nevertheless, Krulak and Dyer went on to note:

> The ability of the helicopter to rise and descend vertically, to hover, and to move rapidly at varying altitudes all qualify it admirably as a supplement or substitute for the slower, more inflexible craft now employed in the ship-to-shore movement. Furthermore, its ability to circumvent powerful beach defenses, and to land assault forces accurately and in any desired altitude, on tactical localities farther inland, endow helicopter operations with many of the desirable characteristics of the conventional airborne attack while avoiding the undesirable dispersal of forces which often accompanies such operations. The helicopter, furthermore, when transported to the scene of operations in aircraft carriers, makes operations possible at ranges

---

[61] LtCol Kenneth J. Clifford, USMCR, *Progress and Purpose: A Developmental History of the United States Marine Corps, 1900–1970* (Washington, DC: Marine Corps History and Museums Division, 1973), 77.

[62] Clifford, *Progress and Purpose*, 77; and LtGen Victor H. Krulak, to the director of the Marine Corps History and Museums Division, 3 August 1970, Victor H. Krulak biographical file, Historical Reference Branch, Marine Corps History Division, Quantico, VA.

which have not yet been achieved by the existing convention-al carriers.[63]

In a follow-up to the Krulak and Dyer study, Commandant Cates tasked Major General Smith, then serving as Assistant Commandant of the Marine Corps, to form a board to "consider measures which the Marine Corps should take in order to fulfill its obligations in maintaining its position as the agency primarily responsible for the development of landing force tactics, techniques, and equipment" in 1949. Around this same timeframe, the 2d Marine Division experimented with helicopter support for amphibious landings in what was known as Operations Packard II and III. The outcome of Packard was so exciting that the board noted that "the helicopter offered the most promising possibilities of being the quantum advance for which the Marine Corps had been searching." The board believed that the helicopter was so transformative for the Marine Corps that "operating helicopter squadrons should be organized and placed in support of FMF maneuvers."[64] The leadership also believed that using the helicopter as an effective vertical assault platform would negate the prevalent belief that the new atomic era had made amphibious assaults too impractical. Furthermore, at this time the U.S. Navy agreed to convert a few of its old World War II-era *Essex*-class fixed wing aircraft carriers such as USS *Valley Forge* (CV 45) into helicopter carriers called landing platform helicopter (LPH) vessels. Out of this, the former World War II-era aircraft carrier *Valley Forge* became LPH 8.

The Smith board, however, noted some serious obstacles that the Marine Corps needed to overcome before fully embracing rotary-wing aircraft. These issues focused on lift capacity, procurement cost, and training. It was also estimated that it would take between 8 and 12 months to get the personnel and mechanical support trained well enough

---

[63] *Amphibious Operations: Employment of Helicopters (Tentative)*, Navy-Marine Corps Instruction 4544 (Washington, DC: Department of the Navy, 1948), 1, copy in Box 14, Historical Amphibious Files no. 243, Marine Corps History Division Archives, Quantico, VA; and Clifford, *Progress and Purpose*, 77.

[64] LtCol Eugene W. Rawlins, *Marines and Helicopters, 1946–1962*, ed. Maj William J. Sambito (Washington, DC: Marine Corps History and Museums Division, 1976), 30.

Figure 15. USS *Valley Forge* (LPH 8)

Source: official U.S. Navy photo.

to successfully operate a rotary-lift aviation squadron. Furthermore, it was also clear that early helicopters had limited lift capacity and that the Navy-Marine Corps team could not afford the necessary amount of rotary-lift platforms to carry significant numbers of Marines as part of a larger amphibious landing operation. Nonetheless, Colonel Dyer noted that Marine Corps helicopters "should be designed for carrier-based operations," be "capable of carrying a payload of about 3,000 pounds (15 combat-equipped Marines)," and have "sufficient fuel for an operating radius of about 100 miles."[65] While these figures might seem extraordinarily low today, they were pressing the extreme edge of the technology's performance at that time. Nevertheless, starting in 1952, the presence of helicopters in the Marine Corps gradually increased to a ra-

---

[65] Rawlins, *Marines and Helicopters, 1946–1962*, 30–31.

Figure 16. USS *Iwo Jima* (LPH 2)

Source: official U.S. Navy photo.

tio of one rotary-wing aircraft to five fixed-wing aircraft. Unfortunate-ly, Congress limited the Marine Corps to a ceiling for total number of aircraft, both fixed- and rotary-wing. Consequently, the Corps had to make the tough decision to cut fixed-wing aircraft in favor of procur-ing more rotary-wing aircraft, which was an unpopular move among the Service's fixed-wing community. By 1967, at the height of combat op-erations in Vietnam, the ratio between the two types of aircraft grew to almost one-to-one.[66]

When the 1st Marine Provisional Brigade, commanded by Briga-dier General Edward A. Craig, was sent to Korea as an emergency stop-gap force in the summer of 1950, the helicopters of Marine Observation Squadron 6 (VMO-6) went along. As part of the fighting along the Pu-

[66] LtCol William R. Fails, USMC, *Marines and Helicopters, 1962–1973* (Washington, DC: Marine Corps History and Museums Division, 1978), 42.

san Perimeter, VMO-6 took part in a "series of improvised mobile operations," during which the helicopters "more than proved their worth." Craig noted that "Marine helicopters have proven invaluable. . . . They have been used for every conceivable type of mission." In fact, Craig wanted as many helicopters as he could procure. During an inspection tour of the war zone as commanding general of Marine Forces Pacific, Lieutenant General Lemuel C. Shepherd Jr. noted, "There are no superlatives adequate to describe the general reaction to the helicopter. Almost any individual questioned could offer some personal story to emphasize the valuable part played by the five [Sikorsky] HO3S [helicopters] available." In 1950, senior Marine Corps leaders, such as Commandant Cates and the director of the Marine Corps Aviation Division, Brigadier General Clayton C. Jerome, all advocated for immediately increasing helicopter production and delivery to Korea as soon as possible.[67]

By the mid-1950s, the lift issue for helicopter squadrons in the Marine Corps had been suitably resolved to the point at which the Service could consider revamping its World War II-era tables of organization and equipment since it was clear that the FMF of 1955 was far too heavy for much of it to be utilized in a vertical lift mode. Consequently, Major General Robert E. Hogaboom was tasked by the Commandant with forming a board "to conduct a thorough and comprehensive study of the entire FMF, including aviation, with the purpose of making recommendations for the optimum organization composition and equipment of the FMF."[68]

What emerged from the board was the most far-reaching change to the organizational structure of the Marine Corps since the publication of the *Tentative Manual of Landing Operations* in 1934. The Hogaboom study recommended the near-total restructuring of the entire FMF. The board divided up the process in phases to lessen the impact on the Marine Corps budget. Most importantly, the board recommended a lighter combat ech-

---

[67] Rawlins, *Marines and Helicopters, 1946–1962*, 43–44.

[68] Rawlins, *Marines and Helicopters, 1946–1962*, 73. Other notable members of the board included BGen Edward C. Dyer, Col Lewis W. Walt, Col William K. Jones, and Col Keith B. McCutcheon. The board heavily relied on testimony from officers assigned to Marine Corps Schools, Headquarters Marine Corps, and Marine Corps Test Unit Number 1 at the Development Center, Quantico, VA.

Figure 17. Two Marine Corps helicopters from Marine Corps Air Station, Tustin, CA

Source: official U.S. Department of Defense photo.

elon and the removal of some specialized forces into a divisional support unit known as Force Troops. The reorganization also cleared the way for the Marine Corps to embrace the vertical assault concept, which meant the heavy use of helicopters in offensive operations. Moreover, the board noted that it gave "special attention to the subject of Atomic Task Forces. The best solution to this problem is to structure the tactical units of the Fleet Marine Force, both ground and air, to fully exploit atomic fire support . . . available in Naval Task Forces."[69]

Vertical assaults via heliborne operations were featured heavily during the Vietnam War. The experiences of this conflict inspired yet another Cold War leader, the 29th Commandant of the Marine Corps,

---

[69] *Report of the Fleet Marine Force Organization and Composition Board* (Washington, DC: Department of the Navy, 1957).

General Alfred M. Gray Jr. While Gray is considered one of the Service's most influential Commandants of the entire Cold War era, most of his impact took place in the halls of Marine Corps Schools or during North Atlantic Treaty Organization (NATO) exercises with the 4th Marine Amphibious Brigade and the 2d Marine Division, both of which Gray commanded prior to his appointment to Commandant.

Gray started his Marine Corps career as an enlisted man. He briefly attended Lafayette College in Pennsylvania, where he was an outstanding baseball player before leaving school for financial reasons. When the Korean War broke out, he unhesitatingly joined the Marine Corps. His recruiting officer was none other than then Major Louis H. Wilson Jr. (Later, as the Commandant of the Marine Corps, Wilson had the honor of promoting his former enlisted protégé to the rank of brigadier general.) After graduating from boot camp, Gray was assigned to an amphibious reconnaissance platoon. By 1952, Gray's platoon commander, Captain Francis R. Kraince, recommended him for an officer's commission, believing that Gray "possesses in an outstanding way the ability to never deviate from his mission no matter how difficult the attainment of final success is."[70]

By the time of the Vietnam War, Gray had matured as an officer and a leader. During the interim between Korea and Vietnam, Major General Raymond G. Davis helped direct Gray toward the signals intelligence field. Gray played a significant role in the formation of the 1st Composite Radio Company, then located at Kaneohe Bay, Hawaii, and the creation of a signals intelligence military occupational specialty within the Marine Corps. In 1964, he received his first assignment in Vietnam, which was to establish a then–classified signals intelligence presence in the country. With the support of Lieutenant General Krulak, the then–commanding general of Fleet Marine Force, Pacific, Gray went on to conduct a series of highly successful cryptologic operations in Vietnam.[71]

---

[70] Scott Laidig, *Al Gray, Marine: The Early Years, 1950–1967*, vol. 1 (Arlington, VA: Potomac Institute Press, 2012), 9–10, 16–18, 22.
[71] Laidig, *Al Gray, Marine*, 191–218.

By 1966, after a brief foray at Headquarters Marine Corps, Gray was back in the artillery with the 12th Marine Regiment, arrayed along the demilitarized zone (DMZ) that demarked the border between the Democratic Republic of Vietnam (North Vietnam) and the Republic of Vietnam (South Vietnam). This tour of combat duty cemented the leadership qualities that made up the future 29th Commandant. During Gray's time in the Marine Corps' I Corps tactical zone along the DMZ, long-range People's Army of Vietnam (PAVN) artillery fire constantly harassed the Marines, so much so that Gray and the Marine leadership decided to do something about it. In Operation Highrise, U.S. Army and Marine Corps artillery units, along with naval gunfire just offshore, fired more than "1000 rounds a day" into North Vietnam and the DMZ to suppress the PAVN harassing fire that was affecting infantry operations in I Corps.[72]

On 8 May 1967, the PAVN launched a major attack against the Marines at the Con Thien firebase. The fighting in and around all the Marine bases near the DMZ remained heavy throughout the summer of 1967. Gray and his artillery poured in supporting fire to suppress that of the enemy. At the time of the Con Thien attacks, Gray was in command of a composite artillery battalion made up of Marine and Army artillerists at Gio Linh. Taking thousands of rounds of incoming enemy artillery fire, Gray's supporting 105-millimeter howitzer battery had "more than 60 wounded; all the wounded occurred in the firing pits as the Marines were firing their guns."[73] During the fighting around Gio Linh, Gray personally rescued two wounded Marines who had inadvertently wandered into a mine field, an action for which he later received the Silver Star.

Every night, the PAVN shelled the base at Gio Linh, hardly missing the well-known location. The effect of such fire, not to mention the casualties it incurred, was becoming too much for Gray's troops. To relieve the pressure, Gray decided on a bold plan to temporarily move out of Gio Linh, fire his artillery from an alternate position at night, and then move back to the base just before dawn and before the PAVN could react to what had just happened. It was an early version of maneuver warfare

---

[72] Laidig, *Al Gray, Marine*, 300–3.
[73] Laidig, *Al Gray, Marine*, 309–11.

in action, and Gray and his men pulled it off perfectly. The bloody May 1967 fighting around the Con Thien firebase was a lesson in static warfare that Gray never forgot.[74]

In many ways, actions in Vietnam caused the Marine Corps to temporarily forget its core mission of amphibious warfare. A good case in point is two of the largest, most ambitious peacetime amphibious landing exercises that the Navy and Marine Corps ever attempted. Known as Operation Steel Pike I (October–November 1964) and Operation Silver Lance (February–March 1965), these maneuvers were conducted to demonstrate that after years of decline following the successful amphibious landing at Inchon in 1950, and with the looming possibility of large-scale deployment to South Vietnam in 1965, the maritime services were still capable of managing such an operation.[75] While a sizeable portion of the 2d Marine Division had deployed to Beirut, Lebanon, in 1958 during Operation Blue Bat, the landing of Marine forces there was largely administrative in nature and were not considered a true test of the amphibious capability of the Navy and Marine Corps.[76]

The larger of the two amphibious exercises, Operation Steel Pike, saw nearly the entire 2d Marine Division transported to the coast of southern Spain, where the maritime services, including Spanish forces, conducted three days of amphibious landing exercises. The exercise had three objectives. Primarily, it was meant to detail the time it would take for the United States to transport a large Marine expeditionary force (MEF) across vast oceanic distances. It took the amphibious task force approximately 10 days to reach Spain. Secondly, once established ashore, the landing force was to exercise its recently developed short airfield for tactical support concept for use by Navy and Marine Corps attack aircraft. Finally, it was also designed "to test for the effectiveness of command

---

[74] Col Gerald H. Turley, USMC (Ret.), *The Journey of a Warrior: General Alfred M. Gray, 29th Commandant U.S. Marine Corps (1987–1991)*, 2d ed. (Arlington, VA: Potomac Institute Press, 2010), 57–59.

[75] LtCol James B. Soper, "Observations: Steel Pike and Silver Lance," *U.S. Naval Institute Proceedings* 91, no. 11 (November 1965).

[76] Jack Shulimson, *Marines in Lebanon, 1958* (Washington, DC: Marine Corps Historical Branch, 1966).

and control of air and ground units while in simulated combat. . . . under warlike conditions."[77]

While the exercise was marred by the unfortunate deaths of nine Marines due to a midair collision between two helicopters from Marine Medium Helicopter Squadron 262 (HMM-262), several operational lessons emerged for Navy and Marine Corps amphibious planners. First and foremost was that even with more than 60 dedicated Navy amphibious ships, 17 Military Sea Transportation Service (MSTS) vessels, and numerous contracted commercial hulls, the 2d Marine Division still had to leave behind a substantial amount of its heavy equipment and logistics at its home base of Camp Lejeune. Moreover, much of the MSTS shipping was found unsuitable for actual amphibious operations. Finally, it was readily apparent that the "naval gunfire and air support were not sufficient for an operation of this dimension."[78] Nevertheless, the overall operation was the first true test of a major long-distance combat-simulated amphibious landing maneuver since the Korean War.

In late February 1965, "with the Vietnamese crisis as an ominous background," the 1st MEF, based at Camp Pendleton, California, conducted Operation Silver Lance, a slightly smaller amphibious landing exercise. Based out of three widely separated West Coast ports (San Diego, Long Beach, and San Francisco), the operation included a large portion of the 1st Marine Division for an 18-day operation that culminated on 5 March 1965 when the division performed an amphibious landing on the beaches of Camp Pendleton.[79] During the workups to the landing, the Navy and Marines Corps experienced several aviation mishaps, including the loss of an aircraft and its pilot due to being mistakenly shot down by a Convair RIM-2 Terrier naval antiaircraft missile. Interestingly, the editors of the Palm Springs, California, *Desert Sun*, a local news-

---

[77] James S. Santelli, *A Brief History of the 8th Marines* (Washington, DC: Marine Corps History and Museums Division, 1976), 74–75; and Sgt Harvey Hall, "Steelpike I," *Leatherneck*, January 1965, 19–20. In an interesting sidebar, the Marines were surprised to find that the Spanish military, especially its Air Force, was still equipped with German World War II-era equipment, such as the Messerschmidt ME-109 attack fighter, which most Marines had "only seen in news reels." See Fails, *Marines and Helicopters, 1962–1973*, 37.

[78] Santelli, *A Brief History of the 8th Marines*, 76.

[79] *Desert Sun* (Palm Springs, CA), 23 February 1965.

paper that covered Silver Lance, stated that the exercise codenames for the notional participant countries bore "a striking resemblance to South Vietnam, North Vietnam, and Red China."[80] Beginning with the involvement of the Marine Corps in Vietnam in 1965, however, the maritime Services never again attempted to conduct amphibious landing exercises on the same scale as Steel Pike and Silver Lance.

This lack of amphibious landing maneuvers did not mean that the Marine Corps ceased conducting contingency operations altogether. During the spring of 1965, the 6th Marine Expeditionary Unit was deployed to quell real-world unrest in the Dominican Republic and assist with the evacuation of endangered U.S. citizens there. Two U.S. Army 82d Airborne battalion combat teams also flew in to secure the San Isidro Airfield and its surrounding environs. Due to the political situation ashore, the Marines and their Army compatriots, all under the command of Army lieutenant general Bruce Palmer Jr., were limited to using small arms to suppress frequent sniper fire coming from rebel forces that were usually hidden in densely populated areas. Due to the increased violence, the decision was made to bring in reinforcements. On short notice, the 2d Marine Division commander, Major General Ormand R. Simpson, activated the 4th Marine Expeditionary Brigade (4th MEB), but by the time the bulk of the 4th MEB was ashore, much of the rebel resistance was already dying down. While the Marines eventually mollified the local population through direct civil affairs support, the 6th Marines had four men killed in action mainly due to sniper fire. Eventually, an Inter-American Peace Force, which consisted of units from multiple Latin American counties, was formally established on 20 May 1965, which "laid the groundwork for the withdrawal of United States Forces from the Dominican Republic."[81]

As was the case with most post-World War II interventions, the Dominican Republic crisis of 1965 also came with its share of increased political activity from Washington, DC, and other locations in Central and

---

[80] *Desert Sun*, 1 March 1965.
[81] LtGen William K. Jones, USMC (Ret), *A Brief History of the 6th Marines* (Washington, DC: Marine Corps History and Museums Division, 1987), 133–38.

South America, which directly affected how operations were conducted on the ground. Once the U.S. military intervention in the Dominican Republic made the news, "there were violent anti-American demonstrations by Argentine students in Buenos Aires. . . . and other demonstrations were reported from Caracas, Bogota, Lima, and Santiago de Chile."[82] Further, the "breakdown of services in Santo Domingo had created many problems for the Dominican people," and U.S. personnel struggled to restore a semblance of order.[83]

The intervention forces clearly needed to provide emergency medical, food, and water services for the entire general population. Tons of food, bedding, blankets, and emergency medical supplies were quickly dispatched to the Dominican Republic via Navy and commercially contracted shipping. For example, the U.S. merchant ship *Alcoa Ranger* (1944) delivered "80 tons of dry milk, 189,000 pounds of flour, 500,000 pounds of vegetable oil, and 1.6 million pounds of cornmeal."[84] One of the most significant lessons learned from the intervention was the need to consider support to the civilian population and its effect on the logistical supply chain to include providing for a storage and distribution plan for the tons of material that rapidly piled up along the runway of the San Isidro Airfield. Communications was also an issue. For several days after the United States decided to intervene, the only way to communicate with the U.S. embassy in the capital of Santo Domingo was "through commercial telephone and telegraph exchanges held by rebel forces." Soon after the Marine Corps and Army ready force had landed, communications with the embassy only slightly improved, thanks to the presence of "one of the Embassy officials who was an amateur radio operator."[85] In fact, interagency communications remained problematic throughout the entire intervention, which was another major lesson from the operation.

---

[82] "U.S. Landings Widely Assailed," in *Dominican Crisis 1965*, ed. Richard W. Mansbach (New York: Facts on File, 1971), 43.

[83] Maj Jack K. Ringler and Henry I. Shaw Jr., *U.S. Marine Corps Operations in the Dominican Republic, April–June 1965* (Washington, DC: Marine Corps History and Museums Division, 1970), 35.

[84] Ringler and Shaw, *U.S. Marine Corps Operations in the Dominican Republic*, 35.

[85] Ringler and Shaw, *U.S. Marine Corps Operations in the Dominican Republic*, 50–51.

Finally, unlike their Army counterparts, the 4th MEB was not as adequately supported by dedicated civil affairs assets. While the Army units had the 42d Civil Affairs Company working for them, the 4th MEB had just a single "legal officer in the G-1 section filling the billet of civil affairs officer."[86] The Army supplied a trained civil affairs officer to the Marines. Most Marine Corps after-action reports on the intervention noted that having a "permanent civil affairs and psychological warfare staff" would have greatly assisted the Marines with anticipating issues they encountered with food and medical support to the population as well as establishing "sufficient distribution points" for such emergency supplies.[87]

Large-scale amphibious warfare had one last minor hurrah before the Vietnam interregnum put an immediate near stop to all consideration of such operations for eight years. In February 1965, due to continued attacks on the U.S. military compounds at Pleiku and Qui Nhon, South Vietnam, by guerrilla units from the National Liberation Front (NLF), more commonly known as the Viet Cong, President Lyndon B. Johnson ordered Navy aircraft to attack targets in North Vietnam.[88] Initially called Flaming Dart II, the bombings eventually blended into the larger and more comprehensive Operation Rolling Thunder air campaign against targets in North Vietnam. Rolling Thunder, however, was delayed until 2 March due to inclement weather. Nevertheless, the overall com-

---

[86] Ringler and Shaw, *U.S. Marine Corps Operations in the Dominican Republic*, 53.

[87] Ringler and Shaw, *U.S. Marine Corps Operations in the Dominican Republic*, 52–53.

[88] "Memorandum from Secretary of Defense McNamara," 30 July 1965, in *Vietnam, June–December 1965, vol. 3, Foreign Relations of the United States, 1964–1968* (Washington, DC: Government Printing Office, 1996), 280. The National Liberation Front (NLF) in South Vietnam emerged as a coalition group against the Ngo Dinh Diem government in 1960. Although Communists in South Vietnam were part of the NLF, the latter organization included a broad spectrum of political, religious, and ethnic groups and ideologies. In this study, references to the Viet Cong are either specifically from a source or to a specific NLF military unit. For more on the creation of the NLF, see Carlyle Thayer, *War by Other Means: National Liberation and Revolution in Vietnam, 1954–60* (Boston: Allen & Unwin, 1989; revised 2021), xxvi–xxx; Robert K. Brigham, *Guerrilla Diplomacy: The NLF's Foreign Relations and the Viet Nam War* (Ithaca: Cornell University Press, 1999), 3–11; Christopher Goscha, *Vietnam: A New History* (New York: Basic Books, 2016), xii, 265–72; Hang T. Nguyen, *Hanoi's War: An International History of the War for Peace in Vietnam* (Chapel Hill: University of North Carolina Press, 2012), 50–53; and Pierre Asselin, "Forgotten Front: The NLF in Hanoi's Diplomatic Struggle, 1965–1967," *Diplomatic History* 45, no. 2 (2021): 330–55, https://doi.org/10.1093/dh/dhaa091.

mander of the U.S. Military Assistance Command, Vietnam (USMACV), U.S. Army general William C. Westmoreland, believed that improved security for the large U.S. airbase near Da Nang was immediately required due to its importance in supporting Rolling Thunder and "the questionable capability of the Vietnamese to protect the base." Fortunately, the 9th MEB, commanded by Brigadier General Frederick J. Karch, a veteran of the Battle of Iwo Jima, was preparing to conduct some exercises in the Philippines near U.S. Naval Base Subic Bay. Westmoreland requested from U.S. Navy admiral Ulysses S. Grant Sharp Jr., the Joint commander of all naval forces in the Pacific, that the 9th MEB be landed and immediately proceed to Da Nang to begin local security operations around the air facility. U.S. Ambassador to South Vietnam Maxwell D. Taylor, a highly regarded retired Army general, quickly gained permission from the South Vietnamese government for the landing to take place.[89]

U.S. leadership in the region felt some urgency in getting credible combat forces near Da Nang but were hesitant over how a sudden buildup of forces might be viewed in the United States. To address both issues, Karch, now located with his staff on Okinawa, was ordered to immediately deploy two of his three battalion landing teams (BLTs). Karch's initial plan was to land BLT 3d Battalion, 9th Marines (BLT 3/9) west of Da Nang on Red Beach 2 on the morning of 8 March 1965 and then use a truck convoy along Route 1 to move the BLT into a defensive position around the airfield. All did not go to plan, however, as poor weather made the surf conditions horrendous that morning, which required a change to the landing's H-hour from 0600 to 0900. It remained tough going all morning long for the gear-laden leathernecks, who were soon met by South Vietnamese schoolgirls who placed leis of flowers around their necks, including Karch. South Vietnamese civilians lined the convoy route and presented colorful banners "in both Vietnamese and English"

---

[89] Jack Shulimson and Maj Charles M. Johnson, *U.S. Marines in Vietnam: The Landing and the Buildup, 1965* (Washington, DC: Marine Corps History and Museums Division, 1978), 6–10, hereafter *The Landing and the Buildup, 1965*.

Figure 18. BGen Frederick J. Karch led through a crowd near Da Nang, Republic of Vietnam

Source: official U.S. Marine Corps photo.

welcoming the Marines. Karch's second BLT, BLT 1st Battalion, 3d Marines (BLT 1/3), was flown in that afternoon from Okinawa.[90]

The landing of the 9th MEB marked the first stage of a massive U.S. military buildup to take place in Vietnam and, in hindsight, proved a watershed moment for the Marine Corps of the early Cold War era and everything that followed. By 1969, the Marine Corps had upward of two entire divisions and aircraft wings ashore—later known as III Marine Amphibious Force (III MAF)—in the I Corps tactical zone of operations in South Vietnam, which contained the provinces of Quang Tri, Thua Thien, Quang Nam, Quang Tin, and Quang Ngai, extending from the DMZ in the north to the central highlands and the beginning of the U.S. Army's II Corps tactical zone in the south. By late 1965, III MAF was anything but amphibious. With the notable exception of Operation Starlite (1965), during which Marine Corps amphibious tractors were used in a waterborne assault mode in combination with other land-based forces, the Marine Corps rarely participated in any amphibious activity for the following eight years of combat operations. Instead, geographical location and timing pushed Marine Corps units to engage in a series of inland conventional ground combat operations against both PAVN regular and NLF irregular forces. Ironically, in the far southern tactical zone of South Vietnam known as IV Corps, which largely covered the watery Mekong Delta region, the U.S. Navy supported U.S. Army forces in numerous small-scale amphibious operations throughout the entire war. The requirement for readily available security forces at Da Nang in 1965 tied the previously amphibiously oriented Marines to the rugged far northern sector of South Vietnam while the primarily land-based Army took on amphibious duties in the Mekong region. Although reversing the two Service's roles would have made much more sense, General Westmoreland was reluctant to replace the Marines in I Corps once they became heavily engaged in region by late 1965.

While amphibious warfare for the Marine Corps clearly languished during the Vietnam War, it is important to note at least one operation-

---

[90] Shulimson and Johnson, *The Landing and the Buildup, 1965*, 9–12, 14.

al innovation that originated with Brute Krulak—the Combined Action Program (CAP). Between 1962 and 1964, Krulak traveled to Vietnam eight times to study the conflict in person and figure out a way to combat the elusive NLF. He studied the works of Mao Zedong to understand the Chinese Communist leader's theories on guerrilla warfare, which carried over into how the PAVN and NLF forces prosecuted the conflict. Similarly, Krulak examined the earlier successes of British irregular warfare expert Sir Robert G. K. Thompson, who fought against a Communist-led insurgency in Malaysia from 1948 to 1960. During the conflict in Malaysia, Thompson argued that the government retained the support of the people, especially those in the "country's rice-growing heartland," by initiating rural development projects, providing humanitarian assistance, and improving the quality of life in rural districts.[91]

In sum, combating any rural Maoist-inspired insurgency started with the government's ability to protect its own population. Krulak believed that the U.S. government and military forces, alongside those of South Vietnam, should do the same thing in Vietnam. When Krulak became commanding general of Fleet Marine Force, Pacific in 1964, he made sure Marines deploying to Vietnam were at least familiar with irregular warfare concepts and willing to apply them once they arrived on the ground. He even ensured that some aspects of irregular warfare were incorporated into the noncombat amphibious exercise, Operation Silver Lance (February–March 1965).[92] The Marine Corps would pioneer Krulak's concepts as part of the CAP during the conflict, and it proved to be a largely successful irregular warfare operational strategy against the PAVN and NLF Communist cadres in the more rural areas of I Corps.

The first test of the CAP concept took place near where the 9th MEB had landed in 1965, in the village of Le My. Slightly more than 12 kilometers northwest of Da Nang, the Marine forces there believed that it was "truly 'enemy country'." They approximated that "this small village alone, consisting of about 700 Vietnamese living in 8 hamlets, sup-

---

[91] MSgt Ronald E. Hays II, USMC (Ret), *Combined Action: U.S. Marines Fighting a Different War, August 1965 to May 1971* (Quantico, VA: Marine Corps History Division, 2019), 2–3.
[92] Hays, *Combined Action*, 2–3.

Figure 19. Map of corps boundaries in South Vietnam

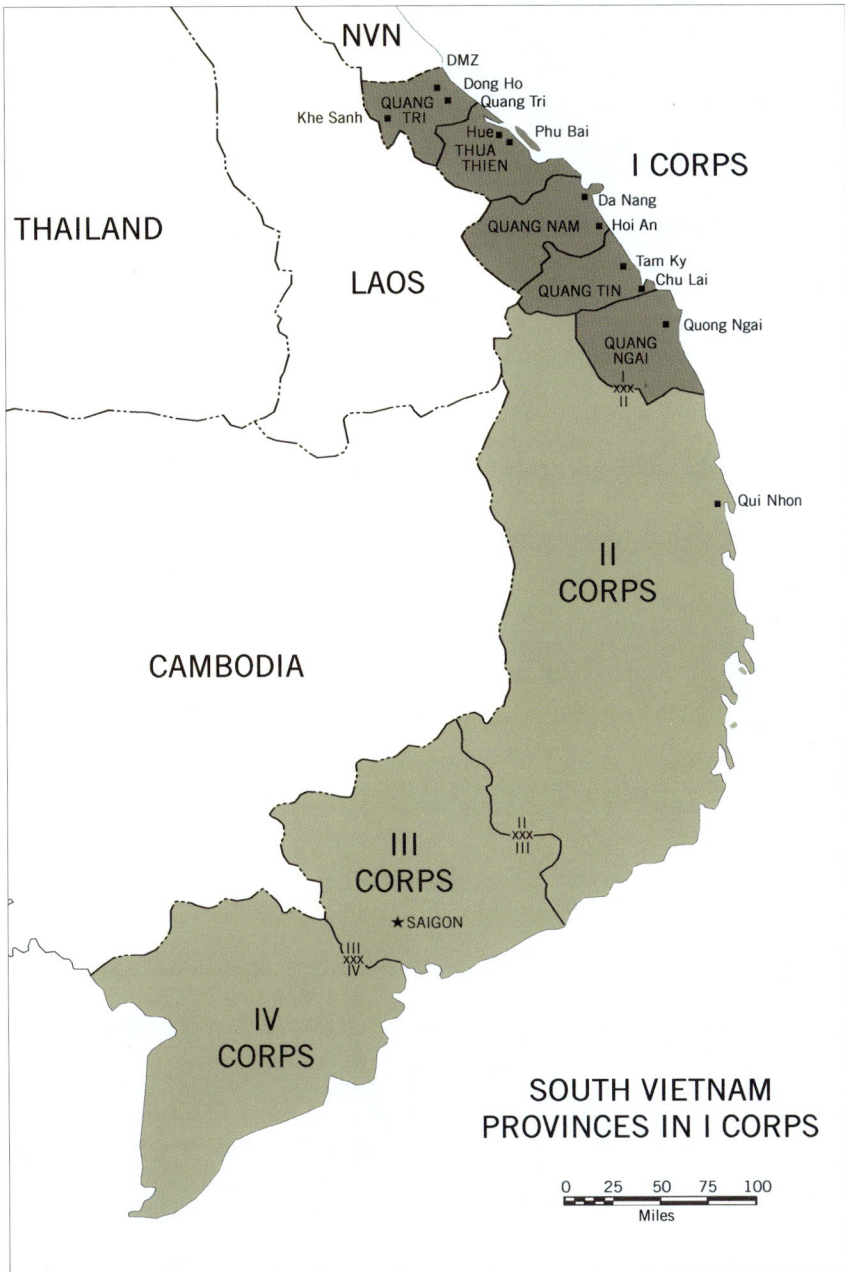

Source: courtesy of Marine Corps University Press.

ported 2 Viet Cong platoons of about 40 men each." BLT 2d Battalion, 3d Marines (BLT 2/3), under the command of Lieutenant Colonel David A. Clement, retook Le My after some significant ground combat. "Instead of moving on as they had in previous operations," one historian noted, "Clement decided to hold the village with its surrounding hamlets after clearing them, a process that had become known as *pacification*." The villagers surprised the U.S. personnel when they "reacted positively to this new development." It was not long before Le My was firmly secure from the effects of the NLF, which ultimately enhanced the overall security of the nearby Da Nang air base.[93]

Marine Corps leaders considered the CAP experiment so successful that they expanded the program throughout the I Corps area. General Westmoreland, however, was not especially impressed with the idea. He believed that the U.S. military effort in Vietnam had neither the time nor patience to properly employ the CAP. Instead, he put more stock in large-unit operations that conducted conventional search-and-destroy missions against suspected enemy troop concentrations. Nevertheless, Krulak remained adamant that the CAP and the security emphasis of the Marine Corps "belonged where the people were, not where they weren't."[94]

Despite this apparent success, multiple issues arose primarily between Marine Corps squads and South Vietnamese local village militia, commonly known as Popular Forces. Ideally, the combined platoons in the CAP would consist of elements from both Popular Forces and Marine Corps forces. In practice, this remained easier said than done. The Popular Forces rarely defended a village successfully entirely on their own. Further, most were armed with antiquated World War II-era U.S. surplus weaponry. To make matters worse, the South Vietnamese government paid the Popular Forces a paltry $19 (USD) a month, which was "less than half of his regular Vietnamese army counterpart." Finally, "graft and corruption" was standard practice for some village leaders

---

[93] Hays, *Combined Action*, 5–7.
[94] Hays, *Combined Action*, 9–10.

Figure 20. 1stLt George S. Dorgatt teaches a class in Vietnam

Source: official U.S. Marine Corps photo.

who "padded the muster rolls of their platoons to extort the salaries of 'ghost' soldiers."[95]

The Marines assigned to the CAP were varied, but all were required to be volunteers and have had at least a few months' combat experience in Vietnam. Often, Marine platoon commanders loathed allowing their best noncommissioned officers to join the program. During the CAP's first two years, Marines in it came predominately from the infantry, but between 1968 and 1970 many did not necessarily have even a background in combat arms. Moreover, the Communist cadres in Vietnam had a significant head start on the Marines in winning the loyalty of the local population. Although the CAP was seen as an overall success in I Corps, USMACV and even the Marines Corps never fully invested in the strategy. CAP forces peaked at 2,220 men, but this "represented only 2.8 percent of the 79,000 Marines in Vietnam." During the program's five-year existence, the "combined units secured more than 800 hamlets in the I Corps area, protecting more than 500,000 Vietnamese civilians."[96] While the Marine Corps could look with pride on the program's achievements, the ultimately unsuccessful outcome of the eight-year effort in Vietnam caused a great many postwar problems for all the armed forces of the United States and American society in general.

While the aforementioned leaders most influenced the Marine Corps early in the Cold War era, many other significant transformational leaders also influenced that period of the Service's history. Senior leaders early in the Cold War, such as Marine aviator General Keith B. Mc-Cutcheon and modern-day FMF architect Lieutenant General Robert E. Hogaboom, achieved distinction during their careers as innovators and operational visionaries. Later, there were others who started their Marine Corps careers in the 1970s at the height of the Cold War. Generals James T. Conway, James F. Amos, James N. Mattis, Joseph F. Dunford Jr., John F. Kelly, Peter Pace, John M. Paxton Jr., and John R. Allen, among many others, all ultimately ended up as twenty-first-century four-star

---

[95] Capt Keith F. Kopets, "The Combined Action Program: Vietnam," *Military Review* 82, no. 4 (July–August 2002): 79–80.
[96] Kopets, "The Combined Action Program," 79–80.

Figure 21. The Marine Corps' six four-star generals on active duty

Source: official U.S. Marine Corps photo.

generals who successfully led Marines and other Joint and combined forces during conflicts in Iraq and Afghanistan. Yet, their connection to the post-Vietnam Cold War-era was unmistakable. Compared to the end of World War II, when the Marine Corps had just a single four-star general who was occasionally shut out of Joint warfare planning, post-Cold War Marine Corps general officers had come a long way. Most can trace their success to the lessons learned during the Vietnam War, the Wilson-Barrow Cold War renaissance in personnel, and operational leadership gained during Cold War-era exercises and various armed interventions. While there can be no doubt that the Marine Corps struggled greatly for relevance throughout the early Cold War era, it was the leadership and skill of key leaders, applied at critical strategic inflection points in history, that enabled the Corps to survive as a Service, evolve its role and mission, and maintain its core ethos that had traditionally inspired its men and women to achieve considerable combat success.

CHAPTER TWO

No More Vietnams

As the 1970s progressed, the Marine Corps faced yet another fight for its institutional survival, but this threat differed from those of the past. This time, Commandants were going to have to rapidly reset the Service from its Vietnam War emphasis on counterinsurgency. Instead, the Marine Corps would be forced to embrace, for the first time in its history, mechanized warfare against a well-armed adversary—the Soviet Union. In 1973, the United States concluded the Vietnam War, which was, at the time, the longest and most divisive foreign expeditionary operation it ever attempted. During this same time frame, some say expressly due to Vietnam, the United States was internally buffeted by its most serious social turmoil since the U.S. Civil War. Writing long after the fall of the Republic of Vietnam (South Vietnam), former President Richard M. Nixon noted:

> Saigon's fall ten years ago was the Soviet Union's greatest victory in one of the key battles of the Third World war. No Soviet soldiers fought in Vietnam, but it was a victory for Moscow nonetheless because its ally and client, North Vietnam, won and South Vietnam and the United States lost. After we failed to prevent Communist conquest in Vietnam, it became accepted dogma that we would fail everywhere. For six years after Vietnam, the new isolationists chanted "No more Vietnams" as the dominos fell one by one: Laos, Cambodia, and Mozambique in

1975; Angola in 1976; Ethiopia in 1977; South Yemen in 1978; Nicaragua in 1979.[1]

The period following the conclusion of active combat operations in South Vietnam was extraordinarily difficult for the Marine Corps. Indeed, the military and social fallout over the unsuccessful conclusion to the Vietnam War affected all the Armed Services. Nevertheless, the Marine Corps experienced some remarkable high points of performance during the eight-year conflict. The Marine advisors who stayed behind after most major U.S. combat units had withdrawn in 1971 were an especially noteworthy group.[2]

When Nixon was inaugurated president of the United States in January 1969, he assumed office with the promise to end substantial U.S. involvement in the Vietnam War by the end of his first term. Nixon's "peace with honor" effort, however, was met with considerable resistance by the Democratic Republic of Vietnam (North Vietnam), which saw an opportunity to drive the United States out of Southeast Asia entirely. To preclude this from happening, the Marine Corps left behind a number of highly qualified officers and senior enlisted personnel as advisors to the South Vietnamese Marine Corps (VNMC) during the last years of the conflict. The U.S. Army did the same for the Army of the Republic of Vietnam (ARVN). The Vietnamese from both sections referred to these U.S. advisors as Co-Vans. Those Americans who survived this experience formed a pool of high-quality leaders for the post-Vietnam-era Army and Marine Corps. No matter how hard they worked or what they did for their ARVN advisees, the Co-Vans could not fully inspire the South Vietnamese to win on the battlefield more than their ideologically driven North Vietnamese Communist opponents.[3]

---

[1] Richard M. Nixon, *No More Vietnams* (New York: Arbor House, 1985), 212. Nixon wrote *No More Vietnams* in an attempt to address popular perceptions of the Vietnam War that had taken root in the minds of most Americans.

[2] Charles D. Melson and Wanda J. Renfrow, *Marine Advisors with the Vietnamese Marine Corps* (Quantico, VA: Marine Corps History Division, 2009), 9–15.

[3] John Grider Miller, *The Co-Vans: U.S. Marine Advisors in Vietnam* (Annapolis, MD: Naval Institute Press, 2000).

Nevertheless, the Co-Vans of the Army and Marine Corps were at the forefront of renewed combat that broke out in the northern tactical zone of operations, known as I Corps, in South Vietnam in April 1972. Known today as the Easter Offensive, because the heaviest fighting occurred during the Easter weekend, this campaign consisted of People's Army of Vietnam (PAVN) units attacking on a broad front across the demilitarized zone (DMZ) and directly targeting Saigon while striking from the Cambodian border with large columns of tanks, combined arms, and more than 300,000 troops. The northernmost tactical zone—or Military Region 1 (MR 1), as it was called by the ARVN—was where the Marine Corps had operated throughout much of the conflict. It was also where most of its VNMC advisors were stationed.[4]

The Easter Offensive and the lack of success of the ARVN should have set off alarm bells in Washington, DC. Yet, the Nixon administration predominately focused on trying to exit Southeast Asia while still emerging with a modicum of success for its years of effort there. The most shocking aspect of the fighting in MR 1 was that despite nearly eight years of sustained U.S.-dominated combat operations, more than 56,000 servicemembers killed, and a robust and dedicated advisory effort, the majority of the ARVN was largely still not willing or even capable of fighting well.[5]

During the early desperate days of the Easter Offensive, the ARVN struggled to stem the PAVN onslaught pouring across the DMZ. In one instance, the North Vietnamese forces quickly overran the large ARVN fire support base at Camp Carroll without the camp's defenders putting up much resistance. Camp Carroll's former commanding officer was later found making broadcasts for the North Vietnamese, advising the ARVN defenders in MR 1 to surrender as soon as possible—and many did. Some South Vietnamese units, such as the rangers, airborne, and marines, however, put up a stalwart defense for as long as they could. The Easter Offensive incursion seemed to have a threefold objective. First, the North

---

[4] Charles D. Melson and Curtis G. Arnold, *The U.S. Marines in Vietnam: The War that Would Not End, 1971–1973* (Washington, DC: Marine Corps History and Museums Division, 1991), 19–34, hereafter *The War that Would Not End, 1971–1973*.

[5] Melson and Arnold, *The War that Would Not End, 1971–1973*, 33–88.

Vietnamese wished to demonstrate that even after nearly eight years of U.S. involvement in the war and suffering from heavy bombing raids, the PAVN was still an entity to be reckoned with. Second, they wished to show that the United States had failed to bring the ARVN up to a level at which it could adequately defend South Vietnam, even against a limited incursion of conventional forces, despite the Vietnamization effort. Finally, the North Vietnamese hoped to overrun the provincial capital at Quang Tri and possibly start the beginning of a South Vietnamese governmental collapse in Saigon.[6]

The commanding general of the U.S. effort in Vietnam in 1972, U.S. Army general Creighton W. Abrams Jr., was shocked by the success of the PAVN, which now looked and acted like a conventional armed force. In Saigon, Abrams held emergency sessions with his senior staff that included the deputy U.S. ambassador to South Vietnam and former Marine Charles S. Whitehouse. During a meeting on 7 April 1972, Abrams described the offensive as a "full court press." The North Vietnamese made an impression on senior U.S. leadership with this assault that Hanoi was trying to win the war. Abrams noted that intelligence gained from captured North Vietnamese soldiers indicated that they were prepared to win at all costs in 1972. Moreover, he saw that all their troops had been given completely refreshed military equipment. Abrams observed that the PAVN soldiers possessed "new everything," including rations and "floating vests" for them to use while crossing rivers. "They are down there," Abrams stated, "and I'll tell you, they're first class equipment, every last man." In an ominous observation, Abrams revealed that North Vietnamese soldier diaries found on the bodies of those killed in action indicated that "they know that 90 percent of them will die" in the offensive.[7] Having recognized that the PAVN seemed highly motivated, Abrams believed that stopping the offensive would take at least a 30-day sustained effort from everyone. He was now concerned that his South Vietnam-

---

[6] Melson and Arnold, *The War that Would Not End, 1971–1973,* 55–88; and Dale Andrade, *America's Last Vietnam Battle: Halting Hanoi's 1972 Easter Offensive* (Lawrence: University Press of Kansas, 2001), 25–26.

[7] Lewis Sorley, ed., *Vietnam Chronicles: The Abrams Tapes, 1968–1972* (Lubbock: Texas Tech University Press, 2004), 811–13.

ese allies did not possess the same fighting spirit or dedication as their North Vietnamese opponents.

In response, Abrams requested that the United States double the number of Boeing B-52 Stratofortress bombing raids and other fixed-wing sorties against the North Vietnamese, although doing so would require far more B-52s than were already on hand. He especially desired them to strike North Vietnamese troop concentrations as soon as possible. Abrams knew that before the White House might approve his proposed bomber offensive, the ARVN would have to hold on as best it could. This fact put those Marine advisors with the VNMC in MR 1 in a tight predicament. They needed to get the VNMC to help stall the PAVN juggernaut, even while much of the ARVN was collapsing around them, so that Abrams could eventually marshal enough air support to crush the offensive and possibly reverse this setback.[8]

One Marine Corps advisor, Captain John W. Ripley, found himself in the middle of the action at the Dong Ha bridge on Easter morning, 2 April 1972. By late morning, armored elements of the PAVN's elite *308th Division* approached the Cua Viet River crossing after having breached the DMZ and attacked down Route 1, the coastal highway in the south. The ARVN's 3d Division was supposed to defend much of the DMZ, but these troops had proven ineffectual in stopping the PAVN advance early in the Easter Offensive.[9] The 3d Division faced significant issues when the Easter Offensive began. First, instead of holding a solid defensive line, the division was spread across a series of loosely supported outposts and defensive strongpoints along the entire DMZ. Second, the 3d Division had allegedly been assigned deserters from other units "because there were

---

[8] Lewis Sorley, *Thunderbolt: From the Battle of the Bulge to Vietnam and Beyond* (New York: Simon and Schuster, 1992), 318–19.

[9] Andrade, *America's Last Vietnam Battle*, 29–36; and Col Gerry H. Turley, USMCR (Ret), *The Easter Offensive: The Last American Advisors, Vietnam 1972* (Novato, CA: Presidio Press, 1985), 53–58.

Figure 22. Capt John W. Ripley

Source: official U.S. Marine Corps photo.

fewer places to desert to" along the DMZ.[10] Despite this latter issue, the division also had been supplemented by excellent ARVN airborne and ranger units as well as two VNMC brigades.

Prior to his second deployment to Vietnam, Ripley had served as a force reconnaissance Marine and been an exchange officer with the British Royal Marine commandos, receiving training in demolitions among other special warfare skills. In Vietnam, he had served as an infantry commander of Lima Company, 3d Battalion, 3d Marine Regiment, along the DMZ in 1967. Now an advisor to the VNMC brigades colocated with the ARVN's 3d Division on 2 April 1972, Ripley heard a column of PAVN armor headed directly toward the Cua Viet and Dong Ha bridge at approximately 1015. His real problem at that moment was that the bridges were intact, due to ARVN lieutenant general Hoang Xuan Lam's order that they re-

---

[10] John Grider Miller, *The Bridge at Dong Ha* (Annapolis, MD: Naval Institute Press, 1989), 8–9. The U.S. Marine Corps advisors in Vietnam were an extraordinary lot, and many achieved the rank of general officer later in their careers—the most notable being Gen Walter Boomer, who had numerous close calls with the People's Army of Vietnam (PAVN) during the Easter Offensive. The South Vietnamese 3d Marine Battalion had been in extensive combat during the months leading up to the Easter Offensive, incurring a nearly 40 percent casualty rate. It had only been recently reconstituted. The PAVN's *308th Division* had had been acting as a guard force around the North Vietnamese capital of Hanoi, avoiding combat in the process, for nearly eight years. Nevertheless, the *308th Division* was clearly one of the PAVN's most prized units. In 1954, the Viet Minh leadership gave the unit the honor of accepting the historic surrender of the French at Dien Bien Phu. Deliberately, the North Vietnamese codenamed their offensive after Nguyen Hue, a legendary eighteenth-century Vietnamese hero who led his forces from the Central Highlands of South Vietnam toward Hanoi, where he ultimately defeated a large force of Chinese invaders.

main standing to further the possibility of a counterattack, which would allow the enemy armor to cross easily. Lam was convinced that his ARVN forces could ultimately contain the PAVN offensive. Later events demonstrated that he was sadly mistaken.[11]

Having confirmed the approaching NVA column, Ripley was convinced that the bridge at Dong Ha eventually needed to be destroyed, and he jumped on the radio to the advisory headquarters to request instructions. Earlier that morning at 0915, Lieutenant Colonel Gerry H. Turley had received a "completely unexpected" order from the 3d Division's senior U.S. military advisor, U.S. Army colonel Donald J. Metcalf, to take charge as senior advisor at the 3d Division's forward tactical operations center (TOC) at the Ai Tu combat base directly south of Dong Ha. The U.S. Army major on duty there needed to be relieved for exhaustion, which made Turley the de facto senior advisor. Metcalf colocated with the 3d Division's commanding general, Brigadier General Vu Van Giai, at its new command post at Quang Tri City at 1900 the previous night.[12]

Turley's main problem at Ai Tu was getting the truly ad hoc TOC up and running amid the chaos breaking out around him. With ARVN units melting away under the weight of the PAVN assault, other leaders, including Metcalf, fell back to new positions, leaving Turley and his small staff as the only command post capable of "controlling all U.S. supporting arms." Moreover, once General Giai and his senior staff displaced to the new division TOC in Quang Tri City, they were temporarily out of communication, which forced Turley and his little band of advisors

---

[11] Miller, *The Bridge at Dong Ha*, 13, 51–52; Norman J. Fulkerson, *An American Knight: The Life of Colonel John W. Ripley, USMC* (Spring Grove, PA: American Society for the Defense of Tradition, Family, and Property, 2009), 67–71; and Turley, *The Easter Offensive*, 173–76.

[12] Turley, *The Easter Offensive*, 139. Turley's situation as a relatively recent newcomer to the regional advisory command found his assignment to be truly difficult due to receiving responsibility for the location and whereabouts of all forward U.S. Army and Marine Corps advisors in the region. For example, Marine Corps majors James Joy and Walter Boomer and captain Raymond Smith—all of whom later became general officers—were forced to escape and evade North Vietnamese forces who had overrun their respective South Vietnamese defended positions. Boomer's situation was especially difficult in that his Vietnamese Marine radio operator became separated from him in the chaos. It would be several days before Turley could gain a full accountability of the entire advisory in MR 1. There is a discrepancy in the actual name of the senior army advisor in Ai Tu. See the more definitive history, Melson and Arnold, *The War that Would Not End, 1971–1973*, 42–43.

to "operate around the clock recommending B-52 Arc Light strikes, directing tactical air support, and adjusting Vietnamese artillery and [U.S.] naval gunfire support. All the fire support coordination in Quang Tri Province for the next few days was carried out by 30 men in one bunker north of the Thach Han."[13]

Turley had received orders from both the U.S. Army advisory command and the ARVN that the bridge at Dong Ha was to remain intact. Seeing the tanks, however, Ripley knew that the bridge needed to come down if the ARVN had any hope of stopping the PAVN offensive there. He just could not get permission to do it. Nevertheless, Turley received consistent reports that the PAVN armored column was getting ready to cross the Cua Viet in force. He also knew that Ripley and his VNMC had few assets to stop or even slow a vigorous armored assault. Along with just two 106-millimeter recoilless rifles, the best antitank weapon available were short-range, shoulder-fired M-72 light antiarmor weapon (LAW) rockets. U.S. destroyer USS *Buchanan* (DDG 14), on station directly offshore in the Gulf of Tonkin, was already firing five-inch rounds into the PAVN column and targeting the nearby railroad bridge. The *Buchanan* was later credited with the destruction of at least four PAVN tanks—certainly a first for the Navy during the war. Other on-station destroyers soon joined in the bombardment. Nevertheless, PAVN infantry had already taken the railroad bridge just west of the Dong Ha span, even raising their national colors from one of the girders. This bridge was partially damaged, preventing armored vehicles from safely crossing there. Still, the VNMC had their hands full fighting off small clumps of PAVN infiltrators.[14]

Eventually, Turley commanded Ripley to figure out a way to "blow up the Dong Ha bridge," with the tone of his voice clearly indicating the "desperate nature of his order." In response, Ripley set off for the bridge aboard some tanks of the 20th ARVN Tank Battalion, along with their advisor, U.S. Army major James E. Smock. On their approach, the battalion came across an ominous sign. All along Route 1, they saw the de-

---

[13] Melson and Arnold, *The War that Would Not End, 1971–1973*, 48.

[14] Turley, *The Easter Offensive, Vietnam, 1972*, 144, 149–52.

tritus of war—"dead animals, broken-down trucks, push carts, helmets, weapons and full ammunition pouches." Refugees also streamed southward in advance of the PAVN vanguard. Mixed in among them were ARVN soldiers fleeing the battlefield. When the unit arrived on their recently received M48 Patton main battle tanks, Ripley and Smock found that a VNMC sergeant had partially damaged the lead PAVN tank with a LAW rocket, bringing the column to a complete halt just before the north side bridge abutment. This action gave them time to prepare the bridge for destruction. Ripley fortuitously located about 500 pounds of explosives that some ARVN engineers had prepositioned in anticipation of needing to blow up the bridge. His real problem was how he could emplace the explosives with the enemy in force on the far shore and, more importantly, how he would set the explosion off.[15]

To get underneath the bridge, Ripley had to clear an antisapper fence topped with razor-sharp wire. Once he did, Smock passed the boxes of explosives to Ripley, who placed them between the steel girders on the bridge, all while under fire. The enemy could not solely concentrate their fire on Ripley alone, as they were more concerned with engaging the VNMC dug in along the south bank of the river. Although not the sole target of the NVA fire, Ripley had to swing his body hand over hand along the steel bridge beams to complete the work, an action Turley later described as "a feat very similar to a high-wire circus act."[16]

After about two hours of intense work, Ripley and Smock had the bridge ready for destruction. Finding some time fuse cord and manual actuators, Ripley improvised a way to clamp the blasting caps onto the end of the fuse using his teeth to crimp the highly explosive blasting cap onto the time fuse cord. Once he returned to the riverbank, he noticed a box of electrical blasting caps nearby and, returning to the bridge span under fire, emplaced this second set of detonators as an insurance policy. Lacking a hellbox actuator for this second set of electrical blasting caps, Ripley tried to set off the charges using a battery from a nearby wrecked jeep, which did not work. Soon after, the time fuses worked, causing the

---

[15] Turley, *The Easter Offensive, Vietnam, 1972*, 150–53.
[16] Turley, *The Easter Offensive, Vietnam, 1972*, 179.

bridge to erupt in a huge explosion. The span splashed into the Cua Viet, shutting the PAVN's direct "gateway to Quang Tri."[17]

While PAVN forces coming from the west eventually took Quang Tri City after a successful crossing of the still intact bridge at Cam-Lo, they could not fully replicate the success of the Tet Offensive in 1968, when PAVN and National Liberation Front (NLF) forces captured the more politically valuable city of Hue farther to the south. Eventually, ARVN and VNMC counterattacks backed by U.S. airpower stabilized the situation and brought some order out of the chaos. In the aftermath, South Vietnamese president Nguyen Van Thieu ordered the well-regarded ARVN lieutenant general Ngo Quang Truong to replace Lam as the commanding general of all South Vietnamese forces in MR 1. Truong noted that the fall of Quang Tri Province "was a serious psychological blow that deeply affected the morale of troops and the local population."[18] Moreover, by 2 May 1972, "throngs of dispirited troops roamed about, haggard, unruly, and craving food" in Hue City. "Driven by their basest instincts into mischief and even crime," Truong later wrote, "their presence added to the atmosphere of terror and chaos that reigned throughout the city."[19]

The ARVN under Truong launched its counteroffensive, Lam Son 72, in late June 1972, which partially restored the situation and retook Quang Tri City. They were never able to fully eject all PAVN forces from MR 1. The fact that PAVN forces retained control over significant portions of South Vietnamese territory following the conclusion of the Easter Offensive, Lam Son 72, and even after the signing of the Paris Peace Accords in January 1973 later proved to be a deadly mistake for the republic's existence. While General Abrams had been less than pleased with

---

[17] Turley, *The Easter Offensive, Vietnam, 1972*, 185–86. For his part during the Easter Offensive, Capt John Ripley was awarded the Navy Cross for his actions at the Dong Ha Bridge. Even today, some people argue that his act of courage was so extraordinary that he should have received the Medal of Honor.

[18] LtGen Ngo Quang Truong, *The Easter Offensive of 1972* (Washington, DC: U.S. Army Center of Military History, 1979), 50. Ngo argued that criticism of the performance of the ARVN 3d Division has been "unduly harsh and unjustified." He believed that the infusion of former deserters and other undesirable elements within its ranks was "never proven." Ngo believed that no ARVN division could have done any better against the North Vietnamese Easter Offensive. See Truong, *The Easter Offensive of 1972*, 165–68.

[19] Ngo, *The Easter Offensive of 1972*, 50.

the events that took place in MR 1 in 1972, he later commented that the 1st ARVN Division and the VNMC Division had "performed well above any performance by that division, or those brigades, in the time I've been here. . . . So the Marine Division and the 1st Division are in good shape. They're stout, morale's good, and the leadership's good." He was even more pleased with his U.S. advisors who were caught in the thick of the offensive, noting that, for them, "it was an all-out thing. In fact, what you've got mostly is the routine performance of miracles."[20]

To pressure North Vietnam into a negotiated settlement, President Nixon ultimately ordered two major bombing offensives in 1972—Operation Linebacker I (9 May–23 October) and the even more devastating Operation Linebacker II (18–29 December). Linebacker I was a Joint air campaign designed to interdict, damage, and isolate North Vietnam from supporting its now overextended and exposed forces in the south. On 16 August 1972, U.S. fighter-bombers flew more than 370 sorties against targets in North Vietnam. Soon after, Nixon ordered the mining of Haiphong Harbor, the most significant port of entry in North Vietnam for military and other aid from the Soviet Union and the People's Republic of China (PRC). To ensure greater isolation, he also had the railroad connections between North Vietnam and the PRC targeted for destruction. At the end of August 1972, Nixon announced the withdrawal of an additional 12,000 U.S. combat forces from Vietnam while staying at his estate in San Clemente, California, known as his "Western White House." This proclamation "was the 10th in a series of withdrawal announcements since June 1969, when the authorized ceiling was 549,000 troops." Ominously for the war effort, the largest contributor of combat forces among the United States' allies also announced it withdrawal. On 7 September 1972, the Republic of South Korea declared that it would begin removing its remaining 37,000 highly effective troops from South Vietnam by the end of the year.[21]

---

[20] Sorley, *Vietnam Chronicles*, 831–33.

[21] Edward W. Knappman, ed., *South Vietnam: U.S.-Communist Confrontation in Southeast Asia, 1972–1973*, vol. 7 (New York: Facts on File, 1973), 158, 162–64.

Before aviation assets from the USS *Coral Sea* (CVA 43) could emplace mines in Haiphong Harbor on 27 August 1972, the Navy needed to suppress the North Vietnamese shore defenses. In one of the few major surface actions of the war, Operation Lion's Den saw U.S. warships boldly steam directly into Haiphong to conduct a night surface attack prior to the minelaying operation. The commanding admiral of the U.S. Seventh Fleet, Vice Admiral James L. Holloway III, was aboard the lead heavy cruiser USS *Newport News* (CA 148) as an observer as the ship entered the harbor. Escorting *Newport News* was the guided missile cruiser USS *Providence* (CLG 6) and two destroyers, USS *Rowan* (DD 782) and USS *Robison* (DDG 12). Once they entered the harbor, the ships fired more than 700 high explosive rounds at enemy shore installations and radar sites. As the naval assault force prepared to depart the harbor, North Vietnamese forces counterattacked in Soviet-built high-speed patrol boats, focusing on the largest warship, *Newport News*. Since the attack was coming from dead ahead, *Newport News* could not fire its large-caliber eight-inch main battery due to the presence of a forward mounted electronics tower. At this point, Holloway called in air support from *Coral Sea*, stating on a special emergency net that "Jehovah himself aboard the USS *Newport News*" was making the request. Meanwhile, *Newport News* turned broadside to the attackers, enabling its main battery to finally engage them. *Newport News* was later credited with the destruction of at least one of the attacking craft while the *Rowan* possibly damaged a second.[22]

The ordeal of U.S. involvement in the Vietnam War did not end with the Easter Offensive. President Thieu demanded that, for political reasons, General Truong's forces "recapture Quang Tri—regardless of the

---

[22] John Darrell Sherwood, *Nixon's Trident: Naval Power in Southeast Asia, 1968–1972* (Washington, DC: Naval History and Heritage Command, 2009), 68–69. Approximately a month later, USS *Newport News* had one of its 8-inch gun barrels explode during gunfire missions south of the DMZ, killing 20 U.S. sailors and wounding 36 others. It was the largest loss of life for any U.S. Navy gunfire squadron during the entire Vietnam War. A later investigation concluded that a "faulty detonation fuse" caused the accident. As a junior officer during World War II, Adm Holloway had been a gunnery officer aboard the destroyer USS *Bennion* (DD 662). During the Battle of Leyte Gulf, *Bennion* torpedoed a Japanese battleship and sank a destroyer. See Sherwood, *Nixon's Trident*, 69.

cost."[23] The North Vietnamese gave similar orders to their now dug-in defenders, setting the stage for a bloody battle of attrition. Although the ARVN finally recaptured Quang Tri by 16 September 1972, it was a Pyrrhic victory because the best forces, including the marines, airborne units, and rangers, had been pinned down and ultimately decimated. These forces also represented the ARVN's last mobile reserve and its best troops. Nevertheless, the final success of Truong's forces at Quang Tri forced the North Vietnamese to reevaluate their negotiating strategy at the Paris Peace Talks in 1972–73. The United States was adamant that any settlement must include the repatriation of hundreds of prisoners of war (POWs) held by North Vietnam and guarantees for the survival of the Thieu regime. North Vietnam wished to retain its Easter Offensive gains, have the United States pull its forces entirely out of South Vietnam—something Nixon was already committed to doing—and bring an end to the bombing campaigns in the north.[24]

In anticipation of a negotiated settlement by the end of October 1972 and, not coincidentally, just in time for the U.S. presidential election on 7 November, the North Vietnamese, at the behest of the Soviet Union, rushed reinforcements and supplies to their beleaguered forces in South Vietnam. As the United States and North Vietnam negotiated the Paris Peace Accords, the language in it ultimately allowed two contending governments and two opposing armies to remain in South Vietnamese territory. The North Vietnamese believed that such an agreement, coupled with the departure of all U.S. forces and military support, would guarantee them "total victory in the end."[25] This prediction was later proven correct.

Amazingly, no one on the U.S. side thought to gain Theiu's acceptance for any sort of negotiated settlement. It would take Nixon and Secretary of State Henry A. Kissinger three additional months of salesmanship to get both Vietnamese governments to sign on to the Paris

[23] Stephen P. Randolph, *Powerful and Brutal Weapons: Nixon, Kissinger, and the Easter Offensive* (Cambridge, MA: Harvard University Press, 2007), 324.
[24] Randolph, *Powerful and Brutal Weapons*, 324.
[25] Randolph, *Powerful and Brutal Weapons*, 325.

Peace Accords. To coerce the North Vietnamese into signing from a position of strength, Nixon ordered the short but highly violent Linebacker II bombing campaign. While it did not change the gist of the original agreement, it brought down a considerable amount of international scorn on the Nixon administration and the United States. In the end, while not ideal for anyone, all parties finally agreed to the accords in January 1973 and the war—at least for the United States—was officially over.[26]

The map of South Vietnam at this time looked like a patchwork quilt, with North Vietnamese forces in control of portions of the country from the Mekong Delta all the way to the former MR 1 territory extending from the DMZ to Quang Tri City. The bloody ground of Dong Ha and the Cua Viet River was now firmly within their grasp. The PAVN's success in the Easter Offensive gave it a large stretch of the central South Vietnamese coast from Quang Ngai to Tuy Hoa and some far western suburbs of Saigon. While a four-party International Commission of Control and Supervision (ICCS), consisting of the nations of Poland, Hungary, Canada, and Indonesia, was ostensibly in charge of overseeing the ceasefire agreement, most observers believed that they would not agree on "any aspect of enforcement," nor that they would have "a real cease-fire to supervise."[27] Additionally, it was not clear to anyone where the territories held by the ARVN and the PAVN began and ended. This situation guaranteed sustained conflict, at least as far as the South Vietnamese government was concerned.

The Paris Peace Accords required that the United States "dismantle all its military bases" and withdraw all its military personnel, including its advisors to the ARVN.[28] By 27 March 1973, the United States had largely complied with its part of the requirements of the accords. However, South Vietnam, North Vietnam, and the latter's NLF allies in the

---

[26] Dale Andrade, *Trial by Fire: The 1972 Easter Offensive, America's Last Vietnam Battle* (New York: Hippocrene Books, 1995), 514–15, 523–26; and Nixon, *No More Vietnams* (New York: Arbor House, 1985), 152–58.

[27] Randolph, *Powerful and Brutal Weapons*, 333–35.

[28] Maj George R. Dunham and Col David A. Quinlan, *U.S. Marines in Vietnam: The Bitter End, 1973–1975* (Washington DC: History and Museums Division, 1990), 2, hereafter *The Bitter End, 1973–1975*.

south were not interested in implementing anything that might give any advantage to their opposition following the withdrawal of U.S. forces. To add to the confusion, NLF forces had seized two neutral Canadian observers of the accords. Although the observers were released on 15 July 1973, Canada gave notice that it was withdrawing as a member of the ICCS by the end of that month. Most significantly, the military arrangement that allowed PAVN forces to remain in place inside South Vietnam had been a major mistake. Thieu argued that "American estimates placing North Vietnamese military strength in the South at 140,000 were 'imaginary and misleading' and suggested that the actual figure was not less than 300,000."[29]

A cascade of political events inside the United States continued to impact the future of South Vietnam as well as the support given to the Lon Nol regime in Cambodia. In June 1973, Senator Frank F. Church III (D–ID) and Senator Clifford P. Case Jr. (R–NJ) cosponsored and passed a bipartisan measure that "reflected the growing disenchantment of Congress with even minimal American involvement in Asian combat." The Case-Church Amendment ended military aid to Cambodia, dooming the pro-U.S. administration of Lon Nol. It also "prohibited the United States, after 15 August 1973, from engaging in any combat activity in Indochina, especially air operations."[30] In fact, throughout most of 1973, Congress passed a series of measures designed to constrain the warfighting prerogative of the executive branch.

The War Powers Resolution, the most famous anti-intervention measure, passed over Nixon's veto in 1973. The resolution greatly limited any U.S. president from unilaterally deploying military forces in any future overseas expeditionary combat operations. Essentially, it called for presidents to seek the "collective judgment" of both the legislative and executive branches "before U.S. troops are sent into combat, especially for long termed engagements."[31] Section 4(a)(1) of the resolution

---

[29] Dunham and Quinlan, *The Bitter End*, 2, 4–5.

[30] Dunham and Quinlan, *The Bitter End*, 5.

[31] Joint Resolution of 7 November 1973, Public Law 93-148, 87 STAT 555, Concerning the War Powers of Congress and the President; and Louis Fisher, *Presidential War Power*, 3d ed. (Lawrence: University Press of Kansas, 2013), 144–45.

stated that once troops were committed to an overseas combat operation, the president had just 48 hours to report their action to the speaker of the House of Representatives and the president pro tempore of the Senate. Within 60 days, Congress would then decide whether the operation should continue or not.[32]

A good example of how a later president assiduously worked to avoid reporting compliance with the War Powers Resolution came during the Beirut crisis of the early 1980s. In this moment, President Ronald W. Reagan tried to avoid the resolution's requirements by stating that the U.S. Marines he sent into Lebanon were not, and never had been, involved as combatants, and therefore the War Powers Resolution did not apply to such forces. Consequently, Congress passed a statute that declared the date of when the War Powers Resolution clock had started in Lebanon—29 August 1983. The statute did not limit the Reagan administration to three months, as the original War Powers Resolution intended. Instead, they had 18 months to resolve matters in Lebanon. Still, Reagan ended up ordering nearly all Marine Corps forces totally withdrawn from Lebanon as early as February 1984.[33]

While no U.S. president since 1973 has ever publicly admitted that the War Powers Resolution limited their inherent constitutional authority to use the U.S. military in overseas expeditionary operations, all have gone to great lengths to avoid a showdown in the Supreme Court about this issue. For example, when President Reagan ordered U.S. airstrikes against key targets inside Libya in 1986, he informed Congress about the attack only after the strike force had taken off and that such a preemptive action was merely an act of "self-defense" designed to further "deter acts of terrorism by Libya."[34] In another case, before beginning Operation Desert Storm in 1991, President George H. W. Bush made sure he gained congressional approval—although by a narrow margin in the Senate—to use force to eject the Iraqi Army from its occupation of Ku-

---

[32] *Background Information on the Use of U.S. Armed Forces in Foreign Countries, 1975 Revision* (Washington, DC: Congressional Research Service, 1975), 73–75.
[33] Fisher, *Presidential War Power*, 149–50.
[34] Fisher, *Presidential War Power*, 163.

wait. In fact, "military operations in Grenada and Panama were conducted as though the 60-day limit was enforceable—if not legally, then politically."[35]

At least since Reagan's presidency, every administration has usually resorted to employing presidential war powers without consulting Congress before or even sometimes after. Nonetheless, every U.S. military activity that portends to last longer than three months have been debated in Congress with the president largely receiving tacit, if not outright, legislative approval of their unilateral decision to involve U.S. forces in overseas expeditionary operations. The War Powers Resolution, while never truly tested or properly implemented, served to do one major thing in the decades after the Vietnam War, as far as the office of the president and the Department of Defense was concerned: it caused presidents and secretaries of defense to rethink the efficacy of overseas contingency operations.[36] The slogan "no more Vietnams" had significant implications for the Marine Corps in the post-Vietnam War years.

From the late 1970s and through the end of the Cold War in 1991, U.S. leaders and war planners did not believe that Congress or the American people would tolerate another conflict that in any way approximated the Vietnam quagmire. As a result, the U.S. military of the later Cold War era focused almost exclusively on the threat that the Soviet Union posed to Europe and the North Atlantic Treaty Organization (NATO) alliance. This transition pushed the relatively light Marine Corps to reinvent a new role and mission for itself for possible participation in the large-scale European theater, which featured a formidable and heavily mechanized potential antagonist who operated quite differently from its most recent opponents in Vietnam. Moreover, a potential war against the Soviet Union required the Marine Corps to embrace the concept of maneuver warfare on a scale that the Service was neither equipped nor trained for. War planners of the 1980s saw the Marines contributing to the de-

---

[35] Fisher, *Presidential War Power*, 150.

[36] Robert D. Clark, Andrew M. Egeland Jr., and David B. Sanford, *The War Powers Resolution: Balance of War Powers in the Eighties* (Washington DC: National War College University Press, 1985), 1–4, 7, 19–37.

fense of NATO largely in far northern Europe or even above the Arctic Circle near the North Cape of Norway. Not since the Korean War had the Marines considered fighting in such a frozen environment.

Several other political events in the United States that took place in late 1973 and 1974 served to further undermine the future security of South Vietnam. Just days prior to President Nixon's veto of the War Powers Resolution, which Congress easily overrode, the Watergate Scandal broke wide open when the president fired Special Prosecutor Archibald Cox Jr., Attorney General Elliot L. Richardson, and Deputy Attorney General William D. Ruckelshaus. To make matters worse, in the face of looming corruption charges, Vice President Spiro T. Agnew was forced to resign from office. In August 1974, in a dramatic historical moment, Nixon resigned as president of the United States as Congress prepared articles of impeachment against him. His newly appointed vice president, Gerald R. Ford Jr., automatically became the new U.S. president without ever having stood for a formal election for either position.

During this period of political turmoil, there was a growing concern in the Marine Corps as to the readiness of III Marine Amphibious Force (III MAF), headquartered on Okinawa. Since the other two MAFs were stationed inside the United States, they had the advantage of possessing Marines who served within these organizations for a longer term. III MAF inherited the old Vietnam-era tradition of having its personnel serve for a single year before being rotated back to the United States for reassignment or discharge, since most billets on Okinawa did not authorize dependents to accompany Marines so assigned. Until the advent of the Unit Deployment Program (UDP) in October 1977, this policy made the stability and training of III MAF combat units highly problematic.

To simplify manning requirements, there were two full infantry regiments, the 4th and 9th Marines, stationed in Camps Hansen and Schwab in the central and northern parts of Okinawa, respectively, while two battalions of the 12th Marine Artillery Regiment were located further south at Camp Foster. Supporting Marine helicopter squadrons were located at Marine Corps Air Station Futenma, just north of the principal Okinawan city of Naha. The 1st Marine Aircraft Wing was stationed at

Iwakuni, Japan. The 3d Marine Infantry Regiment and the 1st Battalion, 12th Marines, were assigned to the 1st Marine Brigade at Kaneohe Bay, Hawaii. In 1975, Japan hosted a major exposition fair called Expo '75 Okinawa on the island. Consequently, they Japanese severely restricted the size and use of III MAF's training areas and, especially, anything that might impact the run-up to the festivities. This decision forced III MAF forces to seek training opportunities at Camp Fuji, Japan, and in the Philippines and South Korea. While the situation was not ideal, it caused most III MAF units to become well acquainted with off-island deployments.[37]

To minimize the consequences of a one-year deployment on readiness and training, III MAF instituted a battalion landing team (BLT) readiness program. Essentially, a battalion would start from scratch, get the unit staffed in the first 60 days on the island, and then progress to a 100-percent manning level when the battalion would proceed to the predeployment, deployment, and post-deployment training programs. III MAF staggered the battalions throughout the four above phases so that the commanding general did not have more than one or possibly two battalions in the initial buildup phase. While full readiness for all units was rarely achieved, the input program reduced the more pernicious effects of a single-year deployment cycle.[38]

The III MAF commanding general further streamlined matters by splitting the Okinawa-based infantry regiments into two large amphibious ready groups (ARGs). ARG Alpha was centered on the 4th Marines, and ARG Bravo was created around the 9th Marines. ARG Alpha contained the 31st Marine Amphibious Unit (31st MAU), including robust helicopter support. These Marines could conduct "sea-based, over-the-horizon, forced, surface, and vertical amphibious entry anywhere in the Western Pacific area."[39] ARG Bravo largely had the same capabilities but was envisioned for a surface amphibious role only.

---

[37] Dunham and Quinlan, *The Bitter End*, 34–36; and "Expo 1975 Okinawa," Bureau International des Expositions, accessed 17 August 2022.
[38] Dunham and Quinlan, *The Bitter End*, 34.
[39] Dunham and Quinlan, *The Bitter End*, 34.

An event taking place in the Middle East in 1973 eventually had a profound effect on all the U.S. Armed Services. In a rare intelligence failure on the part of the Israel Defense Forces (IDF) as well as the United States, the IDF was surprised by Egyptian and Syrian military forces launching a massive and coordinated attack against IDF outposts in the Sinai Desert and the Golan Heights on 6 October 1973, in a conflict known as the Yom Kippur War. In the previous Six Day War (5–10 June 1967), the IDF had captured and subsequently occupied part of the Sinai all the way to the eastern bank of the Suez Canal and a large portion of Syria's strategic Golan Heights. The Golan region was especially critical terrain, representing the only highly defensible ground between Israel and its longtime enemy Syria.[40]

The first day of fighting—the Jewish holy day of Yom Kippur—saw Egyptian and Syrian forces make significant advances against the IDF in both regions. The effectiveness of the Soviet military hardware used by both Egypt and Syria shocked the IDF and many observers. For example, the IDF had long used U.S. attack aircraft, such as the Douglas A-4 Skyhawk, to blunt the numeric armor and infantry advantage that their Arab opponents held. In just a single afternoon of the first day, Soviet-made SA-6 surface-to-air missiles (SAMs) and ZSU-23 antiaircraft batteries shot down 30 Skyhawks and 10 U.S.-provided McDonnel Douglas F-4 Phantom II fighter-bombers over the Golan Heights. According to writers of the London-based *Sunday Times*, the ZSU-23s "chewed up the Skyhawks if their pilots dropped to deck level in an effort to beat the SAMs." One Dutch military observer then working for the United Nations noted that the "Israelis were losing three out of every five aircraft they sent over" the Golan battlefield.[41] The IDF clearly could not sustain such losses, and before the first day of fighting had ended, its senior leadership temporarily suspended air strikes over the Golan, which shifted the advantage to the heavy Syrian armored columns there. Israeli defense minister Moshe Dayan, having visited the Golan battlefield in those crit-

---

[40] *The Yom Kippur War: By the Insight Team of the London* Sunday Times (New York: Doubleday, 1974).

[41] *The Yom Kippur War*, 161.

ical first days, believed that "the fate of the Third Temple [the modern state of Israel] was at stake."[42]

Similarly, in the Sinai, the Egyptians operated under a seemingly impenetrable umbrella of highly sophisticated Soviet-made SAMs. Because the more critical Golan region was drawing the attention of the Israeli Air Force, the IDF's armored forces took on the defense of the Sinai. Here, the Egyptians surprised the Israelis with new Soviet-produced antiarmor technology, the 9M14 Malyutka wire-guided antitank missile, known in the West as the AT-3 Sagger.[43] The Saggers seemed to pop up everywhere on the battlefield and exacted a heavy toll on IDF armor. One Israeli soldier wounded in the fighting remarked that "ordinarily, an infantry platoon would be equipped with one big anti-tank weapon and two smaller ones. But every third Egyptian seemed to be carrying one, and they were the most sophisticated things I've ever seen." Furthermore, the U.S.-built M60 Patton main battle tank, considered the best tank in the entire U.S. arsenal at that time, had serious design flaws, such as the hydraulic fluid having "a flashpoint so low that it would explode into flame from the heat of a missile impact which otherwise would not have killed the tank's occupants." The M60's fuel and ammunition was also stored too close together, which meant that even a missile that did not penetrate its armor could cause "the fuel and ammunition [to] . . . explode, blasting inward, not outward—again killing the tank's crew."[44] Although the Israelis also used large numbers of British-made Centurion main battle tanks, the Soviet T-62 main battle tanks that the Syrian and Egyptian forces operated now seemed superior to both the U.S.- and British-made weaponry that the Israelis employed.

While the IDF eventually reversed these initial setbacks, even crossing into Syria and Egypt in their counterattacks, the impact of the new and highly effective Soviet military technology sent shock waves throughout the national security community in the United States. While the U.S.

[42] Chaim Herzog, *The War of Atonement: October, 1973* (Boston, MA: Little, Brown, 1975), 97.
[43] Chris McNab, *Sagger Anti-Tank Missile vs. M60 Main Battle Tank: Yom Kippur War 1973* (London: Bloomsbury, 2018), 5.
[44] *The Yom Kippur War*, 170–71.

military had largely focused on winning an honorable peace in Vietnam in the early 1970s, the Soviet Union had clearly stolen a technological march in the interim. For the Marine Corps, the severe attrition of the A-4 Skyhawk to increasingly sophisticated Soviet-made SAMs caused the Service to speed up its acquisition of the revolutionary British-built Hawker Siddeley AV-8A Harrier attack aircraft, which possessed the ability to take off and land vertically as well as operate close to the leading edge of the battlefield, where SA-6s were thought to be less prevalent. At the time, the Marine Corps was still using large numbers of Skyhawks in its own attack squadrons, but the Harriers were also thought to be—falsely in some cases—more survivable than the older Skyhawks. For a Service that relied heavily on its organic attack aircraft to support the Marine air-ground task force concept, especially in the early phases of an amphibious landing, an effective SA-6 SAM umbrella and the growing proliferation of shoulder-fired SA-7 Grail heat-seeking missiles caused great concern for Marine Corps amphibious warfare planners throughout the 1970s and into the 1980s. In reality, "the most shocking aspect" of the Yom Kippur War was "the curtailment of Israeli air supremacy."[45] Lacking adequate artillery to make up for the missing Israeli air strikes resulting from the Arab forces' SAM umbrella, the Israelis were forced to employ ad hoc and only partially effective counter measures.

The lethality of wire-guided, portable antitank missiles also garnered significant attention. In sum, the Sagger was a game-changer. Although the U.S. Armed Services developed an equivalent to the Sagger—the BGM-71 tube-launched, optically tracked, wire-guided (TOW) missile—in 1972, it was not widely available for the operating forces until later that decade. Further, while the Marine Corps maintained only three tank battalions in its active inventory, one battalion per division, they all consisted of the flawed M60 Patton. During the Yom Kippur War, which lasted less than three weeks in duration, both sides suffered substantial attrition rates. "To give but a single example of the magnitude of the

---

[45] Maj Bruce A. Brant, "Battlefield Air Interdiction in the 1973 Middle East War and Its Significance to NATO Air Operations" (master's thesis, U.S. Army Command and General Staff College, 1986), 120.

numbers involved, the total count of tanks lost on both sides must have approached 3000 (75 percent of which were Arab)." As military strategist Martin van Creveld noted, this figure "represents fully one-third of all the tanks that the members of NATO—France included—can muster."[46] It now appeared to all observers that Soviet military technology had not only equaled NATO's but in some cases had surpassed it.

At that time, military planners worldwide needed to reassess the impact and increased lethality of this new post-Vietnam-era weaponry. The new armaments seemed to be driven by "high rates of consumption of ammunition and fuel." Further, some military officials thought that the consumption rates were "unprecedented, and very serious in their implications," signifying that the "loss planning factors of all General Staffs must be drastically changed."[47] Finally, the Arab Coalition Forces in 1973 had proven that surprise was still possible on the modern-day battlefield. The U.S. Joint Chiefs of Staff were going to have to think hard about their previous assumption that in the event of a general war breaking out—most likely in the central German plains—they would have time to reinforce their standing forward deployed forces before Warsaw Pact armored columns operating under the secure protection of its SAM umbrella potentially overran NATO forces. Now, they were not so sure.

Meanwhile, the Marine Corps mission, as well as that of the other U.S. Services, in South Vietnam was rapidly drawing to a close. By 1974, only a few Marine officers were assigned to Vietnam in a liaison role. They were mainly there to assist the VNMC in their ongoing fight with PAVN forces still covertly operating inside South Vietnam. For example, Lieutenant Colonel George E. Strickland stated that he spent most of his tour that year "living with the Vietnamese Marine Corps in a bunker. While in Saigon, I maintained a billet at the Brinks Hotel, three blocks from my office."[48] Strickland frequently stared across the Thach Han River near Quang Tri City at PAVN forces maneuvering on the oth-

---

[46] Martin van Creveld, *Military Lessons of the Yom Kippur War: Historical Perspectives* (Beverly Hills, CA: Sage, 1975), 47.

[47] Trevor N. Dupuy et al., *The Middle East War of October 1973 in Historical Perspective* (Falls Church, VA: NOVA Publications, 1976), 178.

[48] Dunham and Quinlan, *The Bitter End*, 38.

er side, but the VNMC appreciated his presence on what was considered the front lines.

By 1974, however, the situation inside Cambodia was going from bad to worse, with Khmer Rouge Communist forces rapidly gaining the upper hand. The Marine Corps embassy detachment in Phnom Penh, Cambodia, was kept on a constant state of alert due to frequent talk of imminent evacuation. Consequently, from mid-1974 on, all the focus of U.S. forces in the region was about contingency planning and potential evacuation from both South Vietnam and Cambodia. The situation in Cambodia seemed especially urgent. The planned evacuation there became known as Operation Eagle Pull.

The contingency planning situation for Vietnam, which involved guidelines for a general evacuation of U.S. nationals from South Vietnam, was in poor condition. Amazingly, when Major General Herman Poggemeyer Jr., the then-commanding general of III MAF, sent his operations officer, Colonel John M. Johnson Jr., to Da Nang, South Vietnam, to "obtain the voluminous detailed information necessary to conduct an evacuation," his efforts, as well as those of the U.S. Navy's 7th Fleet commander, Vice Admiral George P. Steele II, were actually "thwarted by Ambassador Graham A. Martin." Martin's office controlled the movement of all U.S. military forces inside South Vietnam, and he believed that any potential plans to evacuate Americans from Vietnam "would create the very fall of Vietnam that he was sent there to prevent."[49] Subsequently, in-country plans remained sketchy well into early 1975. Fortunately for the Marine Corps, Johnson returned to Okinawa with a renewed sense of urgency, and both III MAF and 7th Fleet planners began to make serious contingency preparations during the winter of 1974-75.

That winter and continuing into the spring of 1975, the PAVN kicked off another offensive against the ARVN, this time in the Central Highlands and other locations in MR 1. The battles of Phuoc Long and Ban Me Thuot were especially devastating, as PAVN forces overwhelmed the ARVN in a series of coordinated combined arms attacks that jump-

---

[49] Dunham and Quinlan, *The Bitter End*, 52.

started the eventual destruction of South Vietnamese military resistance. In fact, the loss of Phuoc Long represented the "first province since 1954 to fall intact into the hands of Hanoi's forces, and its capture exposed the [ARVN's] gravest weakness, the absence of an uncommitted reserve."[50] The North Vietnamese spring offensive rapidly picked up steam as ARVN resistance in MR 1 and other locations began to collapse hastily. The totality of the South Vietnamese defeat in the spring of 1975 was both sudden and unexpected.

Due to the rapid collapse in MR 1 and the loss of Da Nang to PAVN forces, the United States had to modify its plans for the orderly evacuation of South Vietnam. Major General Kenneth J. Houghton, the commanding general of the 3d Marine Division, placed Colonel Dan C. Alexander in charge of the 9th Marine Amphibious Brigade (9th MAB). Alexander decided to place individual rifle companies aboard four Navy amphibious support ships to "serve under the operational command of the ship's commanding officer" to act as "internal security for the ship and to assist in evacuee processing and administration."[51] The operation to evacuate U.S. personnel and selected foreign nationals in Saigon received the code name Frequent Wind.

Journalist Malcolm W. Browne painted a vivid picture of the ARVN collapse, noting that the South Vietnamese government had lost radio contact with its second largest city, Da Nang, by 30 March 1975. An observer stationed on a ship offshore claimed that the only thing they could see was "wall to wall people along the shore." Browne noted that World Airways had attempted to land a Boeing 727 airliner at the Da Nang airport to evacuate some civilians but was instead "met by about 300 South Vietnamese soldiers, armed with rifles and grenades, who forced their way aboard the big jet. Other people, seeking to flee the beleaguered city," Browne reported, "lay in front of and under the plane to keep it from leaving." He went on to write that other soldiers "mobbed" the airplane as "it taxied off the runway to the ramp. At least one soldier," he recorded, "was seen firing his pistol at the cockpit." Believing that the

---

[50] Dunham and Quinlan, *The Bitter End*, 69.
[51] Dunham and Quinlan, *The Bitter End*, 87–88.

soldiers made departing from the runway dangerous, the pilots took off from a taxi way instead, despite the crowds of people around the plane. After making an emergency landing at Saigon, the pilots asserted that "they knew of no deaths resulting from this." After flight experts in Saigon talked with "passengers and stowaways on the plane," they believed that "between 20 and 30 persons had probably been killed—some run over on take-off, some dropping away from the wheel wells and the cargo hold." When the plane arrived, the specialists inspected it and claimed that they found "the body of one soldier" in the landing gear.[52]

Although no other flights landed at Da Nang, some local ships evacuated people from the piers and beaches around the city. According to Browne, at least 9,000 evacuees had been taken aboard the 333-foot U.S. Military Sealift Command (MSC) ship SS *Pioneer Contender* (1963). Numerous other MSC vessels had been purposely kept empty and in standby status in the event that a large-scale evacuation became necessary. These ships became an essential part of the refugee sealift effort. However, sanitation and internal security conditions aboard these overcrowded vessels was horrendous. Soon after taking as many refugees as could possibly be crowded onboard, the commanding officers of many MSC vessels began losing control of their ships and requested security help from the 3d Marine Division.[53]

Earlier, General Houghton had task-organized all his forces into on-call security detachments for the fast-moving evacuation mission. On 4 April 1975, a rifle platoon from Bravo Company, 1st Battalion, 4th Marines, under the command of recent Naval Academy graduate, Second Lieutenant Robert E. Lee Jr., was sent onboard the vessel in the dark of night to restore order. Lee initially reported that 7,000 refugees were "on board, everything under control." As dawn broke, however, Lee came to realize that nearly 16,000 refugees were aboard and that little food or water was available to them. Many onboard were still armed ARVN deserters. Just the day prior, other armed ARVN deserters had taken over

---

[52] Malcolm W. Browne, "Radio Link Fades; Saigon Is Still Talking with Observers on Ships Off the City," *New York Times*, 30 March 1975.

[53] Browne, "Radio Link Fades"; and Dunham and Quinlan, *The Bitter End*, 205–14, 217–19.

the USNS *Greenville Victory* (T–AK 237) and tried to force the ship to land them on the mainland. The teeming horde of humanity on the *Pioneer Contender* was a major concern for Lee and his men. On at least one occasion, the refugees rushed Lee's Marines as they delivered food and water, putting them in danger of being crushed to death.[54] As the ship continued southward, Lee and his Marines resorted to firing warning shots on more than a few occasions. In the end, they successfully completed their mission of delivering the refugees without any casualties. The following day, 5 April 1975, Navy vessels and support ships no longer accepted any more Vietnamese refugees onboard.

Throughout April 1975, with Khmer Rouge units and PAVN and NLF forces threatening the governments of Cambodia and South Vietnam, respectfully, the experiences of the Marine Corps and the Navy were crammed with boatlifts, "vertical envelopments and extractions made under fire and in poor weather and lighting conditions." Now in the midst of evacuating U.S. and allied personnel and citizens, five Marine helicopter squadrons and the security forces of two Marine infantry battalions executed Operation Eagle Pull and Operation Frequent Wind. In the former, these units "rescued 276 people from Phnom Penh." In the latter, they saved "nearly 7000 from Saigon." While Operation Eagle Pull took less than three hours to complete, the evacuation of Saigon and South Vietnam in general proved far more difficult. For example, throughout much of the two–day evacuation of Saigon, nearby Tan Son Nhut Air Base was under PAVN artillery and antiaircraft fire that ultimately caused the death of two embassy Marines. Elements of the 9th MAB were airlifted in to secure the Defense Attaché Office (DAO) compound and annex grounds using BLT 2d Battalion, 4th Marines (BLT 2/4), to facilitate the evacuation of civilians who were pre–staged or had made their way there. On the night of 29–30 April 1975, a final rear guard of Marines was lifted off the embassy roof, and all Marines, U.S. civilians, and select South Vietnamese nationals were on their way to safety offshore. During the entire extraction, the Marine Corps "lost only a sin-

---

[54] Dunham and Quinlan, *The Bitter End*, 90–92.

Figure 23. Refugees evacuate from Saigon, April 1975

Source: photo by Hubert Van Es.

gle helicopter due to an accident."[55] Even so, it represented a sad end to U.S. involvement in Southeast Asia for many years.

One last dramatic act marked the end of the U.S. departure from Southeast Asia. This incident, however, did not have anything to do directly with the victorious North Vietnamese forces that had toppled the Thieu regime in Saigon. Instead, it involved Khmer Rouge forces that had also recently overthrown the pro-Western Khmer Republic of Lon Nol. The Nol government had never been popular in rural Cambodia, and his regime had been kept in place primarily due to the presence of U.S. airpower. By March 1975, Nol controlled Phnom Penh but not much else in Cambodia. Around the same time that the PAVN was overrunning Saigon the following month, Nol fled Phnom Penh and eventually flew to the United States, but his remaining loyal military and governmental offi-

---

[55] Allan R. Millett, *Semper Fidelis: The History of the United States Marine Corps* (New York: MacMillan, 1980), 605.

cials were not so fortunate. Anyone who elected to stay were ultimately massacred by murderous Khmer Rouge forces of dictatorial Prime Minister Pol Pot, which unleashed a horrific orgy of violence against the people of Cambodia shortly after seizing power.[56]

In fact, anyone considered an enemy of the new government was marked for execution. It is estimated that 1.5–2.2 million Cambodians were either killed outright or were worked to death, starved, or died of disease in the killing fields of the Khmer Rouge. Prior to his ousting by Vietnamese forces and local insurgents in 1979, Pot and his Khmer Rouge adherents were responsible for the death of nearly 25 percent of Cambodia's entire population. Finding the Pot regime extremely horrific, Vietnam invaded Cambodia in 1979, overturned Pot's government, and forced him to flee with some of his loyal followers into exile in the Western mountains of Cambodia along the border with Thailand—an exile from which he never returned. Just two days before his death from a heart attack in 1998, his own Khmer Rouge friends, facing tremendous international pressure, finally agreed to turn him over to international authorities, but it was too late for justice to be served.[57]

To stand out as a newly established agrarian-based Communist state, Pot did not allow the former government of Prince Norodom Sihanouk, which Nol had deposed in 1970, to come home. Instead, his regime embarked on a series of aggressive border disputes with neighboring Vietnam and Thailand. Additionally, his forces seized and held disputed islands in the Gulf of Thailand under his orders. Now that the United States was in the process of leaving the region, Pot's regime believed that it had little to fear from potential U.S. retaliation. Moreover, "long simmering territorial disputes, control of potential oil deposits, and concern about American-supported insurgents had pushed the new communist government in Phnom Penh to defend its oceanic sovereignty around Cambodia." Furthermore, Pot was convinced that the U.S. Central Intelligence Agency (CIA) might be using internationally flagged vessels

---

[56] Dunham and Quinlan, The Bitter End, 100–24; and David P. Chandler, Brother Number One: A Political Biography of Pol Pot, rev. ed. (Boulder, CO: Westview Press, 1999), 102–12, 123–36.
[57] Chandler, Brother Number One, 155–64, 183–86.

to secretly send weapons to insurgents opposed to his regime. Inherent xenophobia, paranoia, concern with the clandestine activities of the CIA, control of its perceived maritime economic exclusion area, border disputes, and a desire to consolidate its power caused the Khmer Rouge Navy to begin seizing commercial vessels in the Gulf of Thailand as early as 1 May 1975.[58]

On the afternoon of 12 May, two Cambodian swift boats stopped and boarded the U.S. merchant vessel SS *Mayaguez* (1944) in international waters. The *Mayaguez* carried a crew of 40 people and was loaded with "107 commercial, 77 military, and 90 empty containers." The vessel was due to pass the island of Poulo Wai, which the Khmer Rouge forces had only recently seized from Vietnam. Charles T. Miller, the captain of the *Mayaguez*, had not expected any trouble and steamed his aged vessel at a leisurely 12.5 knots as it headed for nearby Thailand. Earlier that month, however, the Khmer Rouge Navy had seized numerous Thai fishing boats, fired on the South Korean freighter *Masan Ho*, captured boats full of desperate refugees from South Vietnam, and, most seriously, boarded a Panamanian freighter that they detained for nearly two days while they interrogated the crew and inspected its cargo. Before the Cambodians boarded the *Mayaguez*, the ship's radio operator transmitted a mayday call that the ship was under attack.[59]

Miller's radio mayday was heard by John Neal of the Delta Exploration Company in Jakarta, Indonesia. He immediately notified the U.S. embassy in Jakarta, which in turn notified the White House, U.S. intelligence agencies, the Pentagon, and the headquarters of the commander in chief of the U.S. Pacific Fleet (CINCPAC) in Honolulu, Hawaii. The CINCPAC staff immediately suggested organizing a search for the *Mayaguez* using Navy and Air Force reconnaissance planes. Fortunately, the container ship was not too difficult to find because it had not moved far from its last known location.[60]

---

[58] Clayton K. S. Chun, *The Last Boarding Party: The USMC and the SS* Mayaguez, *1975* (Oxford, UK: Osprey, 2011), 9–11.
[59] Chun, *The Last Boarding Party*, 11–12.
[60] Chun, *The Last Boarding Party*, 12.

Discounting the agony of Eagle Pull/Frequent Wind, the *Mayaguez* incident in May 1975 was the Ford administration's first real foreign policy crisis. Although Nixon's resignation had created a temporary vacuum in the executive branch, Ford had retained most of Nixon's foreign policy and defense team, keeping James R. Schlesinger as secretary of defense, Donald H. Rumsfeld as the president's chief of staff, and the influential Henry Kissinger as secretary of state. On 12 May 1975, Ford convened an emergency session of his National Security Council (NSC) to consider options to respond to the seizure of the *Mayaguez*.[61]

To Ford and some of his national security team, the *Mayaguez* incident looked like another USS *Pueblo* (AGER 2) incident, in which North Korean patrol boats had seized a small and poorly armed U.S. Navy intelligence ship in international waters and quickly towed it into port in 1968. North Korea proceeded to publicly berate the United States for allegedly violating its territorial waters and held the *Pueblo's* captain and crew prisoner for more than a year. It was a humiliating experience for the United States, President Lyndon B. Johnson, and the Navy. At that time, Ford had been the minority leader in the House of Representatives and was extremely critical of the Johnson administration for its slow response to the seizure of the *Pueblo*. He believed that if Johnson had acted quicker and in a more forceful fashion, the North Koreans would not have been given such a propaganda advantage. At the very least, under those circumstances, the Navy could have used carrier aircraft to destroy the sensitive intelligence vessel before the North Koreans made much use of it.[62]

At the 12 May NSC meeting, the participants assumed that the Cambodian captors would take the *Mayaguez* to Kampong Som despite a lack of hard evidence.[63] Ford and his advisors were working off a faulty analogy and poor intelligence from the start. The only thing that the *Mayaguez* and *Pueblo* incidents had in common with each other was that the

[61] Greene, *The Presidency of Gerald R. Ford*, 143–51.
[62] Richard E. Neustadt and Ernest R. May, *Thinking in Time: The Uses of History for Decision Makers* (New York: Free Press, 1986), 58–66.
[63] John Robert Greene, *The Presidency of Gerald R. Ford* (Lawrence: University Press of Kansas, 1995), 144.

Figure 24. SS *Mayaguez*

Source: photo by Hubert Van Es.

ships floated on the water and that they and their crews were illegally detained by a Communist power vehemently opposed to the interests of the United States. Due to Ford's insistence on rapid action as well as the fact that the administration did not possess accurate or adequate intelligence related to the fast-breaking situation, the U.S. government nearly created a major military disaster. Nevertheless, Ford considered the *Mayaguez* affair "one of the successes of his short presidency."[64]

Throughout the crisis, Ford and his advisors believed that North Korea was closely watching the incident for a sign of further U.S. weakness, fearing that it might launch another invasion of South Korea if it was clear that the United States was no longer capable of decisive ac-

---

[64] Neustadt and May, *Thinking in Time*, 58. This seminal book is a "must read" for all practitioners of national security decision making. The authors correctly pointed out that Ford's faulty analogy between the *Mayaguez* and *Pueblo* incidents was the basis for his demand for a quick military response that nearly ended in a major disaster.

tion in the Western Pacific. Consequently, Ford and his national security group were strongly committed to swift and decisive action against the Khmer Rouge, which the United States did not recognize as the legitimate government in Cambodia. Incredibly, CINCPAC was not initially involved in any rescue planning, as most of it took place at the White House. The evening after Ford met with his NSC, a Navy Lockheed P-3 Orion reconnaissance aircraft "reported that the *Mayaguez* had weighed anchor and appeared to be headed for Kompong Som."[65] The great fear of Ford's administration at this point was that if the *Mayaguez* reached the Cambodian mainland, the chances of getting the crew released quickly would become far less likely.

There were other issues to consider as well. For example, the NSC had to deliberate if the use of force against the Khmer Rouge could potentially lead to the killing of the detained crewmembers. Ford told his deputy assistant for national security affairs, U.S. Air Force lieutenant general Brent Scowcroft, that he was worried about losing Americans if forced to conduct a kinetic operation against the Cambodians. Kissinger, the most overt hawk in all the NSC meetings, saw the crisis being about North Korea. He believed that comparisons with the *Pueblo* incident were incorrect. He saw a greater connection with U.S. inaction when North Korea shot down one of the Navy's Lockheed EC-121 Warning Star aircraft in international airspace on 14 April 1969, killing all 31 crewmembers onboard. Kissinger later recounted in his memoirs that the nonresponse from the United States and the "leisurely process of decision making" evident in the Nixon White House at the time "create[d] a presumption in favor of inaction."[66] Meeting privately with Ford, Kissinger was determined that the process must work better in the *Mayaguez* affair, telling the president, "This is your first crisis. You should establish a reputation for being too tough to tackle." Kissinger even went so far as to publicly

[65] Greene, *The Presidency of Gerald R. Ford*, 145.

[66] Christopher J. Lamb, *The* Mayaguez *Crisis, Mission Command, and Civil-Military Relations* (Washington, DC: Joint History Office, Office of the Chairman of the Joint Chiefs of Staff, 2018), 70–71. This superb book should be required reading for anyone interested in the background to the *Mayaguez* affair. It also provides an excellent behind-the-scenes look at the nuanced world of the Ford administration's National Security Council.

suggest that the administration pondered using B-52 bombers against the Cambodians, although he, Ford, and Scowcroft eventually agreed with Schlesinger that doing so would be an unnecessarily risky move. According to Joint History Office historian Christopher J. Lamb, while it appeared that Schlesinger and the Joint Chiefs of Staff were the most hesitant about conducting any kinetic operation, nearly all the members of the NSC principals committee argued that demonstrating an overt response of some kind in the *Mayaguez* affair was more important than the possible loss of life of among the detainees or even by forces conducting the rescue. Demonstrating rapid resolve was not just important, it was imperative.[67]

The worst scenario for Ford and the entire NSC was the distinct possibility that the *Mayaguez* detainees might be rapidly taken to the Cambodian mainland for a long internment. Kissinger, as noted, was for immediate forceful action. He stated at the NSC meeting that "at some point, the United States must draw the line. This is not our idea of the best such situation. It is not our choice. But we must act upon it now, and act firmly."[68] Kissinger wished to send a clear message to all nations in the Western Pacific that although the United States had departed Southeast Asia suddenly and rather ignominiously, they were not a soft target for Communist-inspired regimes in the region. He wanted the United States to respond forcefully—but not too forcefully—to limit negative political fallout coming from either Congress or the media if there was extensive loss of life.

Despite the initial report that the *Mayaguez* was potentially headed for the port of Kampong Som, Scowcroft informed Ford early the morning of 13 May that the vessel was anchored off Koh Tang Island, just offshore of the Cambodian mainland, but intelligence sources remained unsure of the exact location of all the *Mayaguez* crew. Part of the ship's crew was believed to have been taken to the island. The president believed that the Cambodians had possibly executed some of the crewmembers already, an assessment that was later proven incorrect. The following day, Ford

---

[67] Lamb, *The* Mayaguez *Crisis, Mission Command, and Civil-Military Relations*, 70–73, 91–92, 102.
[68] Greene, *The Presidency of Gerald R. Ford*, 144.

ordered Schlesinger to "quarantine the *Mayaguez* and in effect, by making sure that no Cambodian vessels moved between Tang island and the mainland."[69] U.S. Air Force assets based in Thailand would be required to conduct the quarantine.

Although the Thai government was increasingly sensitive to U.S. forces operating from its territory, especially after the conclusion of Eagle Pull and Frequent Wind, the Air Force still maintained two squadrons of long-range Sikorsky CH-53 Sea Stallion and HH-53 Pave Low helicopters. On the afternoon of 13 May, after debating and ultimately rejecting the use of B-52s, Ford ordered the Joint Chiefs of Staff to deploy an Okinawa-based Marine battalion to the Utapao airbase in Thailand. He commanded that no Cambodian vessels be allowed to leave Koh Tang for the mainland. Finally, he wanted the U.S. military to make immediate plans to retake the *Mayaguez* in a boarding action—something that had not been done since the nineteenth century—and prepare to take Koh Tang Island by vertical assault using the CH-53s and HH-53s. Ford and his advisors were dismayed to learn that a Thai fishing vessel, possibly carrying *Mayaguez* crewmembers, had been allowed to depart Koh Tang unscathed, although Ling-Temco-Vought A-7 Corsair II attack aircraft of the U.S. Air Force's 388th Tactical Fighter Wing sunk the Cambodian patrol boats escorting it with a precision bombing attack. Even so, the Cambodians forced the fishing boat to press on. Captain Miller later recounted that the Air Force did a superb job of dropping bombs close to the boat without hitting it. When this failed, however, "two jets overflew the boat from bow to stern and tear-gassed us."[70] During the attack on the patrol boats, "the pilot of the aircraft was patched through directly

---

[69] Lamb, *The* Mayaguez *Crisis, Mission Command, and Civil-Military Relations*, 19–21, 31–33.
[70] Capt John B. Taylor, "Air Mission Mayaguez," *Airman Magazine*, February 1976, 39–47; and Lamb, *The* Mayaguez *Crisis, Mission Command, and Civil-Military Relations*, 24–25, 31–32.

to the NSC meeting" in Washington, DC.[71] This bizarre communications arrangement emphasized the "confusing, convoluted, and overlapping chain of command."[72]

Unknown to U.S. intelligence, the crew had not disembarked at Kampong Som. Their captors had placed them onboard the fishing trawler heading for the destination of the "nearby island of Rong Som Lem." Just before the Ford administration received the news of the attack on the patrol boats, it learned that 18 members of the Air Force's 56th Marine Security Police Squadron who were en route to Utapao, intended to be part of a larger 125-man landing zone security force on Koh Tang, and 5 crewmembers had been lost aboard one of the HH-53 helicopters called Knife-13 that crashed, most likely due to a mechanical failure, near the Thai-Laotian border. Even before the start of the actual operation, the United States had already suffered 23 casualties. Vice Admiral Steele later commented that "the idea that we could use air police and Air Force helicopters as an assault force appears as ridiculous today as it did then."[73] Such a consideration is perhaps understandable due to the demand by the White House for operational speed, as various commanders of different Services struggled to get the appropriate forces prestaged for a larger assault coming from forces based in Thailand, and doing it all without the Thai government noticing the rapid military buildup.

A seaborne operation may have been more palatable. However, the Southeast Asia-based World War II-era attack aircraft carrier USS *Hancock* (CVA 19), which had acted as an ad hoc helicopter carrier in 1975, and

---

[71] James E. Wise Jr. and Scott Baron, *The 14-Hour War: Valor on Koh Tang and the Recapture of the SS* Mayaguez (Annapolis, MD: Naval Institute Press, 2011), 8. Told largely from the point of view of participants, this book provides the best account of the fast-moving and often confusing series of events that surrounded the entire *Mayaguez* event. Another excellent account is John F. Guilmartin Jr., *A Very Short War: The* Mayaguez *and the Battle of Koh Tang* (College Station: Texas A&M Press, 1995). Guilmartin, who flew 119 combat mission with the U.S. Air Force in Southeast Asia, later went on to a career as a distinguished historian at the Ohio State University. Robert J. Mahoney does the best job incorporating the confusion resident in the Ford administration during this time, and he discusses in detail the numerous intelligence failures that surrounded the entire operation. See Mahoney, *The* Mayaguez *Incident: Testing America's Resolve in the Post-Vietnam Era* (Lubbock: Texas Tech University Press, 2011).
[72] Wise and Baron, *The 14-Hour War*, 8.
[73] Wise and Baron, *The 14-Hour War*, 7-10.

the large-deck amphibious assault ship USS *Okinawa* (LPH 3) were both unavailable for the operation due to temporary mechanical problems. As a result, any sort of near-term operation against Koh Tang could only take place via vertical assault using the long-range Air Force helicopters at Utapao. Without an available landing deck, the long-distance flight to and from Utapao would put the initial assault force at serious risk, especially if the Khmer Rouge put up significant resistance.[74]

Nevertheless, the director of the CIA, William E. Colby, and the deputy secretary of defense, William P. Clements Jr., supported launching an earlier assault because they believed that the initial Koh Tang assault force could defend itself while they waited for further reinforcements to arrive. Interestingly, the hawkish Kissinger strongly advised Ford to wait for a larger force to conduct the operation because "if anything goes wrong, as it often does, I think against 100 [Khmer Rouge troops], you would lose more Americans because you do not have overwhelming power. . . . On balance, I would like to get a more reliable force."[75]

The Marine Corps' operational plan for boarding the *Mayaguez* and possibly assaulting Koh Tang fell to the overall ground force commander, Colonel John M. Johnson of III MAF, and Lieutenant Colonel Randall W. Austin, who commanded BLT 2d Battalion, 9th Marines (BLT 2/9). This unit was selected for the mission because it was "one of six infantry battalions of the 3rd Marine Division home-based on Okinawa."[76] Moreover, BLT 2/9 was designed as one of the division's two air contingency battalions, meaning that it could be called on to deploy on short notice. Typically, air alert battalion Marines could not be away from their quarters for any significant time while the unit was assigned to alert duties. Austin's battalion had been together for several months on Okinawa. Captain Walter Wood of Delta Company, 1st Battalion, 4th Marines, was ordered by his battalion commander, Lieutenant Colonel Charles E. Hester, to organize a 120-Marine detachment for the actual boarding op-

[74] Guilmartin, *A Very Short War*, 47; Chun, *The Last Boarding Party*, 15; and Mahoney, *The Mayaguez Incident*, 69–73.

[75] Mahoney, *The Mayaguez Incident*, 71–72.

[76] LtCol Randall Austin, quoted in Wise and Baron, *The 14-Hour War*, 91.

eration. Major Ray E. Porter, the 1st Battalion's executive officer, would accompany the *Mayaguez* assault element. Six Navy sailors and six civilian volunteers from the MSC cargo ship *Greenville Victory* accompanied the force as well. Because Air Force sensors had indicated that the *Mayaguez* was no longer under its own power, U.S. leadership believed that, if needed, the sailors and civilians would need at least three hours after the Marines retook the ship to get the *Mayaguez* underway on its own power. The major issue concerning the boarding operation was the location of any remaining crew, which no one seemed to be able to positively ascertain, and the possibility of at least 30 Khmer Rouge soldiers onboard. Even more ominously, it was noted that a "helicopter flight would have to cover some 270 miles from Utapao to the *Mayaguez* and would take approximately two hours."[77] Consequently, Wood's Marines could not expect much in the way of reinforcements.

Believing that the forces may have to assault Koh Tang, the senior command elements of the operation envisioned deploying an assault force against the island. U.S. intelligence for that location was even worse. No credible maps of the island existed. To rectify this issue, the senior commanders including key Marine Corps leaders used an Army Beechcraft U-21 Ute airplane to conduct an overflight of the entire 5.5-kilometer-long island at about 1,375 meters just 14 hours before the assault was scheduled to begin. Captain James W. Davis of Golf Company, BLT 2/9, "snapped photos using his Minolta 35-mm camera." Intelligence believed that the *Mayaguez* crew was possibly on the northern part of Koh Tang, with the ship resting at anchor about a kilometer north of the island. Subsequently, the decision was made "to assault the two northern beaches of the island."[78] Two narrow beaches on the east and west sides of the northern tip of the island were ideal landing sites for the Marines, who could land simultaneously on both beaches, drive inland a short distance, and capture the small fishing village. It was the only location that had improved buildings that could possibly hold the *Mayaguez* crewmembers.

---

[77] Wise and Baron, *The Mayaguez Incident*, 12–13.
[78] Wise and Baron, *The Mayaguez Incident*, 14–19; and Dunham and Quinlan, *The Bitter End*, 251.

Meanwhile, approximately 50 Marines of Wood's Delta Company would make up the boarding party that would attack the ship at nearly the same moment elements of BLT 2/9 were assaulting the beaches at dawn on 15 May 1975. Initially, the *Mayaguez* boarding force was going to go in via three Air Force Sikorsky HH-53 Jolly Green Giant helicopters. This changed the evening before due to the timely arrival of the Navy's *Knox*-class frigate USS *Harold E. Holt* (FF 1074). Instead, the leadership revised the plan to embark the Delta Company Marines onto *Harold E. Holt*, which would then pull directly alongside the *Mayaguez* at the arranged time and the leathernecks, selected sailors, and MSC volunteers, including a Cambodian linguist with the Army, would simply jump across the deck from the frigate—like it was the War of 1812 again—and seize key operational points on the merchant ship, including the bridge, while hopefully minimizing any collateral damage to any crew who might still be onboard. *Harold E. Holt* could also take the *Mayaguez* under tow, meaning that the boarding force would not have to worry about refiring the aged merchant ship's boilers. The Air Force aircraft from Thailand and naval aviation assets aboard *Coral Sea* were to provide air cover for the entire operation, including strikes on Kampong Som Harbor and the nearby Ream Naval Base in Cambodia. The Air Force also planned to disperse a riot control agent—commonly known as tear gas—onto the deck of the *Mayaguez* to immobilize any Khmer Rouge soldiers located topside. This decision required the Marines to board the vessel wearing gas masks.[79]

At least eight CH-53 and HH-53 helicopters were needed for the first part of the Koh Tang aspect of the operation. These aircraft, along with the assault echelon of Marines from BLT 2/9, were located at the Utapao air base approximately 305 kilometers north of Koh Tang. While two Navy surface vessels were in the vicinity of the island, *Harold E. Holt* would be busy initially with the *Mayaguez* boarding operation. The other ship, USS *Henry B. Wilson* (DDG 6), with its powerful 5-inch-gun main

---

[79] Wise and Baron, *The* Mayaguez *Incident*, 14–17. USS *Harold E. Holt* was named for the Australian prime minister who disappeared while swimming and was presumed lost in December 1967. A strong ally of the United States who advocated for greater Australian participation in the Vietnam War, Holt served only 22 months as prime minister before his untimely demise.

battery, was still a couple of hours away at launch time. As a result, the 180-man Marine Corps assault force, loaded onto the eight helicopters, was to storm ashore without *any* naval gunfire support. Preliminary bombing of the landing sites was also out of the question due to the faulty belief that the Mayaguez crew was being held nearby. The U.S. leadership planned for two of the helicopters to land on the western beach while the other six landed on the east.[80]

Intelligence provided to the assault force stated that while "initial sharp resistance" was anticipated, it was believed that only "20-40 irregulars" were positioned on the island and "little or no opposition was expected." Some reports placed the number of Khmer Rouge soldiers on the island at less than 20 fighters. The Washington, DC-based Defense Intelligence Agency, however, estimated a force of 150-200 while intelligence sources at CINCPAC in Hawaii believed that enemy strength on Koh Tang was around "90-100 men plus a heavy weapons squad," since U.S. aircraft had sporadically experienced fire from machine guns on the island, from the *Mayaguez*, and from other small boats. Operations experts with the Center for Naval Analyses later wrote that "U.S. forces did not know it at the time, but something went wrong with the dissemination of intelligence estimates of enemy forces."[81]

Three helicopters delivered the Marine Corps boarding party to *Harold E. Holt* around 0600 on 15 May 1975. The eight remaining Utapao-based helicopters—referred to with the callsigns "Knife" or "Jolly Green"—began the initial assault on Koh Tang island soon afterward. Within 30 minutes, the Marines were landed in three different places on the island—"20 in the east zone, 60 in the west zone, and 29, including the Command Group, were 1200 meters south of the west zone."[82] Almost immediately on their approach to Koh Tang, the lead helicopters—Knife

---

[80] Urey Patrick, *The Mayaguez Operation* (Washington, DC: Center for Naval Analyses-Marine Corps Operational Analysis Group, 1977), 5-6. Helicopters designated as "HHs" were generally used for long-distance search-and-rescue missions and were considered a bit more survivable than their CH-53 cousins, which were required to carry dangerously exposed external fuel tanks for such tasks.

[81] Patrick, *The Mayaguez Operation*, 7-8.

[82] Patrick, *The Mayaguez Operation*, 8.

21 and Knife 22 to the east; Knife 23 and Knife 31 to the west—came under heavy fire from Khmer Rouge forces dug in near the fishing village. As Marines began unloading on the west beach from the lead helicopter, Knife 21, it came under "heavy automatic weapons fire," including mortar bursts in the landing zone (LZ).[83] Knife 21 received heavy damage but managed to stagger off to the west for about a mile before setting down in the water, where it eventually rolled over and sank. Knife 22 was forced to abort its approach, taking heavy fire that partially damaged its fuel tanks. This helicopter limped back to the Thai coast, where Air Force search and rescue aircraft picked up the embarked Marines, including the assault company commander Captain James H. Davis, and aircrew and took them back to Utapao.[84]

The units landing on the east beach faced a far worse situation. Intense small arms and machine gun fire hit Knife 23, which caused its pilot to struggle to maintain control, but they managed to set down the helicopter, and Second Lieutenant Michael A. Cicere's 3d Platoon from Golf Company, 2d Battalion, 9th Marines, could tumble down its ramp and find cover in a nearby tree line. Knife 31 had all the operation's forward air controllers onboard. Its copilot, Air Force second lieutenant Richard Vandegeer, was killed as the aircraft approached the LZ. The pilot, Air Force major Howard A. Corson Jr., decided to abort near the beach, but as he pulled away, "the aircraft exploded in a ball of fire fueled by JP-4 jet fuel from a ruptured 650-gallon external fuel tank." Even with the damage, Corson was able to "put the helicopter down in some four feet of water."[85] The casualties for this aircraft were horrific, with eight of the occupants dying.[86]

As the helicopters scrambled to land or aborted their landing in the first few minutes of the operation, an Air Force Lockheed EC-130 Compass Call electronic attack aircraft with the call sign "Cricket" took over the duties of vectoring in the inbound helicopter lifts and possibly coor-

---

[83] Guilmartin, *A Very Short War*, 87; and Dunham and Quinlan, *The Bitter End*, 248.
[84] Guilmartin, *A Very Short War*, 86–88; Wise and Baron, *The Mayaguez Incident*, 25–26; and Dunham and Quinlan, *The Bitter End*, 248–49.
[85] Guilmartin, *A Very Short War*, 88.
[86] Wise and Baron, *The 14-Hour War*, 29.

Figure 25. The wreckage of two U.S. Air Force helicopters at Koh Tang Island

Source: official U.S. Air Force photo.

dinating airstrikes from LTV A-7 Corsairs. Seeing the chaos on east beach and in consultation with Austin's inbound command group, Cricket directed follow-on flights to the less lethal but still dangerous west beaches. Because of the fireball that the destruction of Knife 31 created, two trailing helicopters—Knife 32 and Jolly Green 41—aborted their approach to the east beach before receiving any damage. "With smoke rising from the primary LZ and at least three HH-53s down, the helicopter crews tried to sort out what had happened" and how they might get their Marines ashore in one piece.[87] The lack of an effective and trained airborne command and control center greatly affected the forces' effectiveness throughout the entire operation. Austin understood that the airborne mission commander was to place an airborne command aircraft above

---

[87] Guilmartin, *A Very Short War*, 88–89, 91.

the island that was ostensibly in control of all the aviation assets. This command structure proved to not be the case, or it simply was not working as intended. Austin later noted that "to this day, I do not know who the [airborne mission commander] was, exactly what he knew or didn't know about the landing force, nor what his authority and orders were. I do know that the [Airborne Battlefield Command and Control Center] was a continuing frustration to us on the ground throughout the day."[88]

The command group with Austin became separated from the rest of the assault force because their helicopter was forced to land approximately 1 kilometer south of the west beach LZ. By approximately 0900, Austin only had about 131 Marines, or just "73 percent of the first wave," successfully inserted.[89] With the loss of the forward air control team on Knife 31, communication between separated Marine Corps elements and overhead air assets became difficult. There was concern over the fate of Cicere's 20 Marines and the five Air Force helicopter crewmembers on the east beach. While the Marines had very high frequency radios that could communicate with some of the aircraft orbiting overhead, the aircrews could not locate any meaningful targets nor the position of friendly forces on Koh Tang due to the confusing situation on the ground. For some time on the west beach, First Lieutenant James D. Keith, the executive officer of Golf Company, 2d Battalion, 9th Marines, and in temporary command of around 60 Marines, fought on against still-unknown odds. In the midst of the confusion, one junior Marine officer observed the intrepid Keith "on three radios, trying to communicate with multiple commanders simultaneously."[90]

Meanwhile, a little after 0715, Marines wearing gas masks cross-decked from *Harold E. Holt* to the *Mayaguez* as planned, successfully regained control of the merchant ship without any further mishap, and found no one on board. At around 0830, a Lockheed AC-130 gunship began providing suppressing fire to cover for the partially armored helicopter Jolly Green 13, which was sent in to rescue Cicere's isolated force.

---

[88] Austin, quoted in Wise and Baron, *The 14-Hour War*, 96–97.
[89] Patrick, *The Mayaguez Operation*, 9.
[90] Wise and Baron, *The 14-Hour War*, 8–9.

The attempt went poorly, as Khmer Rouge gunners poured heavy fire into the aircraft, setting a flare case on fire and destroying the cockpit control panel. Jolly Green 13's pilot, Air Force first lieutenant Charles R. Greer Jr., got the damaged helicopter to stagger away from the hot LZ, leaving Cicere and his troops stranded.[91]

The Marines faced a critical situation ashore. The initial wave was in serious trouble, and its support helicopters had either been destroyed or so severely damaged that the operation's overall ground force commander, Colonel John M. Johnson, was concerned that he would not have enough flyable helicopters available for the second wave of critically needed reinforcements. This complication left only the 131 already landed Marines as the insertion force rather than the originally planned 180 Marines.[92] To make matters worse, Johnson did not have direct communication with Austin and could only respond to emerging crises retroactively.

The Khmer Rouge was clearly in force on the island, but the landing units had difficulty determining their exact location in the thick jungle foliage. A small 10-man patrol under Second Lieutenant James V. McDaniel attempted to suppress a Cambodian machine gun and possibly link up with Austin's still separated command group. McDaniel's group did not make it far, as hidden Khmer Rouge soldiers ambushed the Marines. Lance Corporal Ashton N. Loney was killed in the first minutes of the ambush. McDaniel, along with several others, were wounded. In the face of a cascade of enemy hand grenades, McDaniel ordered his Marines to withdraw back to the west beach perimeter. He recalled that as they moved back, he could hear "the enemy laughing in derision all around me."[93] McDaniel was awarded the Navy Cross for his gallantry in action on Koh Tang.

Later that morning, another orbiting helicopter successfully landed additional Marines on the west beach. They made another attempt to link up with Austin's still separated command group, connecting with

---

[91] Guilmartin, *A Very Short War*, 100–1.
[92] Wise and Baron, *The 14-Hour War*, 37.
[93] Wise and Baron, *The 14-Hour War*, 39–40.

it around noon just as the second wave arrived.[94] Unifying the units relieved some enemy pressure on the LZ, allowing the Marines to land three more helicopter loads of reinforcements from E Company, 2d Battalion, 9th Marines, which further suppressed the Cambodians' fire. The respite also gave Austin time to consider whether to sustain the attack or plan an extraction.

Instead, the Joint Chiefs of Staff ordered a cessation of all offensive operations. What was not known to those hotly engaged on Koh Tang was that earlier that morning, Khmer officials released all the *Mayaguez* crew after learning that the Ford administration offered to end the operation as soon as they freed the crew. The international press relayed the message because the United States had no direct connection with the Khmer Rouge government. The Cambodians quickly agreed. By mid-morning, a Thai fishing vessel filled with white flag-waving "Caucasians" was picked up by *Henry B. Wilson*. Photographs of the national security team at the White House show Ford's advisors congratulating an ebullient Ford for gaining the release of the *Mayaguez* crew. Even so, Ford ordered the continuation of airstrikes against the Cambodian mainland. At a press conference afterward, Ford's press secretary, Ronald H. Nessen, tried to downplay the intelligence debacle at Koh Tang. In a true understatement, he remarked, "All's well that ends well."[95]

Colonel Johnson lamented, "Let's see us get off a piece of property we don't even own," when hearing the order to disengage being sent down from above. Likely due to the extreme uncertainty on Koh Tang or perhaps because of Johnson's response, the cessation order was swiftly rescinded and the second wave of reinforcements continued toward Koh Tang.[96] The formerly aggressive mood at the White House had completely changed, however. Now that the ship and crewmembers had been recovered, most of the NSC advocated ending the Koh Tang operation as soon as possible.

---

[94] Wise and Baron, *The 14-Hour War*, 43.
[95] Greene, *The Presidency of Gerald R. Ford*, 149.
[96] Guilmartin, *A Very Short War*, 111.

Figure 26. President Gerald R. Ford and his national security team in a midnight meeting

Source: Gerald R. Ford Presidential Library.

The Marines still engaged in deadly combat on Koh Tang had none of this information. The earlier failure to rescue the east beach group left "only three of the second-wave helicopters flyable," a number that was "insufficient for either reinforcement or extraction" based on the day's occurrences. In sum, Austin had to consider whether to dig in for the night or plan for an extraction before running out of daylight and, most importantly, flyable helicopters. Around this time, Austin finally received some good news. *Henry B. Wilson* and *Harold E. Holt* both received permission to provide direct support fire of the Marines ashore, which started to affect the Cambodians. Next, a trained forward air controller and fighter pilot, Air Force major Robert W. Undorf, arrived over the island in a North American Rockwell OV-10 Bronco observation aircraft, which had the radio capacity to directly communicate with the Marines on the ground. Given the callsign "Nail 68," Undorf restarted the close air support of the A-7s and properly coordinated rescue operations. A second OV-10, "Nail 47," soon joined Undorf. Finally, *Coral Sea*, steaming at nearly 30 knots, closed the range on the island and quickly rigged its massive flight deck for helicopter operations. The Air Force helicopters no longer had to make the long transit to and from Utapao. Undorf also made it a priority to get Cicere and his Marines off the east beach. *Henry B. Wilson* even put its captain's gig in the water as a backup to yet another planned east beach rescue attempt.[97]

One of the problems Undorf had to consider was a half-sunken swift boat just off the east beach that the Cambodians had reboarded to use its machine guns. *Henry B. Wilson* moved in and blasted the boat with 22 5-inch rounds. Meanwhile, mechanics quickly repaired the damaged Air Force helicopters aboard *Coral Sea*, making them ready to rescue the east beach refugees. Two more AC-130 gunships also arrived on the scene. Employing heavy suppressive fire against the Khmer Rouge forces, Jolly Green 11 landed and picked up Cicere and his Marines, including the aircrew of Knife 23. As Jolly Green 11 lifted off, a Cambodian machine gun in the tree line fired on it.[98] At this point, without coordinating with the Marines on the ground, the Air Force dropped a 1,500-pound BLU-82 bomb in the center of Koh Tang. All noted the shock effect of the explosion, as it burst even friendly force eardrums and blew Undorf's Nail 68 "1000 feet higher and jolted Jolly Green 11's automatic flight controls off-line." The Marines "requested that BLU-82 not be used again."[99] Soon afterward, Undorf contacted Austin to ask him if he would rather have Undorf coordinate further offensive airstrikes or help with an extraction. Now aware that the *Mayaguez* hostages had been released, Austin chose the extraction option. The issue was whether they still had enough flyable helicopters to recover all the Marines and downed aircrews before it got too dark.

Once Cicere's isolated east beach force had been safely rescued, Davis and Austin immediately began contracting the west beach perimeter to

---

[97] Guilmartin, *A Very Short War*, 117–121.

[98] Guilmartin, *A Very Short War*, 121–26; and Lamb, *The* Mayaguez *Crisis, Mission Command, and Civil-Military Relations*, 62–63.

[99] Lamb, *The* Mayaguez *Crisis, Mission Command, and Civil-Military Relations*, 62–63.

prepare for the final extraction. With approximately two rifle companies still ashore, Davis and Austin planned for Echo Company, 2d Battalion, 9th Marines, to pass through Golf Company, 2d Battalion, 9th Marines, in preparation for extraction. Having far fewer Marines now ashore, Davis was worried about being overrun. After numerous attempts, Knife 51 finally landed under heavy fire. Davis and his company gunnery sergeant, Gunnery Sergeant Lester A. McNemar, collapsed the rest of the tight perimeter and tried to ensure that all their Marines made it aboard. McNemar and Air Force technical sergeant Wayne L. Fisk went to extraordinary lengths to ensure that no one was left behind. In doing so, Fisk nearly lost his own life. As soon as he hit the helicopter ramp, the aircraft suddenly lifted off with the ramp still down. Fisk would have slid off if not for the timely intervention of Davis and McNemar, who physically held onto him until they leveled off.[100]

In the aftermath of Koh Tang, it became apparent to one and all that the operation was nearly a substantial military disaster. If Ford and his advisors would have waited for half a day for the highly capable *Coral Sea* to get into position, it would have significantly helped matters, likely helping the Marines and Air Force avoid the 18 personnel lost during the operation's combat phase. Coupled with the 23 security police and air crewmembers killed in the crash of Knife 13 in Thailand, the single-day operation was tragically costly, made even more so because the Khmer Rouge would have likely released the *Mayaguez* crew without an assault. Still, the battle courage of the Marines and the Air Force pilots and aircrews was certainly noteworthy. Nevertheless, the Ford administration's lack of intelligence and micromanagement of the operation was a major issue.

To make matters worse, the Marine Corps reported that it had three Marines missing in action from a machine gun team of Echo Company, 2d Battalion, 9th Marines. Team leader Lance Corporal Joseph N. Hargrove, Private First Class Gary L. Hall, and Private Danny G. Marshall had somehow been left behind during the extraction. A Marine Corps inves-

---

[100] Guilmartin, *A Very Short War*, 142–43.

tigation found that it was not likely that the three Marines tried to swim out to sea. Both Hall and Hargrove were unqualified as swimmers, and Marshall only made it to the third class level. No one had seen them injured or wounded during the firefight, and they were not noticed in the final extraction that occurred around 2200 on 15 May. The three men had allegedly been told to relocate to a position on the left side of the perimeter near Davis's location. Yet, the team did not belong to Davis's company. The investigating officer, Major Peter C. Brown, did not believe that the team ever linked up with anyone and, in the last-minute confusion, failed to get on board the last helicopter.[101] The story relating to the unfortunate demise of Lance Corporal Hargrove and his team, and the ultimate location of their remains, has been frustratingly difficult to determine to this day.[102]

Just two months after the Battle of Koh Tang, the Vietnamese retook the island from the Cambodians. The *Mayaguez* was sold for scrap four years later. Soon after the battle ended, the Marine Corps commissioned the Center for Naval Analyses to review the entire operation to mine it for lessons. First and foremost, the study revealed that the short time to plan the operation, coupled with built-in command and control issues as well as extremely poor intelligence on enemy intentions and capabilities on Koh Tang, all combined to make the entire affair a near disaster. Historian and Koh Tang Air Force veteran John F. Guilmartin Jr. was more to the point when he stated that "if ever an infantry unit was set up by circumstances for failure, 2/9 was it. Yet 2/9 did not fail. The same point applies to the two air force helicopter squadrons."[103] Guilmartin

---

[101] Wise and Baron, *The 14-Hour War*, 67–71.

[102] Wise and Baron, *The 14-Hour War*, 69–70, 76–77. The search for Hargrove and his teammates has been made even more frustrating due to the continually shifting accounts told by Em Son, the commander of the Khmer Rouge forces on Koh Tang Island. Historian Ralph Wetterhahn did the deepest dive on the issues surrounding the final moments of the Hargrove machine gun team. See Wetterhahn, *The Last Battle: The Mayaguez Incident and the End of the Vietnam War* (New York: Carroll and Graf, 2001). Baron and Wise noted that as late as 2011, no conclusive DNA results have positively identified the remains of LCpl Hargrove, PFC Hall, or Pvt Marshall. The lack of confirmed information has been exceptionally frustrating for the families of the deceased Marines.

[103] Guilmartin, *A Very Short War*, 42. Guilmartin was a U.S. Air Force major during the Battle for Koh Tang Island and flew the last flyable rescue aircraft the Air Force had on hand.

believed that if not for the innate bravery of nearly everyone involved in the fight, it would have turned out much worse.

Joint History Office historian Christopher J. Lamb provided the best post-operational analysis of the *Mayaguez* incident. He noted that recently declassified NSC messages demonstrated that the White House was fully in charge of decision making throughout the entire operation. Most importantly, they made deliberative decisions without rushing to judgement. They did take risks, such as their failure to wait for the arrival of *Coral Sea* or the possibility that they might injure or kill the detainees whose liberation was, after all, part of the overarching objective by sinking vessels around Koh Tang. Lamb makes it clear that the rescue of the detainees was clearly secondary to a "demonstrative use of force for geostrategic reasons."[104] Lamb believes that the preponderance of archival evidence illustrates that the president "was *not* influenced by domestic political considerations" and was determined to go after the Cambodians regardless of the costs.[105] He further argued that faulty intelligence, which was indeed bad; runaway emotions at the White House during the new administration's first international crisis; or even concern about the welfare of the crew inadequately explains the reason for the Ford administration's decision to order the operation. Rather, Lamb was convinced that "concern over declining U.S. credibility" was "the overriding concern of U.S. decision makers and their controlling objective."[106]

Many mistakes were made during the *Mayaguez* incident, starting with Ford, who possessed a misplaced sense of urgency over the entire affair, down to the courageous Air Force pilots and aircrew who were more familiar with rescue operations than tactics necessary for a successful vertical assault. Moreover, the Marines fed their forces into the fight piecemeal, due to the distance between Koh Tang and Utapao as well as not having enough long-range helicopters on hand. In truth, they had been lucky to not lose more personnel to the intense and greatly underestimated Khmer Rouge resistance.

---

[104] Lamb, *The Mayaguez Crisis, Mission Command, and Civil-Military Relations*, 122.
[105] Lamb, *The Mayaguez Crisis, Mission Command, and Civil-Military Relations*, 113.
[106] Lamb, *The Mayaguez Crisis, Mission Command, and Civil-Military Relations*, 131.

Congress also investigated the crisis. A few members seemed more concerned about Ford's orders to conduct airstrikes against the Cambodian mainland in violation of the 1973 Church-Case Amendment than details related to what happened on Koh Tang. One member of the House Committee on International Relations, Donald W. Riegle Jr. (D-MI), a former Republican who had switched parties in 1973 over his differences with Nixon's handling of the Vietnam War, believed, "Had this not occurred at a time and under circumstances—post-Vietnam when we were feeling a sense of frustration and national humiliation, we would not have felt such a strong need to assert ourselves militarily to prove, in the words of former President Nixon, that we are not a pitiful, helpless giant."[107] Most of the committee agreed that the loss of military personnel was highly regrettable, but the United States was also extremely fortunate to get the *Mayaguez* and its crew back so soon after its seizure.

The United States and its Marine Corps eventually moved on from the *Mayaguez* affair and focused on other missions around the globe. Following Vietnam, largely owing to its experience in Southeast Asia, all the U.S. military Services faced increased micromanagement when involved in future contingency operations. The political stakes at home and abroad were just too high. At the beginning of Operation Rolling Thunder, President Johnson famously boasted that the military "can't hit an outhouse without my permission."[108] This situation was just the beginning of what later military commanders sardonically referred to as the 10,000-mile screwdriver effect, as robust instantaneous worldwide communications created a greater tendency for Washington-based national security leaders to tinker with details for far-flung military operations in real time. Moreover, the U.S. media started to play a greater role in reporting on U.S. overseas contingencies. Operational commanders believed that television becoming an increasingly important source of information for the public by the mid-1960s made their jobs more difficult.

---

[107] *Hearings before the Committee on International Relations and Its Subcommittee on International Political and Military Affairs*, 94th Cong. (15 May 1975) (testimony of Col Zane E. Finkelstein, legal advisor and legislative assistant to the Chairman of the Joint Chiefs of Staff), 60.
[108] Michael Beschloss, "LBJ and the Descent into War," HistoryNet, 5 February 2019.

By that decade, "92 percent of American homes had a television set." From then on, Americans received wartime news nearly as it was taking place.[109] They would no longer experience the homespun World War II-era fireside chats broadcast directly from the White House. Vietnam was the first conflict in which a diverse, ubiquitous, and largely adversarial media pool provided live detailed coverage to the United States.

For the next two decades, graphic scenes of warfare were usually transmitted directly into people's homes in time for nightly newscasts within several days of the events occurring. The introduction of the Cable News Network (CNN) in the late 1980s changed this paradigm even further. With CNN now broadcasting news in real time 24 hours a day, other news networks raced to catch up. CNN's coverage of combat operations during the Gulf War (1990–91) riveted viewers. At the time, opponents of the United States even took this CNN factor into account. Sitting administrations found that this new real-time media scrutiny was difficult to control and heightened political stakes.[110]

Based on events and the operations that the Marine Corps conducted near the end of the Vietnam War and later against the Khmer Rouge portended just as much confusion and death in any future campaigns. These last events in Cambodia represented a sad ending to decades-long U.S. military involvement in Southeast Asia. Afterward, the United States turned its focus fully to NATO and its mission to defend Western Europe against potential Soviet aggression. For the next 18 years, the mantra for those in Congress and with the American public was indeed "no more Vietnams."[111]

---

[109] James Wright, *Those Who Have Borne the Battle: A History of America's Wars and Those Who Fought Them* (New York: PublicAffairs, 2012), 177–78. Wright served as an enlisted Marine and later became an historian and academic administrator, eventually serving as president of Dartmouth College from 1998 to 2009.
[110] Steven Livingston, *Clarifying the CNN Effect: An Examination of Media Effects According to Type of Military Intervention* (Cambridge, MA: Joan Shorenstein Center, Harvard University, 1997).
[111] Nixon, *No More Vietnams*, 212.

CHAPTER THREE

Trials of the 1970s

On 12 October 1972, the U.S. Navy attack aircraft carrier USS *Kitty Hawk* (CVA 63) was on station off the coast of North Vietnam assisting with operations related to Operation Linebacker I, the intense bombing campaign designed to convince the Democratic Republic of Vietnam (North Vietnam) to negotiate an end to the Vietnam War with the United States. The ship was scheduled to resume its role in combat air operations the next day. For months, however, racial tension aboard *Kitty Hawk* had been building.

Kitty Hawk had deployed from San Diego, California, in February 1972. The success of People's Army of Vietnam (PAVN) forces in Military Region 1 during the Easter Offensive forced the Navy to extend the ship's Western Pacific deployment by several months. To make matters worse, the crew had been long overworked. In the engineering department alone, 600 sailors "worked eight hours on and four hours off for 60 percent of the cruise, while in other departments sailors worked six hours on and six hours off. Each crewman in engineering, furthermore, received on average only six days off during the 247 days of *Kitty Hawk*'s deployment."[1] An act of sabotage delayed USS *Ranger* (CVA 61), which had been scheduled to relieve *Kitty Hawk*, when a member of its crew jammed an 18-inch steel rod into the vessel's main reduction gear.

---

[1] John Darrell Sherwood, *Black Sailor, White Navy: Racial Unrest in the Fleet During the Vietnam Era* (New York: New York University Press, 2007), 57.

An earlier relief attempt by USS *America* (CVA 66) had also met with misfortune when a faulty main feed pump of its engine forced it into the dockyards for repairs. Consequently, *Kitty Hawk* was required to remain on deployment for an excessive amount of time.[2]

Kitty Hawk's commanding officer, Captain Marland W. Townsend, had unintentionally exacerbated the racially tense situation aboard by allowing White and Black sailors to create informal separate berthing areas. Additionally, after having spent nearly eight months stationed off Vietnam, the crew received a short six-day liberty port visit to Subic Bay in the Philippines, expecting that they would be heading home shortly after. Instead, they were informed they would be returning to further operational assignments off the Vietnamese coast. The crew had no indication as to when the deployment might end. To further compound matters, while ashore, sailors smuggled copious amounts of drugs onboard the ship. While Townsend later believed that overall drug use was not pervasive aboard his ship, other anecdotal evidence has indicated that it was widespread. The commanding officer of the ship's Marine detachment, Captain Nicholas F. Carlucci, believed that "heavy users" would smoke either heroin or hashish "twice a day in laced cigarettes which were virtually undetectable."[3]

Further, many of *Kitty Hawk*'s younger Black sailors entered the Service following exposure to the highly charged political and racial environments pervasive throughout the United States in the late 1960s and early 1970s. In such an atmosphere, even a mild disagreement between White and Black crewmembers could easily spin into a major conflict. After the sailors worked long hours on "a seemingly endless cruise," one naval historian writes, the sailors were "then allowed to congregate" in separate groups "during their off-duty hours to hold gripe sessions, often under the influence of drugs and alcohol." These sessions led to an atmosphere where "a handful of perceived injustices on the ship could spark a major riot."[4]

---

[2] Sherwood, *Black Sailor, White Navy*, 79.
[3] Sherwood, *Black Sailor, White Navy*, 59.
[4] Sherwood, *Black Sailor, White Navy*, 60.

On the evening of 12 October 1972, a dispute about a sandwich between a White messman and a Black crewmember served as that spark. Several fights broke out between Black and White crew on the mess deck. The ship's Marine detachment was called out to restore order, and some arrived carrying nightsticks. Many of the young Black sailors saw the Marines as a police force, similar to one they might have encountered back home, causing the situation to escalate. Both Townsend and Commander Benjamin W. Cloud, a Black naval officer serving as the ship's executive officer, later arrived on the scene and attempted to defuse the situation. Instead, further violence broke out between the Marines and sailors on the hangar deck.[5]

This tension spilled over into the rest of the ship. Black crew attacked White colleagues as random targets of opportunity in other locations on the ship. While Townsend, Cloud, and Carlucci diligently worked throughout the night to calm the situation, it only seemingly got worse. Soon White sailors began arming themselves with lengths of cable or various maintenance tools and threatened retaliation. The following day, relative order was restored aboard Kitty Hawk, and the crew resumed active combat operations, springing into action without any further disruption. The ship's staff recorded only three serious injuries, but Townsend took immediate steps to break up the segregated berthing spaces and transferred the principal riot ringleaders off the ship.[6] In hindsight, the Kitty Hawk riot was a harbinger of things to come in the early 1970s. Much of this tension was related to a combination of institutional prejudice, detritus from the Vietnam War, increased drug use, and social fallout on the home front. Both the Navy and the Marine Corps need-

---

[5] Sherwood, Black Sailor, White Navy, 88. Capt Carlucci was a combat-wounded Marine who served with distinction earlier in the Vietnam War. His Marine detachment was acting in accordance with a centuries-old tradition that permanently assigned Marines aboard U.S. Navy vessels. These Marines were, in addition to their other duties, to provide shipboard security on behalf of its captain.

[6] Sherwood, Black Sailor, White Navy, 90, 94, 97–98. Sherwood believed that both Capt Townsend and Cdr Cloud showed "tremendous restraint" throughout the riot, which served to stop the escalating violence. The fact that Kitty Hawk could conduct combat operations the day after was a testament to the willingness of these two officers to "improvise" solutions to the problem, often at great personal risk.

ed to address the institutional injustice within their Services, and they had to do it quickly.

Consequently, an even more sensational incident occurred on 3 November 1972, when Black sailors aboard USS *Constellation* (CVA 64) staged a protest over their captain's alleged intent to involuntarily discharge 250 Black sailors. This allegation turned out to not fully be the case. Performance, not race, was the primary factor in who was selected for discharge. Nevertheless, the rumor that only Black sailors were being discharged spread quickly among the crew. Although *Constellation* had formed an active human-relations council on the ship earlier that year, it clearly had not worked well based on the level of dissatisfaction coming from the Black sailors.[7]

The situation grew more serious. The protesters demanded that the ship's captain, Captain John D. Ward, meet with them in-person on the mess deck to discuss their grievances—the discharge issue was just one of several of their problems. The captain refused, but the protesters continued to demand that he meet with all of them. This exchange went on for days. Eventually, Ward ordered *Constellation* to return to its pier at San Diego, California. He let the Navy leadership ashore know about the situation and ordered the dissidents ashore as part of a beach detachment so that they would have access to legal counsel, hopefully isolating the growing disaffection. He expected only about 90 sailors to go ashore, but more than 144, including a handful of White sailors, disembarked. Most of them refused to return until their grievances were heard.[8]

All this activity caught the attention of Admiral Elmo R. Zumwalt Jr., the Chief of Naval Operations. Zumwalt ordered Ward to dissolve the beach detachment. If the sailors still refused to return to duty after a short cooling-off period, Ward was to place them in an unauthorized absence status and get the ship underway without them. The next morning, the detachment showed up in uniform at pier side, but most still refused to board. Eventually, with the concurrence of Secretary of the Navy John

---

[7] Henry P. Leifermann, "The Constellation Incident: A Sort of Mutiny," *New York Times*, 18 February 1973, 17, 22–23, 26–30.
[8] Sherwood, *Black Sailor, White Navy*, 160–61.

W. Warner III, 120 of the dissidents were placed under disciplinary status and transferred to a barracks on Naval Air Station North Island in San Diego. Out of the 46 eventually discharged, only 10 were under less than honorable conditions. The fact that the *Constellation* protest took place in full view of the American media and the Washington political establishment made things worse for the Navy by seemingly confirming a pattern of behavior taking place on their ships.[9]

During this same time frame, the Marine Corps was facing similar issues. Historian James E. Westheider notes that at Marine Corps Base Camp Lejeune in North Carolina, the Service "began tracking racial assaults" in August 1968 and "recorded 160 of them before the end of the year." As conditions became increasingly intense, a "biracial committee of seven officers concluded that racism on the base and in the community contributed to an explosive situation."[10] The committee's findings proved prescient when a full-blown race riot broke out among Black and White servicemembers of the 1st Battalion, 6th Marine Regiment, at the enlisted club on 20 July 1969. The incident started at a crew party that took place just prior to their scheduled six-month Mediterranean cruise. Tensions were already high due to the pending deployment, with heavy drinking making matters worse. Dozens were injured in the brawl and Corporal Edward Bankston, a White servicemember, "lay dead from a fractured skull."[11] The significance of the Camp Lejeune riot became clear soon after and "prompted the creation of a Special House Armed Services Committee to investigate racial tension on United States military bases." The committee recommended that the U.S. Department of Defense (DOD) "institute a program of education in race relations at all levels of command with an emphasis on the platoon and company lev-

---

[9] Sherwood, *Black Sailor, White Navy*, 160–63.

[10] James E. Westheider, *The African American Experience in Vietnam: Brothers in Arms* (Lanham, MD: Rowman & Littlefield, 2008), 87–88.

[11] James E. Westheider, *Fighting on Two Fronts: African Americans and the Vietnam War* (New York: New York University Press, 1997), 94–95, 112–13.

els."[12] In response, the DOD created a Defense Race Relations Institute (DRRI) to address the problem. Following even a larger riot at Travis Air Force Base, California, in 1971, Deputy Undersecretary of Defense David Packard established a Race Relations Education Board.

Deployed Marines seemed especially susceptible to the growing unrest. Private James E. Raines, a member of Battalion Landing Team 3d Battalion, 2d Marines (BLT 3/2), was embarked aboard the Navy's USS *Trenton* (LPD 14), which was participating in a routine Mediterranean deployment out of Camp Lejeune. Raines had recently spent time in the ship's brig and was already in the process of being discharged for staging a small revolt in the crew mess area. Upset with Raines's pending situation, Black Marines and sailors rioted on 18 November 1972 over his alleged mistreatment. An ensuing investigation conducted by Colonel Alfred M. Gray Jr., the then-commanding officer of the 2d Marine Regiment, suggested that the real cause of the disagreement had more to do with a "lack of middle level leadership than race."[13] Many felt that recent publicity given to events on *Kitty Hawk* and *Constellation*, coupled with distrust of leadership among Black servicemembers and the broader issue of institutional racism, played a role in the unrest.

Soon after the *Trenton* affair, another incident took place aboard USS *Inchon* (LPH 12). This time, however, the events on that ship on the night of 26 January 1973 was "more than just a scuffle; it was a full-scale riot."[14] Earlier that day, *Inchon*, with more than 1,500 Marines of the BLT 1st Battalion, 9th Marines (BLT 1/9), onboard, had departed the Gulf of Tonkin for Okinawa, Japan. As with any packed ship, spaces were hot and crowded and boredom was common for the Marines. That evening, a movie was being shown on the ship's hangar deck. Being a Western, the film happened to attract mostly White Marines from BLT 1/9. At some point, a group of about 20 Black sailors and Marines stood up and

---

[12] Isaac W. Hampton II, "Reform in the Ranks: The History of the Defense Race Relations Institute, 1971–2014," in *Integrating the Military: Race, Gender, and Sexual Orientation since World War II*, ed. Douglas W. Bristol Jr. and Heather M. Stur (Baltimore, MD: Johns Hopkins University Press, 2017), 123–24.

[13] Sherwood, *Black Sailor, White Navy*, 194–99.

[14] Sherwood, *Black Sailor, White Navy*, 216.

rendered a Black power clinched-fist salute. They then began fighting with the other Marines. The violence spilled out into different locations on the ship and ended only when *Inchon*'s commanding officer, Captain John K. Thomas, sounded General Quarters, forcing the ship's personnel to occupy their battle stations. Several Marines were treated for minor injuries. At least six Marines received some sort of disciplinary action—including one who was discharged for bad conduct—for their participation in the affair. Of even more significant concern, the riot occurred on a vessel that was known to have had a robust human relations council. Nevertheless, all these programs failed "miserably on the night of 26 January 1973."[15] Outbreaks of race-based violence among the Marine Corps occurred throughout the Service, including instances at Kaneohe Bay, Hawaii, and Millington, Tennessee.

In late August 1972, USS *Sumter* (LST 1181), carrying a cadre of Marines, was steaming off the coast of Vietnam during its regularly scheduled Western Pacific deployment. Events that took place on this cruise made things anything but routine, however. While at sea, the Marines and sailors were allowed to take turns playing music over the ship's loudspeakers to break up the hot and monotonous hours on the tightly packed vessel. In one moment, the interchange of music had far-reaching consequences for at least three Black Marines, especially Private First Class Alexander Jenkins Jr., a 19-year-old from Newport News, Virginia, "whose outgoing personality had earned him a turn as the ship's D.J." Jenkins recalled playing a variety of music but noted that when he played, "White Man's Got a God Complex" by the Last Poets, a Harlem-based Black music group known for their activism during the civil rights movement of the late 1960s, it "really set the white guys off."[16]

Shortly after playing the record, some White Marine officers confronted Jenkins, accusing him of trying to incite a riot. Following this altercation, Jenkins and 63 of the 65 black Marines on the ship request-

---

[15] USS Inchon *Command History of Calendar Year 1973*, Operation Naval Report (OPNAVREP) 5750-1 (Washington, DC: Chief of Naval Operations, 1974); and Sherwood, *Black Sailor, White Navy*, 216–19.

[16] John Ismay, "At War: The Untold Story of the Black Marines Charged with Mutiny at Sea," *New York Times Magazine*, 19 August 2020.

ed to speak with their battalion commander, located on another ship, about being prevented from playing the records of Black artists, especially those that addressed Black culture and issues, but his request was denied. Early in September, fistfights between Black and White Marines broke out all over the ship. Marine leadership, however, zeroed in on Jenkins, Private First Class Roy L. Barnwell, and Lance Corporal James S. Blackwell, considering these three Black Marines the " 'ringleaders' who were instigating general unrest and resistance to their orders." Later put ashore at Da Nang, Vietnam, and eventually put in the brig back in Okinawa, all three Marines faced one of the most serious charges in the military judicial system: "mutiny at sea during a time of war," a charge that came with a possible death sentence.[17] Fortunately, saner heads eventually prevailed. After a long and arduous legal process, the Service dropped the mutiny charges and the three Marines faced lesser charges. Jenkins eventually got an honorable discharge while Blackwell and Barnwell both received undesirable discharges. The trauma of the Last Poets incident on *Sumter* and the actions of the Marine Corps legal authorities toward Jenkins, Blackwell, and Barnwell haunted them for years afterward. The event also demonstrated just how easily tensions between Black and White servicemembers could boil over in 1972.

Racial unrest in Marine Corps units seemed especially pronounced in III Marine Amphibious Force (III MAF) organizations on Okinawa. In 1971, Captain Anthony C. Zinni, who later became the commanding general of U.S. Central Command, was a distinguished combat veteran of two tours of duty in Vietnam, including experience as a Vietnamese Marine Corps (VNMC) advisor during his first tour in 1967. As an advisor, Zinni saw significant combat action. After 10 months in the field advising the VNMC, he had lost 40 pounds and developed severe dysentery and hepatitis, among other diseases. Much to his chagrin, a brief visit to a U.S. Army medical aid station ended this tour in Vietnam, and he was medically evacuated back to the United States. In 1970, during his second Vietnam tour, Zinni was given command of Company A, 1st Battalion,

---

[17] Ismay, "At War."

5th Marines. During one early security mission, he was gravely wounded. Several hours later, he and other wounded Marines were finally evacuated via helicopter. His wounds were so severe that he was airlifted to a hospital on Guam. Once he had mostly recovered, Zinni, hoping to get back to Vietnam, tried to convince his doctors to return him to full duty status. His surgeons refused to allow a return trip to combat and had him reassigned to an apparently less stressful and physically taxing job on Okinawa. Consequently, Zinni spent the last eight months of his second Vietnam tour in the 3d Force Service Regiment, a logistics unit located at Camp Foster.[18]

Zinni's new assignment turned out to be anything but less stressful. Soon after his arrival at Camp Foster, he "noticed units practicing riot control formations and the use of special riot control equipment." The camp appeared to him to be "under siege" and he believed that he was "sitting on a powder keg."[19] This siege mentality revolved around the growing racial unrest apparently prevalent in every Marine Corps unit on the island. Zinni eventually got assigned to command the underperforming guard company at Camp Foster, which placed him in the unenviable position of quelling any future riots or altercations that might break out. The camp commander gave Zinni carte blanche to recruit 100 Marines for his guard force. Zinni was adamant that the guard company should consist of the most imposing Marines available and, more importantly, it should be racially diverse. This decision was a brilliant move on his part, because it was not long before Zinni and his guard company had to deal with several racially motivated riots. Zinni firmly believed that in responding to any incident, "no matter what happened, we would get a black Marine, a white Marine, a Hispanic Marine, and a Samoan Marine—a rainbow detail—going out to handle it." Zinni's policy of firm but fair leadership worked well. Still, he and his Marines had

---

[18] Tom Clancy with Gen Tony Zinni and Tony Koltz, *Battle Ready* (New York: G. P. Putnam's Sons, 2004), 91–108. Zinni was blessed in his career with two important mentors—LtCol Bernard E. Trainor, his former battalion commander in Vietnam, and Gen Alfred M. Gray Jr., who, as the G–3 (operations officer) of the 2d Marine Division, made Zinni his assistant and helped create a new infantry school syllabus at Camp Lejeune.

[19] Clancy with Zinni and Koltz, *Battle Ready*, 117.

to find innovative ways to resolve the highly charged and often racially motivated anger that was prevalent in most Okinawa-based tenant commands more than once. Due to his success, Zinni soon had plenty of racially diverse volunteers for his guard company.[20] The III MAF commanding general at that time also established courtesy patrols in liberty locations across the island. These patrols were only partially effective in tamping down racial confrontations in a liberty spot located just outside Camp Hansen, Okinawa, that Marines called "Kinville."

The Marine Corps sensed that these rising racial tensions were connected to the detritus of the Vietnam War and the growing discontent of the Black community in the United States. Even before the fighting ended in Vietnam, the Service had established 190 human relations committees. By 1971, the Marine Corps established a Human Relations Institute with a mission similar to the DRRI at the Recruit Depot at San Diego, California. This program created mandatory race relations classes for all Marine Corps personnel. Within a year, "every Marine, regardless of rank, had to complete a twenty-hour course."[21] The program, which continued for several years, had limited success due to a lack of educated trainers and basic command apathy. Basically, it was useful when the command emphasized the program, but it failed when they did not.

Several issues seemed to cause the anger among Black servicemembers within the Marine Corps. First, many Black Marines believed that the military justice system practiced systemic discrimination, punishing Blacks more frequently and harshly than Whites. This issues especially pertained to cases that involved nonjudicial punishment, also known within military circles as "Article 15." In these cases, commanding officers had broad discretion over the frequency and punishment levels for Marines who committed minor offenses against the Uniform Code of Military Justice. Statistics from that time seem to support their stance. Black servicemembers throughout the U.S. military were more frequent-

---

[20] Clancy with Zinni and Koltz, *Battle Ready*, 121–27.
[21] Westheider, *Fighting on Two Fronts*, 134. As a second lieutenant at Camp Lejeune, the author cotaught, alongside a highly capable Black staff sergeant, several of the required human relations classes in 1977. These classes were part of the division's annual training requirements.

ly charged with offenses under the uniform code and were more likely to receive an administrative discharge than their White counterparts, as the *Sumter* incident demonstrated in 1972. Further, Black servicemembers were more likely to undergo a court-martial, be "confined before trial," receive a conviction, and receive "long sentences."[22]

Nathan R. Packard, a professor at the Command and Staff College at Marine Corps University and an officer in the Marine Corps Reserve, has done the most recent work on race relations and disciplinary issues in the Marine Corps during the 1970s. Packard notes that "in 1969, African Americans made up only 9.3 percent of the armed forces but nearly 50 percent of the confined population." Throughout the Vietnam War, "there was not a single black Judge Advocate officer on active duty in the Marine Corps." Consequently, Black defendants went to trial with only White defense counselors representing them and having their fates "decided by an all-white jury in a courtroom presided over by a white judge."[23]

According to a 1972 study of how Marines were treated by courts-martial, conviction rates did not differ significantly for defendants regardless of race. At the less frequently convened general courts-martial, the Service convicted 100 percent of the 27 Marines—11 White Marines and 16 Black Marines—charged. For the mid-level special courts-martial, White Marines had an 89.68-percent conviction rate, slightly lower than the 90.79-percent rate for Black Marines—still an extremely high conviction rate for both races. Based on the conclusions of this study, the Marine Corps clearly held various level courts-martials "more frequently than the navy." Similarly, in lower courts, these cases ended in either not guilty or dismissed verdicts more often.[24] In sum, the higher the level of military court, the more likely a defendant would receive a guilty verdict no matter the race. The study also determined that the rate

[22] Bernard C. Nalty, *Strength for the Fight: A History of Black Americans in the Military* (New York: Free Press, 1986), 329.

[23] Nathan R. Packard, " 'The Marine Corps' Long March': Modernizing the Nation's Expeditionary Forces in the Aftermath of Vietnam, 1970–1991" (PhD diss., Georgetown University, 2014), 60–62.

[24] Ronald W. Perry, *Racial Discrimination and Military Justice* (New York: Praeger, 1977), 23–24.

of incarceration for enlisted Marines who were convicted at a court-martial declined as their education rose.[25]

Black servicemembers, Marines included, experienced a distinct lack of equal opportunity in the 1960s and early 1970s. This matter caused so much concern that as early as 1962, President John F. Kennedy formed a committee—the President's Committee on Equal Opportunity in the Armed Forces—to examine the subject. Commonly known as the Gesell Committee for its chairman, Gerhard A. Gesell, its members worked to find out the depth of the problem of discrimination in the armed forces. Because Gesell graduated with both a bachelor's and law degree from Yale University, many fellow committee members had a similar background. Other members had been active in the civil rights movement, including Burke Marshall, the head of the Department of Justice's Civil Rights Division; Nathaniel S. Colley, a civil rights lawyer from California who fought for fair housing laws; and Whitney M. Young Jr. of the National Urban League, among other notable civil rights advocates. Although a presidential committee, the members reported their findings directly to Secretary of Defense Robert S. McNamara.[26]

The Gesell Committee visited numerous defense establishments, usually conducting interviews of various personnel via two-person biracial teams. The committee concluded that "serious discrimination against black service men and their families existed at home and abroad within the services and in the civilian community, and that this discrimination affected black morale and military efficiency." For example, as outrageous as it seems today, Naval Air Station Pensacola, Florida, did not allow Black servicemembers in the early to mid-1960s to guard the main gate out of fear of offending the local community. Other base commanders segregated Black servicemembers and their families into substandard housing complexes. The committee was especially critical of the tendency of Navy and Marine Corps leadership to place Black enlistees in the "supply and food services" occupational specialties. The committee

---

[25] Perry, *Racial Discrimination and Military Justice*, 36.
[26] Morris J. MacGregor Jr., *Integration of the Armed Forces, 1940–1965* (Washington, DC: U.S. Army Center of Military History, 1981), 535–37.

members argued that the Navy and Marine Corps lagged "far behind the Army and Air Force, particularly in the area of community relations."[27] In sum, it "confirmed the presence of discrimination against black servicemen both on and off military base and effectively tied that discrimination to [lowered] troop morale and military efficiency."[28]

By 1972, General Robert E. Cushman Jr., the 25th Commandant of the Marine Corps, was moved to "declare that racial discrimination had no place in the Corps." Cushman was determined to put the Gesell Committee's recommendations—now nine years old—into practice, banning "race as a consideration in the assignment of quarters, the selection of individuals for work details, or the administration of justice." Doing so would not mean that the Marine Corps fully eliminated all vestiges of institutional racism, however. Rather, his "directive was going to require careful and persistent monitoring."[29]

In the 1970s, things were finally beginning to change for women in the military. Since the end of World War II, the 1948 Women's Armed Services Integration Act governed female service in the U.S. armed forces. The legislation essentially sought to balance the growing requirement for women in the ranks with what Service leaders considered the "prevailing need on proper women's roles" in the military. This included "consistently structuring women's military careers on the fundamental belief that women's service should not compromise their femininity or their future role as wives and mothers." Female Marines received coaching from "grooming experts used by Pan American Airways to train stewardesses."[30] By 1967, 70–80 percent of servicewomen "did not complete their first enlistment; on average, women spent 14 months in military service before leaving altogether." Some of this enlistment turnover, however, could be traced directly to motherhood. Prior to the mid-1970s, pregnant military members were "automatical-

---

[27] MacGregor, *Integration of the Armed Forces*, 538–39.

[28] MacGregor, *Integration of the Armed Forces*, 554–55.

[29] Nalty, *Strength for the Fight*, 328.

[30] Tanya L. Roth, " 'An Attractive Career for Women': Opportunities, Limitations, and Women's Integration in the Cold War Military," in *Integrating the U.S. Military*, 75, 82.

ly discharged."[31] After 1975, the DOD revised its policy so that only ser-vicemembers could request a pregnancy or parenthood discharge rather than receiving an automatic dismissal.

During this same time frame, all the U.S. Service academies were fi-nally opened to women. Prior to the Vietnam War, women had decidedly limited Service opportunities. This traditional approach changed in the late 1970s and continued throughout the 1980s as the demand for wom-en to serve in a larger variety of military occupational specialties rose. While women still faced exclusion from certain combat roles through the beginning of the twenty-first century, the number of occupational spe-cialties accessible to women rapidly expanded throughout the 1980s and 1990s. By the late 1980s, women were authorized to fly military aircraft and were eventually allowed to transition into combat fighter wings. In looking back on Operation Desert Storm (1990–91), Secretary of De-fense Richard B. "Dick" Cheney confessed that "we could not have won" without women pilots. More than two decades later, in 2013, Secretary of Defense Leon B. Panetta admitted to the obvious and "announced the end of the combat exclusion policy that applied to women specifically."[32]

Another major problem that the Marine Corps contended with in the mid-1970s was its scarcity of Black officers. The previous decade, the Ge-sell Committee noted the lack of Black officers across the Services, find-ing this situation especially shocking because President Harry S. Truman had first ordered the Services to fully integrate in 1948. They believed that the situation meant that few Black officers would have the oppor-tunity to serve on promotion boards, which could become easily biased against Black servicemembers in the competition for promotion, some-thing that seemed particularly troubling in the Marine Corps. Nearly ev-ery Marine officer in the early 1970s was White. While the enlisted ranks were far more diverse, with Black Marines making up approximately 13 percent of the force during the Vietnam War and rising to as high as 19 percent by the late 1970s, the highest-ranking Black officers in the Ma-rine Corps at that time were Lieutenant Colonels Frank E. Petersen Jr. and

---

[31] Roth, " 'An Attractive Career for Women'," 85.
[32] Roth, " 'An Attractive Career for Women'," 86–89.

Kenneth H. Berthoud Jr., and no Black woman held a field-grade commission. "As late as 1973," one historian notes, "the highest-ranking African American woman officer in the Corps was Captain Gloria Smith."[33] Even the Service academies proved deficient in this regard. Despite having accepted a few Black cadets at the U.S. Military Academy at West Point, New York, following the U.S. Civil War, at the height of the Vietnam War in 1968 there were only "seventeen black cadets at West Point and only ninety-seven African-Americans out of ninety-eight hundred underclassmen at all three service academies."[34]

The Marine Corps did not have a quick solution for its lack of Black officers, and the national Service academies at the time could only take a minor step toward resolving the problem. In the early to mid-1970s, the Marine Corps decided to follow the model of the Navy's successful Naval Reserve Officers Training Corps (NROTC) program of the mid-1960s at Prairie View A&M in Texas. Prairie View was a historically Black college/university (HBCU) with a strong academic reputation, which enabled the Navy to increase its percentage of qualified Black officers in just a few short years. While such a program did not remedy the overarching dearth of Black officers in the military, it gave the Navy and Marine Corps an idea for rapidly closing the gap between Black and White officers.[35]

By the mid-1970s, the Marine Corps was actively recruiting officer candidates from several highly respected HBCUs, such as Savannah State University (SSU) in Georgia and Southern University and A&M College in Louisiana that had established robust NROTC programs on their campuses. Students at these schools competed for acceptance into the NROTC scholarship program. Some of these cadets were labeled "Marine option" candidates, meaning that after graduation they would be commissioned

---

[33] Westheider, *Fighting on Two Fronts*, 122. Westheider believed that part of the scarcity of Black Marine officers during the Vietnam War era had to do with the segregated and biased U.S. educational system and the fact that officers were generally required to have attended college. Few Black Americans at the time could meet this requirement "since only about 5 percent of the adult African American male population held degrees."

[34] Westheider, *Fighting on Two Fronts*, 123. Black candidates who attended the Service academies graduated at the same rate as their White counterparts.

[35] Charles Johnson Jr., *African Americans and ROTC: Military, Naval and Aerospace Programs at Historically Black Colleges, 1916–1973* (Jefferson, NC: MacFarland, 2002), 198–203.

as second lieutenants in the regular Marine Corps. At the time, the most extensive Marine Corps officer candidate program (not counting Service academy accessions), the Platoon Leaders Class (PLC), led the way toward improving officer representation. The PLC started in 1934 as an option for officer candidates who did not attend a university or college with a NROTC program, as the PLC could admit any qualified student. The significant difference between the NROTC and PLC was that successful PLC candidates were only initially offered a reserve commission. Nonetheless, the Marine Corps found the PLC program beneficial during World War II, when it required large numbers of temporary officers.[36]

Colonel Fred L. Jones recounted an excellent example of how bad things were related to minority officer recruiting. Jones initially grew up in segregationist Mississippi in the late 1940s. His father got a government job in Hawthorne, Nevada, and moved his entire family there in the 1950s, when Jones was starting the 6th grade. Traveling on a Greyhound bus for the move, Jones recalled that his family had to sit in the segregated rear section of the bus while it traveled through the Deep South. Once they got past east Texas, such bigoted restrictions were lifted. Jones vividly recalled that all the kids in the Hawthorne government-run "Babbit Housing" complex openly played and went to school together, but the Black families were still housed separately on one side of the complex and White families were situated on the other. In the 1960s, Jones became a standout college football player at Oregon State University. His coach was a Reserve Marine captain named Robert O. McKittrick. McKittrick recommended that Jones sign up for the Marine Corps Officer Candidates School (OCS) program once he graduated.[37]

After arriving for training at Marine Corps Base Quantico, Virginia, Jones noticed that he was only one of two Black OCS candidates. He was commissioned in December 1964, although he had no desire to make the Marine Corps his career and envisioned himself as a civilian in just

---

[36] Johnson, *African Americans and ROTC*, 122, 198–201.

[37] Fred H. Allison and Col Kurtis P. Wheeler, USMCR, eds., *Pathbreakers: U.S. Marine African American Officers in Their Own Words* (Quantico, VA: Marine Corps History Division, 2013), 4–6, 44–46.

three years. While at Quantico, Jones recalled being invited to an on-base party at the home of Major Hurdle L. Maxwell, a highly regarded Black Marine officer. He noted that many other prominent Black Marine officers were there, including then-Major Frank Petersen. After Maxwell informed Jones that he was probably only the 45th Black officer currently on active duty, Jones felt that his original decision to do the minimum amount of time was a good one. He also noticed that the 14 Black officers at Maxwell's party likely represented "one-third of all active-duty black officers in the entire Corps." Despite Jones's concerns, Maxwell believed that Jones had the makings of a career officer and "bet him a steak dinner" that he would opt to stay in the Service. Jones disagreed and told him that "there is absolutely no way in the world I could stay in an organization like this, when I look around and there is nobody that looks like me."[38]

Jones, however, did stay in the Marine Corps for a full and distinguished career. Later assigned to a Marine officer instructor billet at SSU in the early 1970s, he reluctantly accepted it. Nevertheless, once there, he hit the ground running and turned a heretofore struggling NROTC unit into a high-performing program. Jones recounted that it was not long before SSU would "really become a Marine unit. Everybody wanted to be a Marine." Jones taught and trained Walter E. Gaskin Sr., a future lieutenant general and II Marine Expeditionary Force commander. Gaskin was, according to Jones, "by far, my best student." Gaskin was also "the very first Marine officer to be commissioned" from SSU. "From then on," Jones remembered, "there were a bunch of them that came after."[39] Jones went on to instruct many high performing NROTC students, including Donnie L. Cochran, the first Black officer to command the Navy's elite Blue Angels flight demonstration squadron.

---

[38] Fred H. Allison and Col Kurtis P. Wheeler, USMCR, eds., *Pathbreakers: U.S. Marine African American Officers in Their Own Words* (Quantico, VA: Marine Corps History Division, 2013), 5, 44–46. Jones later went on to a highly successful career as a Marine officer, serving in both war and peace. He retired as the chief of staff to the commanding general of Marine Corps Combat Development Command. There is no indication if Hurdle Maxwell ever collected on his steak dinner.

[39] Allison and Wheeler, *Pathbreakers*, 93–95.

Figure 27. MajGen Charles F. Bolden Jr.

Source: official U.S. Marine Corps photo.

Even after receiving assignments that took him away from SSU, Jones never stopped mentoring Black Marine officers, such as future Lieutenant General Ronald S. Coleman. As a young captain, Coleman recalled Jones telling him to "take the job that nobody else wants" and then do it well.[40] Coleman took Jones's advice to heart and went on to enjoy a highly successful career. Other significant Black officers who started during the late 1960s and early 1970s—such as U.S. Naval Academy graduate, airwing commander, astronaut, and eventually administrator of the National Aeronautics and Space Administration, Marine Corps major general Charles F. Bolden Jr.—were also assigned to the officer selection program.[41]

Although the number of Black officers in the Marine Corps grew throughout the 1970s, recruitment remained an ongoing challenge. Major General Leo V. Williams III joined the Marine Corps after graduating from the Naval Academy in 1970. He likened Black officer recruiting to "a sine wave and that it has been very, very much dependent upon the leadership at the moment."[42] Williams believed that Black officer recruitment faltered any time senior Marine Corps leadership lost focus on its officer diversity effort. Despite this vacillation, HBCUs in the 1970s produced several high-performing Marine officers who later made flag rank. The aforementioned Coleman enlisted in the Navy before joining the Marine

---

[40] Allison and Wheeler, *Pathbreakers*, 154.
[41] Allison and Wheeler, *Pathbreakers*, 231–32.
[42] Allison and Wheeler, *Pathbreakers*, 211.

Corps after graduating from Cheyney State College, an HBCU in Philadelphia, Pennsylvania. Lieutenant General Willie J. Williams graduated
from Stillman College, a HBCU with a secondary mission of preparing
young men for the ministry in Tuscaloosa, Alabama, in 1974. Both Major General Arnold Fields and Major General Clifford L. Stanley graduated from South Carolina State College, yet another HBCU with an excellent
reputation, in 1969.[43] Many others came through these same programs.

In 1968, the Naval Academy appointed Marine Corps major Edward L.
Green to the faculty, where he taught leadership and military law, making him the first Black Marine officer to hold an instructor billet there.
Green strongly believed that until the Marine Corps reached "an adequate
black officer distribution throughout the command and policy-making
levels, the basic fairness of the entire institution will remain in doubt."
Thanks to efforts of Green and others, "eight of the 12 black midship-
men in the class of 1972 joined the Marine Corps." While the number of
Black officers in the Marine Corps by 1973 increased to "367 men and 11
women," these numbers still only represented "2.03 percent" of the total officer corps. Despite the low percentage, this growth indicated a "far
greater awareness" throughout the Service about the importance of having a diverse officer corps and how such a force could be useful in creating even more effective combat power in the ranks.[44]

Not all Black Marine officers who entered the Corps at this time came
from HBCU programs. Major General Jerome G. Cooper was born in 1936
in the segregated southern town of Mobile, Alabama. He graduated from
a Catholic high school in 1953 and, even in those difficult times, gained
acceptance to the University of Notre Dame in Indiana on an academic scholarship. He joined Notre Dame's NROTC unit, becoming the first
Black Marine Corps officer commissioned from it. Having received orders
to report to Marine Corps Base Quantico, he noted that the train between
Washington, DC, and the Virginia base was "completely segregated."[45]

---

[43] Allison and Wheeler, *Pathbreakers*, 9–12, 15.
[44] Henry I. Shaw Jr. and Ralph W. Donnelly, *Blacks in the Marine Corps* (Washington, DC: Marine
Corps History and Museums Division, 1988), 74–75.
[45] Allison and Wheeler, *Pathbreakers*, 3–4.

Figure 28. Astronaut Charles F. Bolden Jr.

Source: National Aeronautics and Space Administration.

Cooper later served as an infantry officer with the 9th Marines during the Vietnam War, during which he saw significant combat action and was awarded two Purple Hearts and the Bronze Star.

Cooper left active duty while still in his 30s, though he remained an officer in the Marine Corps Reserve, and returned to Mobile. While there, he became a business entrepreneur and eventually won a seat in the Alabama state legislature in 1974. Five years later, he was appointed to a state cabinet position by Governor Forrest H. James Jr. Promoted to flag rank in the Marine Corps Reserve, Cooper later commanded the 4th Force Service Support Group in Atlanta, Georgia. Cooper went on to even greater heights when President George H. W. Bush appointed him as an assistant secretary of the Air Force. In 1991, while serving as assistant secretary, Cooper spotlighted the incredible contributions of the Tuskegee Airmen of World War II to push senior DOD leadership to recognize the group's importance to the larger Air Force. In May 1991, he brought together Lieutenant General Frank Petersen, Navy vice admiral Samuel Gravely, and Air Force general Benjamin O. Davis Jr. at the officer's club at Bolling Air Force Base in Washington, DC, for an "informal session" on what it meant to be a Black servicemember and a flag officer. Junior officers who attended that historical event included Ronald L. Bailey, a future commanding general of the 1st Marine Division; Willie Williams, a future Marine Corps chief of staff; and Clifford Stanley, who was later appointed to the position of undersecretary of defense for personnel and readiness during the administration of President Barack H. Obama after retiring from the Marine Corps as a major general. President William J. "Bill" Clinton appointed Cooper as the U.S. ambassador to Jamaica near the end of his professional career. Cooper noted that, incredibly, the predominately Black nation of Jamaica had never had a Black American as the U.S. ambassador. Despite the racism he occasionally encountered, Cooper remained a steadfast and positive mentor to the younger Black officers who joined the Marine Corps in his wake.[46]

In 1973, President Richard M. Nixon abruptly ended the national draft and declared that going forward, the U.S. military would be an all-volunteer force (AVF). It was a popular move. During the Vietnam War, all the Services were heavily dependent on a widely reviled national draft due to the decision by the National Command Authority to maintain its substantial commitment to the North Atlantic Treaty Organization and not engage—for the most part—the military's reserve establishment or the National Guard in any overseas contingency operations. Furthermore, the National Guard often found itself deployed at home dealing with rising social unrest in many U.S. cities and on college campuses across the country. In 1970, one such intervention on the campus of Kent State University in Ohio resulted in the deaths of four student protesters at the hands of an improperly trained National Guard unit.[47] Military personnel expert Bernard D. Rostker noted that "conscription" was never "the norm" in the United States. In fact, until World War II and the ensuing Cold War, Americans generally did not prefer to maintain a large standing military. Rostker noted that while the United States largely ran a universal draft with few exemptions allowed for draftees during World War II, the pre-Vietnam War draft "was a poor substitute for universal service," with the Selective Service System allowing increasingly higher

---

[46] See, Kendal Weaver, *Ten Stars: The African American Journey of Gary Cooper—Marine General, Diplomat, Businessman, and Politician* (Montgomery, AL: NewSouth Books, 2016), 115, 120, 154, 168, 195, 205, 216–18, 249–50.

[47] Eliot A. Cohen, *Citizens and Soldiers: The Dilemmas of Military Service* (Ithaca, NY: Cornell University Press, 1985), 166–70; Beth Bailey, *America's Army: Making the All-Volunteer Force* (Cambridge, MA: Belknap Press of Harvard University Press, 2009), 58–59; and Michael Corcoran, "Why Kent State Is Important Today," *Boston (MA) Globe*, 4 May 2006.

numbers of deferments by the 1960s.[48] However, resistance to the draft during an unpopular war in Vietnam "led to a re-evaluation of the expanded State" and what the nation could demand of its citizenry following the conclusion of military activity in Southeast Asia in 1973.[49]

The troubles that emerged in Vietnam from U.S. forces being "composed largely of short-term draftees" fighting an extended conflict in "a region unrelated to American security" led the United States to embrace a larger standing AVF.[50] The question of whether military recruiters could find and retain the personnel necessary for such a force remained unanswered when the draft ended. Alongside this uncertainty, two military experts, Rostker and Richard V. L. Cooper, both believed that "one of the major effects of the draft was to remove the military from the marketplace" by compelling the U.S. armed forces to "compete in the civilian marketplace for qualified personnel."[51] None of the Services were prepared to contend for people at that moment.

Before ending the draft, Nixon set up a commission to study the impact of transitioning a draft-dependent force into an AVF. Prior to this action, the president stated, "We have lived with the draft so long that too many of us accept it as normal and necessary," and that he wanted the commission to address how to move forward.[52] Chaired by Thomas S. Gates Jr., who had served as one of President Dwight D. Eisenhower's secretaries of defense, the committee also included Alan Greenspan, famed economist and future chairman of the Federal Reserve; Jerome H. Holland, president of the renowned HBCU Hampton University in Virginia; Father Theodore M. Hesburgh, president of the University of Notre Dame; Milton Friedman, a Nobel Prize laureate and an economics profes-

---

[48] Bernard Rostker, *I Want You!: The Evolution of the All-Volunteer Force* (Santa Monica, CA: Rand, 2006), 2–3.

[49] John Whiteclay Chambers II, *To Raise an Army: The Draft Comes to Modern America* (New York: Free Press, 1987), 264–65.

[50] Chambers, *To Raise an Army*, 272.

[51] Richard V. L. Cooper and Bernard Rostker, *Military Manpower in a Changing Environment* (Santa Monica, CA: Rand, 1974), 2.

[52] *The President's Commission on an All-Volunteer Armed Force* (Washington, DC: White House, 1970), 11.

sor at the University of Chicago; and Roy O. Wilkins, the executive director of the National Association for the Advancement of Colored People.[53]

The commission reported its conclusions in February 1970. Significantly, it found that the system of a mixed force of conscripts and recruits during the Vietnam War required the Services to find more volunteers each year than they would have if the forces were solely volunteers:

> To judge the feasibility of an all-volunteer force, it is important to grasp the dimensions of the recruitment problem in the next decade. If conscription is continued, a stable mid-range force of 2.5 million men (slightly smaller than pre-Vietnam) will require 440,000 new enlisted men per year. To maintain a fully voluntary stable force of the same effective strength, taking into account lower personnel turnover, we estimate that not more than 325,000 men will have to be enlisted annually. In recent years about 500,000 men a year have volunteered for military service. Although some of these volunteered only because of the threat of the draft, the best estimates are that at least half—250,000 men—are "true volunteers." Such men would have volunteered even if there had been no draft, and they did volunteer in spite of an entry pay that is roughly 60 percent of the amount that men of their age, education, and training could earn in civilian life.[54]

The key assumption in this declaration depended highly on all the Services achieving a lower turnover rate, which would require many first-term volunteers electing to stay. This prospective retention also required new conditions in the military that would attract volunteers, something that had not taken place even during the years before Vietnam. Low pay, racial unrest, disorderliness, and austere working and living conditions combined to defeat the AVF before it got started. The

---

[53] *President's Commission on an All-Volunteer Armed Force,* viii–ix. Wilkins's participation on the committee was affected due to ill health.
[54] *President's Commission on an All-Volunteer Armed Force,* 6.

commission candidly admitted that improvements in pay and conditions were overdue for a force that had long relied on poorly paid conscripts who left the Service as soon as possible. Paradoxically, it also cautioned against quick conclusions based on the apparently higher budget for an AVF. Although seemingly more expensive due to planned improvements in pay and living conditions, the actual price for the volunteer military would be much lower because most expenses to maintain the armed forces are hidden costs that extend over years and recruiting cohorts. Moreover, the commission reported, "Men who are forced to serve in the military at artificially low pay are actually paying a form of tax which subsidizes those in the society who do not serve."[55] On 28 September 1971, Nixon signed amendments to the Military Selective Service Act of 1967. This legislation extended the draft for two years but also indicated a commitment to the suggestions from the Gates Commission, including recruiting an AVF.[56] After the DOD suspended the draft in January 1973, the Nixon administration allowed the Selective Service Act to expire that summer, bringing an end to the longest peacetime draft in U.S. history.[57]

The creation of the AVF placed tremendous pressure on the recruiting establishments of all the Services. During any given year in the 1970s, Marine Corps recruiters needed to generate approximately 50,000 volunteers. As a result, close to 80 percent of enlistees in the Marine Corps were under the age of 25. Comparatively, personnel under 25 represented 61 percent and 50 percent of the Navy and Air Force, respectively.[58] These percentages indicated that the Marine Corps was not successful at getting its first-term enlistees to reenlist despite the offered incentives to stay. Subsequently, Marine Corps recruiters faced even higher pressure to find youthful replacements for those who declined to reenlist.

---

[55] *President's Commission on an All-Volunteer Armed Force*, 8–9.

[56] An Act to Amend the Military Selective Service Act of 1967; to Increase Military Pay; to Authorize Military Active Duty Strengths for Fiscal Year 1972; and for Other Purposes, Pub. L. No. 92-129 (1971).

[57] David E. Rosenbaum, "Nation Ends Draft, Turns to Volunteers," *New York Times*, 28 January 1973, 1.

[58] Martin Binkin and Irene Kyriakopoulos, *Youth or Experience?: Manning the Modern Military* (Washington, DC: Brookings Institution, 1979), 7.

One of the most controversial issues that emerged during the Vietnam War focused on recruit quality. During the struggle, Secretary of Defense Robert S. McNamara instituted a social experiment program called Project 100,000. In October 1966, McNamara ordered the DOD to lower the standards for minimum acceptability for new enlistees in relation to mental aptitude and physical capabilities. Prior to that month, the armed forces disqualified any recruits who failed the Armed Forces Qualification Test (AFQT) by scoring in the "10th to 30th percentile," referred to as Category IV. Under Project 100,000, the Services accepted those who placed in Category IV, referring to these combatants as "new standards men."[59] By 1969, more than 300,000 new standards men were sent to Vietnam.[60]

Due to their low aptitude scores, these servicemembers were most likely to be assigned to the combat arms, making them a higher risk to become a potential casualty. According to one study, these enlistees were "unlikely to qualify for technical training that would otherwise keep them off the front lines." Subsequently, the new standards men were "killed in disproportionate numbers," with one estimate stating that they were "three times more likely to be killed in action." During their service, these soldiers were "also reassigned 11 times more often than their peers and were between 7 and 9 times more likely to require remedial training. Project 100,000 recruits were more likely to be arrested, too." Once they returned to the United States, these veterans fared worse than "comparable men who did not join military service." As one author argues, these results from Project 100,000 made it "one of the biggest—and possibly cruelest—mistakes of the Vietnam War."[61] Most commissioned and noncommissioned officers who served in Vietnam saw McNamara's program as an abject failure that unfairly targeted the less physically and mentally qualified, forcing these soldiers to serve in

---

[59] Capt David A. Dawson, "The Impact of Project 100,000 on the Marine Corps" (master's thesis, Kansas State University, 1994), 5–7; and Matt Davis, "Project 100,000: The Vietnam War's Cruel Experiment on American Soldiers," BigThink, 14 November 2018.
[60] Davis, "Project 100,000."
[61] Davis, "Project 100,000."

a place where they were not going to succeed, possibly get killed, or potentially get somebody else killed.

Project 100,000 had various effects on the Marine Corps. Without hard data, many people came to believe that the project adversely affected the Corps' 1970s disciplinary rates. Further examination of the project's post-Vietnam War ramifications illustrates that they were not as severe as initially assumed. Historian David Anthony Dawson discovered that the Marine Corps, due to a prolonged personnel crisis during the conflict, would have likely lowered its acceptable AFQT scores even without the program. The growing unpopularity of the struggle within the American public made attracting high-quality volunteers nearly impossible. The draft only partially ameliorated the overall problem.[62]

In comparison, Vietnam was not unusual for the Marine Corps when examining the new standards men. Dawson notes that the Marine Corps used a higher number of low-score men during World War II and in Korea than in Vietnam, yet there were "no reports of rampant disciplinary problems in 1945 or 1953."[63] Still, the Marine Corps initially opposed Project 100,000 because it would cause the Service to turn away better-qualified recruits. A Marine Corps-wide study promulgated in 1967, however, claimed that the Service was "sympathetic to the program and its purpose." Additionally, the report noted, the Marines would "continue to respond willingly" provided that doing so did not adversely affect overall "combat readiness." Even so, the records demonstrate that new standards men did not perform as well in combat as higher-scoring enlistees.[64]

It was commonly believed that the military justice system more frequently punished new standards men. Statistics related to this process indicates that they were more frequently punished for minor infractions, but not significantly so.[65] For example, the nonjudicial punishment rate

[62] Dawson, "The Impact of Project 100,000 on the Marine Corps," 152.
[63] Dawson, "The Impact of Project 100,000 on the Marine Corps," 5.
[64] Dawson, "The Impact of Project 100,000 on the Marine Corps," 88–89. Dawson uses hard data and incorporates some excellent oral histories of senior Marines on active duty in the 1990s who had experienced the turmoil within the Marine Corps during the mid-1970s.
[65] Dawson, "The Impact of Project 100,000 on the Marine Corps," 5.

for new standards men was 23.8 percent, while the punishment rate for all others not in this category was 18.2 percent. While 5.3 percent of the new standards men were "convicted by courts-martial," at least 4.7 percent of all other Marines had a similar experience.[66] While many Marines were convinced that the high rate of unruliness in the post-Vietnam era was attributable to the new standards men, Project 100,000 ended in December 1971, and those enrolled in this program had been limited to a two-year enlistment contract since 1967. Although some new standards men may have been allowed to reenlist, few of these recruits, most likely, remained in the Marine Corps by 1975.

Still, the Marine Corps had the worst disciplinary rate among all the Services in the mid-1970s. Dawson believes this issue arose because "after the Marine Corps withdrew from Vietnam, the quality of new recruits remained low as a result of Marine Corps policies implemented during the transition to the All-Volunteer Force." According to Lieutenant General Samuel Jaskilka, the deputy chief of staff for manpower, the recruiting market of the immediate post-Vietnam era was the real issue. The Marine Corps elected to select potential recruits based on their "trainability" rather than "stick-to-itiveness." The Service also decided to focus on AFQT scores as opposed to a young person's possession of a high school diploma. By the early 1970s, however, the AFQT was likely compromised. Some recruiters coached their potential enlistees on how to pass the test or, in extreme cases, even instructed them how to falsify the actual test score. Nevertheless, while all the Services were quite aware that "high school graduation had been shown to be a much better predictor of successful performance than the AFQT score," only half of all incoming Marine Corps recruits possessed a high school diploma between 1971 and 1973.[67] Until General Louis H. Wilson Jr. took over as Commandant of the Marine Corps in 1975, the Service never approached more than 60 percent of its recruits being high school graduates prior to enlistment. Concerned about this situation, Congress required in 1975 that at least 55 percent of all new recruits in every Service must possess

---

[66] Dawson, "The Impact of Project 100,000 on the Marine Corps," 149.
[67] Dawson, "The Impact of Project 100,000 on the Marine Corps," 155–56.

a high school diploma. Among the nation's armed forces, only the Marine Corps needed to request a waiver for this stipulation.[68]

A highly influential member of the Senate Committee on Armed Services, Senator Samuel A. Nunn Jr. (D–GA), offered some shocking disciplinary statistics on the Marine Corps in the mid-1970s that seemed to bear out the performance difference between servicemembers without a high school diploma compared to those with one. Nunn noted that from 1973 to 1975, the absent without leave (AWOL) rate in the Marine Corps "increased by 28 percent and desertion rates increased by 66 percent."[69] Nunn pointed out that the desertion rate for the Navy had doubled during this same time frame as well. Nevertheless, the desertion rate for the Marine Corps in 1975 was "10 times the rate during World War II, three times the maximum in Korea, over twice what it was during the height of the Vietnam War, and about seven times the rate prevailing in the peacetime years of the early 60s. The Corps 1975 rates of courts-martial, [AWOL], and desertion incidents far exceeded the combined rates of the Army, Navy, and Air Force."[70]

Based on a Headquarters Marine Corps study on desertion in the ranks, defense analysts Martin Binkin and Jeffrey Record noted some disconcerting trends about the Service in the mid-1970s:

Of all the males who entered the Marine Corps in 1972, 19.0 percent of those who had not completed high school had deserted by April 1974, compared to 5.5 percent of those who had a high school diploma. In terms of mental category, the rate of desertion was inversely proportional to the score attained on standardized tests, ranging from a desertion rate of 5.6 percent among those scoring above average, 10.8 to 13.8 percent among those of the average group, and 17.1 percent among those scor-

---

[68] Allan R. Millett, *Semper Fidelis: The History of the Marine Corps* (New York: MacMillan, 1980), 619.

[69] Edward W. Brooke and Sam Nunn, *An All-Volunteer Force for the United States?* (Washington, DC: American Enterprise Institute for Public Policy Research, 1977), 15.

[70] Brooke and Nunn, *An All-Volunteer Force for the United States?*, 15; and Martin Binkin and Jeffrey Record, *Where Does the Marine Corps Go from Here?* (Washington, DC: Brookings Institution, 1976), 62–63.

ing below average. Virtually no difference was detected between racial groups; 12.3 percent of all whites and 12.9 percent of all blacks deserted. The study concludes that education level, in combination with any or all of the other factors studied (mental group, race, age, and term of service) was the best discriminator between deserters and non-deserters.[71]

During the 1970s, author Bruce Bliven Jr. estimated that the U.S. military needed approximately 2.1 million people on active duty to maintain the nation's various security commitments. To do so, the armed forces had to recruit "365,000 men and women a year," or "about 1000 volunteers per day." Maintaining such a substantial force was something that "no nation in history has attempted without relying on conscription."[72]

Yet, during the first two years of the AVF experiment, military recruiters consistently met this requirement. This success may have emerged from both a serious economic recession in 1973 that resulted from an unexpected oil embargo against the United States by the Organization of the Petroleum Exporting Countries, which pushed unemployed Americans to the armed forces, and that the Services no longer needed to maintain the high personnel numbers from the Vietnam War. While the number of lowest-scoring recruits had begun to decrease by 1975, the number of those scoring highest on the AFQT also declined. Consequently, military leadership was concerned that this situation could result in "shortages of people for the ever-growing number of military specialties that require advanced technical skills and superior intelligence."[73] In the AVF era, recruiters competed with the private sector for high-quality recruits, which was particularly problematic as the military still offered little incentive for a young person to join in the post-Vietnam War era.

Because of the negative fallout from Vietnam and congressional tampering with the G.I. Bill, the AVF could not have come along at a worse time. As early as 1972, General Robert E. Cushman Jr., the 25th Com-

---

[71] Martin and Record, *Where Does the Marine Corps Go from Here?*, 63.
[72] Bruce Bliven Jr., *Volunteers, One and All* (New York: Reader's Digest Press, 1976), 3.
[73] Bliven, *Volunteers, One and All*, 10.

mandant of the Marine Corps, noted that "today's volunteers are real ones—no longer motivated by the draft. We are in a recruiting scramble of major proportions."[74] As an example of how seriously Marine Corps leadership took the transition to an AVF is seen in how its lead recruiting advertising agency, J. Walter Thompson, a firm that had long handled the Service's public recruiting messaging, increased from just five publicists to more than a dozen to handle this work before the end of the 1970s. This move reflected that "the new world of consumer advertising would be significantly more intense, and significantly more expensive."[75] Throughout the decade, Marine Corps leadership recognized that it was in a competition for recruits like never before.

Consequently, Marine Corps leaders placed tremendous pressure on recruiting officers and stations throughout the 1970s to meet its enlistment demands at a time when national service for American youth was in low regard. To maintain certain critical occupational specialties, the DOD strongly advocated for the Uniformed Services Special Pay Act of 1972. This bill offered significant bonuses to volunteers who enlisted for at least three years in a combat arms specialty. Secretary of Defense Melvin R. Laird Jr. stated that preliminary results suggested that the bonus effectively attracted people for "a longer enlistment period" while also raising "the quality of people entering an occupational field."[76] Laird urged all the Services to apply variable reenlistment bonuses and to offer "Shortage Specialty (Proficiency) Pay" for certain critically needed enlisted skills.[77]

Pressure on the recruiters for the armed forces was so intense that Senator Nunn held hearings on the matter before the Senate Armed Ser-

---

[74] Maj Harold M. Owens, "The All-Volunteer Force: Reaction or Alternative?," *Marine Corps Gazette* 57, no. 10 (October 1973): 30. Cushman was implying that the possibility of being drafted was the driving force behind the decision of many young men to "volunteer" for a specific Service during Vietnam. "True volunteers" were those who joined after the threat of being drafted had ended.

[75] Beth Bailey, *America's Army: Making the All-Volunteer Force* (Cambridge, MA: Belknap Press of Harvard University Press, 2009), 82.

[76] Melvin R. Laird, *Progress in Ending the Draft and Achieving the All-Volunteer Force: Report to the President and the Chairmen of the Armed Services Committees of the Senate and the House of Representatives* (Washington, DC: Government Printing Office, 1972), 35.

[77] Laird, *Progress in Ending the Draft and Achieving the All-Volunteer Force*, 35–36.

vices Committee in October 1978. Nunn was concerned that the Services were falling below strength. He also wished to address some sensational recruiting malpractice that had been uncovered in Ohio. The recruiters blamed some of this misconduct—which Commandant Wilson promised would be immediately corrected—on the intense pressure they faced in finding enough qualified volunteers. A few recruiters cut some serious ethical corners to meet their goals. A number testified that recruiting for the AVF required them to work at least 15 or 16 hours per day, seven days a week. Such an operational tempo, they asserted, took a heavy toll on their personal well-being and family life. Wilson did not fully agree with this assessment. He told Nunn that while " an individual desire to succeed" in the recruiting process certainly created pressures, "professional pressure is no stranger to Marines in the execution of their assigned duties."[78] Wilson reported that he had received 523 volunteers for the recruiting service in 1977, but the Service selected only 232 "of the best." He also stated he would personally ensure that the quality of future recruiters would remain high.[79]

Despite these intense efforts, all the Services experienced recruiting shortfalls toward the late 1970s. Indiscipline created a prominent problem because a significant number of first-term recruits did not finish out their enlistment contracts. In 1978, for instance, "41.4 percent of those who were high school dropouts did not survive (for various reasons) the first three years of their enlistment, compared with 22.7 percent of diploma holders, a pattern that has prevailed at least since the end of the draft."[80] Numerous early discharges in the Navy and Marine Corps at that time were of an adverse nature and usually for character or behavioral disorders. The Government Accountability Office "estimated that attrition of those who entered the services during fiscal year 1974

---

[78] *Hearings on Military Recruiting Practices, before the Subcommittee on Manpower and Personnel of the Committee on Armed Forces, United States Senate* (10–11 October 1978) (letter from Gen Louis H. Wilson to Senator Sam Nunn, 6 September 1978), 3–4, hereafter Hearing, 10–11 October 1978.

[79] Hearing, 10–11 October 1978, 3–4.

[80] Martin Binkin, *America's Volunteer Military: Progress and Prospects* (Washington, DC: Brookings Institution, 1984), 5.

through 1977 cost the Government an estimated $5.2 billion USD in veterans benefits and unemployment compensation."[81]

According to one expert in the mid-1970s, Gary R. Nelson, the armed forces went with a sort of "volunteer-in, volunteer-out philosophy." If any servicemember was "unhappy and not performing," Nelson noted, "we let him out." This approach caused attrition to skyrocket from "a base of about 25 percent of people not completing their first terms prior to the All-Volunteer Force, to 37 percent with the 1974 cohort." He believed that higher reenlistment and lower attrition rates would reduce pressure on the recruiting service. At the same time, Nelson argued, the Services could increase their base by "bringing in people into the supply side of the market that weren't there previously." His suggestions were to recruit women, to relax physical and mental standards, and to increase the age requirement.[82] Interestingly, the Services at this point all saw the benefits of recruiting highly qualified women and expanding their eligibility for serving in certain military occupational specialties. Previously, the Gates Commission overlooked the possibility of recruiting qualified women to support the post-Vietnam AVF. By 1981, the Army hinted that the DOD may need to reinstate the draft to adequately maintain a viable and broadly representative force.[83]

The demographic makeup of the AVF changed in 1973 and continued well into the twenty-first century. For example, that year, women constituted approximately 3 percent of the U.S. armed forces. Twenty-one years later, women then composed 12 percent of it. Today, women comprise between 15 and 18 percent of the total AVF. Similarly, White enlistees in the Services dropped from a high of 82 percent in 1973 to 70

---

[81] *Attrition in the Military—An Issue Needing Management Attention* (Washington DC: General Accounting Office, 1980), i.

[82] Richard W. Hunter and Gary R. Nelson, "The All-Volunteer Force: Has It Worked, Will It Work?," in *Registration and the Draft: Proceedings of the Hoover-Rochester Conference on the All-Volunteer Force*, ed. Martin Anderson (Stanford, CA: Hoover Institution Press, 1982), 17.

[83] Binkin, *America's Volunteer Military*, 4. See George C. Wilson, "Army Hints Draft May Be Required; Manpower Demands for Reagan Policy Detailed in Report," *Washington Post*, 9 July 1981, A1. Gen Louis H. Wilson Jr., along with all the other Joint Chiefs of Staff, believed that registration for the draft should be made mandatory at least for all 18-year-old males across the nation.

percent by 1994, and Black and Hispanic representation in the ranks rose significantly. At one point in the 1990s, African-American volunteers made up nearly 37 percent of the Army. Additionally, in the two decades between 1974 and 1994, the amount of married recruits in the AVF grew from about 40 percent to more than 60 percent.[84] Despite these trends, the armed forces still had issues with some of the AVF enlistees. In 1980, the DOD reported a problem with the AFQT to Congress, acknowledging that all the Services had enlisted far more low-scoring recruits than they had previously between 1976 and 1980. DOD leadership estimated that as many as "360,000 recruits—over one-quarter of all male recruits" who volunteered in that time scored "below the minimum supposedly required for enlistment" on the aptitude test.[85]

By 1976, a number of Gallup polls considered the transition to volunteerism a disaster.[86] Further, with military pay getting buffeted by double-digit inflation and with a parsimonious Congress sharply cutting educational benefits, it is little wonder that many believed that a renewed peacetime draft would produce a better military than that of the mid-1970s AVF.

By the end of the decade, there was a growing realization that the level of military compensation was inadequate to attract and retain quality personnel. Nunn and a colleague, Senator John W. Warner III (R-VA), cosponsored an amendment that would immediately raise military pay. The Joint Chiefs of Staff strongly supported the amendment, but President James E. "Jimmy" Carter Jr.'s administration initially opposed it. After much back-and-forth testimony between Warner and officials at the Pentagon, Carter eventually acquiesced to improving benefits and pay compensation for the AVF. General Maxwell R. Thurman, the commanding general of the Army Recruiting Command and one of the prin-

---

[84] Edwin Dorn, "Sustaining the All-Volunteer Force," in *Professionals on the Front Line: Two Decades of the All-Volunteer Force*, ed. J. Eric Fredland et al. (Washington, DC: Brassey's, 1996), 19–20.

[85] Mark J. Eitelberg, "The All-Volunteer Force after Twenty Years," in *Professionals on the Front Line*, 67.

[86] Eitelberg, "The All-Volunteer Force after Twenty Years," 67. For more on the issues that the U.S. defense establishment faced after the 1970s, see "Is America Strong Enough?," *Newsweek*, 27 October 1980, 12–26.

cipal architects of the AVF, was ecstatic. "If Congress comes through with promised military pay raises and restores the GI bill," Thurman reportedly stated, "a quality All-Volunteer Force is definitely 'recruitable'."[87] Amazingly, all the Services exceeded their enlistment objectives in fiscal year 1980, exceeding the previous year by more than 52,000 recruits.[88] In sum, better military pay attracted high school graduates who evidently performed better during their first-term enlistments, which resulted in higher performing servicemembers opting to remain in the Service and reducing further pressure on the recruiting service to find replacements.

However, everything still was not entirely well with the AVF. In the aftermath of the Vietnam War, rising drug and alcohol abuse in all the Services emerged as a major problem. At that time, Marine Corps leadership appeared torn over whether to assist confirmed drug users with some sort of rehabilitation program or just to punish and discharge those who they caught. Moreover, there was no access to a generalized or an easily administered drug test. Unless someone directly caught a Marine using drugs, the Service had no quick way to identify and discharge them. Instead, the Marine Corps occasionally allowed a Marine to request voluntary enrollment in a drug treatment program, which permitted them to avoid being charged with drug use under the Uniform Code of Military Justice. Most Marines, however, chose to dodge the stigma of going through a rehabilitation program and decided to take their chances with not getting caught.[89]

Many Americans today suppose that rising drug use in the military was directly related to ill effects extending from the Vietnam War. In 2007, historian Jeremy Kuzmarov conducted a comprehensive study of the influence that mass media had on this perception of the connection between service in Vietnam and increasing drug use. Kuzmarov claims that "the mass media played a critical role in shaping the sustained 'drug panic' of the late 1960s and early 1970s." With governmental hearings

---

[87] George C. Wilson, "General Favors Volunteer Army over Draftees," *Washington Post*, 8 August 1980, A7.
[88] Rostker, *I Want You!*, 407.
[89] Cosmas and Murray, *Vietnamization and Redeployment*, 359–64.

creating "maximal public exposure" on the subject, "newspapers, magazines, and television created the impression" that drugs had caused a breakdown in the military while also supposedly becoming a "full-fledged 'epidemic' " in the United States.[90] Although Kuzmarov contends that the nation overhyped the connection between the Vietnam War and postwar drug use in the military, the use of recreational drugs, especially marijuana, was clearly on the rise across the United States in the 1970s, which was reflected in the armed forces as well. Still, the linkage between the last years of the Vietnam War and this postwar drug use remains specious. Brigadier General Edwin V. Simmons, the assistant division commander of the 1st Marine Division, argued against the connection between drugs and Vietnam War service, claiming that the Marine who arrived in Vietnam in 1971 was "probably only 13, 14, or 16 years old" when the conflict started. "So he grew up in a different high school environment than his predecessor did, five or six years ago," Simmons reasoned, "and he brought many of the attitudes of that environment into the Marine Corps with him."[91] This outlook included an increased overall tolerance for personal drug use.

For Marine Corps leadership, the Service's policies related to drug usage caused confusion. Many Marines caught with drugs in the 1970s admitted to having consumed the same substance before they joined the Service. The 24th Commandant of the Marine Corps, General Leonard F. Chapman Jr., expressed significant concern about substance abuse in the Service. He sent a message to all Marine Corps commands noting, "The Marine Corps is neither funded nor equipped to carry out the burden of noneffective members for the inordinate length of time that civilian institutions are finding necessary to achieve the rehabilitation of addicts." After remarking that reversion rates also made this untenable because they have been "discouragingly high," he stated that the Marine Corps' "medical resources are sufficiently taxed by duty-connected

---

[90] Jeremy Kuzmarov, *The Myth of the Addicted Army: Vietnam and the Modern War on Drugs* (Amherst: University of Massachusetts Press, 2009), 55.

[91] Graham A. Cosmas and LtCol Terrance P. Murray, *U.S. Marines in Vietnam: Vietnamization and Redeployment, 1970–1971* (Washington DC: Marine Corps History and Museums Division, 1986), 353.

physical problems" without adding care for substance abusers. With the reduction in strength after Vietnam, he emphasized the need for professionalism in the Service, requiring the enlistment and retention of "those who will conscientiously meet and maintain high standards," a category in which "drug users do not fit." Despite his concern, he gave "all commanders exercising general court-martial authority" the power to direct the retention of or the discharge of "any enlisted man involved with narcotics use or possession" in February 1970.[92]

Without a specific direction from the top, subordinate Marine commanders were often conflicted as to precisely how to deal with drug offenders, especially when it came to highly skilled personnel. One Marine communications officer noted that anyone reported for drug use would lose their clearance, meaning that the Service would have "one less worker." This situation, the officer stated, was "very painful to us" when "a high skilled kid [was] busted." Other commanders were confused about who to consider a first-time minor offender or an experimental user. Yet, Lieutenant General Keith B. McCutcheon stated that the Commandant's policy was simply a restatement of what Marine Corps leaders were already doing. Punishment for minor and inconsequential usage was at the discretion of the unit commander, and most followed the pattern that if the enlistee "does straighten up, he stays, and if not, then he goes out."[93] The leeway given to officers meant that their personal beliefs on whether drug use was a sickness or a crime influenced how they applied those sanctions during the 1970s. By 1980, the Marine Corps estimated that "37 percent of all Marines" were consuming illegal substances at some point.[94] Future Commandants who were armed with better and more universal methods of detection tightened up on Chapman's strict approach, much to the ultimate benefit of the Marine Corps.

The Marine Corps faced one other significant trial in the mid-1970s with the death of Private Lynn E. McClure in December 1975 while training at the Marine Corps Recruit Depot San Diego, California. McClure's

---

[92] Cosmas and Murray, *Vietnamization and Redeployment*, 360–61.
[93] Cosmas and Murray, *Vietnamization and Redeployment*, 361–62.
[94] Thomas E. Ricks, *Making the Corps* (New York: Touchstone, 1997), 22.

demise was not the first time the Service's training resulted in the death of an enlistee. In 1956, a platoon of Marine recruits at Marine Corps Recruit Depot Parris Island, South Carolina, suffered a tragedy known as the Ribbon Creek disaster. On the night of 8 April 1956, an inexperienced drill instructor, Staff Sergeant Matthew C. McKeon, "led seventy-four recruits of Platoon 71 from their barracks at the rifle range to Ribbon Creek."[95] Unhappy with what he considered his platoon's lackadaisical performance, McKeon plunged his men into the creek at night. Not realizing that the marsh was tidal, meaning that its depth depended on the time of day, he pushed his trainees into water that was over their heads. Some of them began to thrash about in a panic. Most of the exhausted and sodden platoon managed to struggle up the embankment of Ribbon Creek but, to everyone's horror, six of their comrades drowned.

All these memories came flooding back following McClure's mortal injury. For some time, incidents of recruit abuse seemed to be building, especially during the final years of the Vietnam War. For instance, in June 1968, Private Thomas Bartolomeo failed to fill his canteen as instructed and was, in turn, punched several times by a drill instructor. The punches apparently caused massive internal bleeding and led to Bartolomeo's death. His drill instructor pled guilty to maltreatment and assault but not murder. He received a short prison sentence and left the Service after his term ended.[96] However, Bartolomeo's death appeared to be just the tip of a potential iceberg. In two years, from 1964 to 1966, "120 Drill Instructors were relieved of duty at Parris Island, 73 of them for recruit maltreatment or abuse." Moreover, in a year and a half from January

---

[95] Millett, *Semper Fidelis*, 528.

[96] H. Paul Jeffers and Dick Levitan, *See Parris and Die: Brutality in the Marine Corps* (New York: Hawthorn, 1971), 4–7. This book was a harsh attack on the Marine Corps recruit training process that allegedly existed in the late 1960s and early 1970s. The foreword was written by Representative Mario A. Biaggi (D-NY), who concurred with the author's strong stance against the contemporary training methods at Parris Island. Biaggi's claims were strongly disputed by Representative L. Mendel Rivers, (D-SC), who was the chairman of the House Committee on Armed Services at the time. As was the situation with SSgt McKeon in 1956, SSgt Bronson was a good drill instructor, but he had allowed a loosely supervised motivational pugil stick session to get way out of hand. Referred to a court-martial by MajGen Kenneth J. Houghton, Bronson was later acquitted of all charges.

1968 to September 1969, "seventeen Marine recruits died at Parris Island," though the Service attributed all of these to accidents or disease.[97]

As the Marine Corps approached its 200th birthday in 1975, incidents at the recruit depots seemed to be on the rise once again, even though the vast majority of drill instructors served with professionalism and distinction and trained their recruits without incident. On 6 December 1975, however, an incident between McClure and his drill instructor, Staff Sergeant Harold L. Bronson, exposed issues within the Marines' practices. Up to this point in his training, McClure had disciplinary issues, including being absent without leave, which resulted in his transfer to a motivational platoon within the Special Training Branch (STB) at San Diego in December 1975. The Marine Corps created the STB following the Ribbon Creek tragedy to handle recruits who seemed to have the most disciplinary issues and difficulty with completing training. The worst recruits assigned to the STB were sometimes discharged as unsuitable. Those who were retained were encouraged by their drill instructors to show proper motivation so that they could be later reintroduced into the mainstream training cycle and proceed to graduation as full-fledged Marines. The whole point of being in a motivation platoon was for the recruit to work at getting out of it.[98]

It was later revealed that McClure had been recruited under suspicious circumstances. He had initially attempted to enlist in both the Army and the Air Force but was rejected due to exceptionally low aptitude test scores. He later took the AFQT for the Marine Corps and scored a 59, which was quite low but acceptable for induction at that time. McClure also lied to his recruiters about having never been in trouble with the law. Most likely, these issues led to McClure's struggles with training.[99]

---

[97] Jeffers and Levitan, *See Parris and Die*, 11–12.

[98] *Hearings on Marine Corps' Recruit Training and Recruiting Programs, before the House Armed Services Subcommittee on Military Personnel*, 94th Cong., 2d Sess. (24–26 May; 2–3, 9, 23, 29 June; and 9 August 1976), hereafter *Hearings on Marine Corps' Recruit Training and Recruiting Programs*; and LtGen Jack W. Klimp, USMC (Ret), and Col Warren Parker, USMC (Ret), "Lessons Forgotten: Marine Corps Recruit Training," *Marine Corps Gazette* 101, no. 11 (November 2017): 23–27.

[99] *Hearings on Marine Corps' Recruit Training and Recruiting Programs*.

During a poorly supervised pugil stick fighting session, Bronson allowed recruits in the platoon to forcefully strike McClure in the head despite his refusal to actively participate. Although he was wearing full protective gear, a thoroughly frightened McClure did not defend himself. Bronson ordered other recruits to continue attacking McClure even after he "fell to the ground screaming for mercy." Having received repeated blows to his head, McClure suffered a hematoma, lapsed into a coma, and died 14 weeks later. It was a tremendously ugly and brutal event that had been entirely preventable.[100]

Only weeks after the McClure incident hit national news, a recruit at Parris Island, Private Harry W. Hiscock, lost the use of one of his hands due to a hazing stunt by his drill instructor, Sergeant Robert F. Henson. In January 1976, Henson sought to scare Hiscock, who was seen as underperforming. After dumping powder out of a live round, Henson loaded it into his weapon, pointed it at Hiscock and pulled the trigger. The round, however, still contained enough powder to discharge, hitting Hiscock in the hand and causing permanent injury. To make matters worse, Henson and one fellow drill instructor then attempted to bury the episode and never reported it. In the aftermath, Henson and five other sergeants received various levels of punishment of the stunt. Later, the now-permanently disabled Hiscock detailed a culture of abuse that drill instructors, including Henson, created for the recruits of the motivational platoon. Hiscock reported that drill instructors encouraged recruits to punch and kick others for what they considered poor performance. General Wilson, having just taken over as the 26th Commandant of the Marine Corps six months earlier, was determined to end these negative practices related to recruit training as soon as possible. In a September 1976 interview with *People*, Wilson lamented that he was "embarrassed and disheartened that such things can happen. I am determined

---

[100] "The Corps on Trial," *Time*, 12 July 1976; and *Hearings on Marine Corps' Recruit Training and Recruiting Programs*. The *Time* article detailed multiple other incidents that led to serious or life-threatening injuries at both Parris Island and San Diego around the same time as McClure's death.

that they stop; we're going to conduct our training with firmness, fairness, and dignity."[101]

To implement his planned corrections, Wilson asked his most trusted senior staff officer, Lieutenant General Robert H. Barrow, the deputy chief of staff for manpower, to come up with a way for the Marine Corps to immediately fix its recruit abuse problem that had resurrected itself despite the positive changes that followed the Ribbon Creek affair. Additionally, Barrow was convinced that the Service needed to address recruiting irregularities that allowed McClure to enlist despite his issues. Testifying before the House Committee on Armed Services in June 1976, Barrow stated that the "objective of our recruitment effort is the enlistment of high-quality men and women. Unfortunately, and in spite of built-in safeguards, some individuals who did not meet our quality standards, principally in terms of mental and moral qualifications, have been enlisted." Barrow recommended that going forward the Marine Corps needed to focus on better screening and training of recruiting service personnel, closer scrutiny of the management of the recruiting service, and a reduction of quota pressure that seemed to be driving recruiters to cut corners.[102]

In addition to increasing the quality of recruits, Wilson and Barrow implemented changes specifically to reduce the abusive actions of drill instructors. Barrow also insisted on improved training and screening for all drill instructors, including implementing a psychiatric screening for all instructor candidates and the "execution of a sound training syllabus." Drill instructors were no longer allowed to "put [their] hands on a recruit in any way." Additionally, the hours of training and hours of drill instructor interaction with the recruits was reduced. Most significantly, the Marine Corps leadership was "strengthening our supervision

---

[101] *Hearings on Marine Corps' Recruit Training and Recruiting Programs, before the House Armed Services Subcommittee on Military Personnel* (25 May 1976) (testimony of Pvt Harry Hiscock); Clare Crawford-Mason, "Boot Camp Should Be Tough, but Never Brutal: Gen. Wilson Tells that to the Marines," *People*, 13 September 1976; and James P. Sterba, "Marine Recruit Abuse Continues," *New York Times*, 7 March 1976, 1.

[102] *Hearings on Marine Corps' Recruit Training and Recruiting Programs, before the House Armed Services Subcommittee on Military Personnel* (26 May 1976) (statements of Gen Louis H. Wilson, LtGen Robert H. Barrow, and BGen Richard C. Schulze), hereafter Hearing, 26 May 1976.

at all levels," ensuring that at least two officers were in place to watch over individual groups of enlistees and their treatment.[103]

Barrow announced that from June 1976 onward, recruit depot commanders would have full control of the operational aspects of recruiting and recruit training as well as the responsibility for recruits to successfully make it through boot camp. If a recruit was discharged before graduation for reasons of unsuitability, false enlistment, or other recruiter wrongdoing, then the recruiting station did not receive credit and would, in turn, bring scrutiny on whoever had sent that recruit to boot camp. Barrow also promised to implement more thorough background checks on the moral fitness of potential recruits and, most importantly, the shutting down of all the motivation platoons. Further, Barrow reported that there would be at least 2 general officers assigned to each recruit depot and an additional 84 junior officers added to "perform duties as assistant series commanders and company executive officers."[104]

During the last week of May 1976, Wilson also testified before the House Committee on Armed Services. He made specific reference to the McClure and Hiscock incidents, stating that, as Commandant, he took "full responsibility for the unacceptable actions of a few Marines." He was also convinced of the need for changes in both the recruiting service and training. In Wilson's view, "There are no more demanding, challenging, or important assignments for Marines, both officers and NCOs, than those associated with recruiting and recruit training. They are in every sense the lifeline of the Corps." Wilson boldly announced that henceforth recruits were going to meet "quality instead of quantity," committing to reducing the Service's end strength to "meet our quality goals." In doing so, he commented, the Marine Corps would require that recruits with a high school diploma should count for "three out of every four" enlistees.[105]

---

[103] Hearing, 26 May 1976; and Bernard Weinraub, "The Marine Corps Is Softening the Role of the Hard-Boiled Drill Instructor," *New York Times*, 11 May 1977.
[104] Hearing, 26 May 1976.
[105] Hearing, 26 May 1976.

Figure 29. Marine Corps recruiting poster from the 1980s

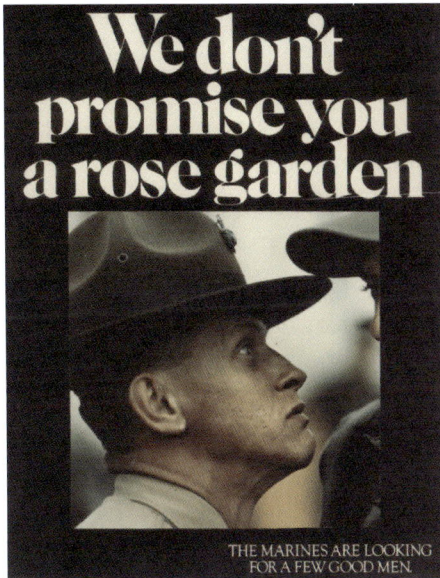

We don't promise you a rose garden

THE MARINES ARE LOOKING FOR A FEW GOOD MEN.

Source: Marine Corps History Division.

Wilson was adamant about recruit quality. In his first year as Commandant, he stated to the committee that the ratio of high school graduates in the Service rose to 67 percent and that he set the goal of 75 percent for fiscal year 1977. He forcefully informed the entire Marine Corps that these figures were a hard and fast requirement for the recruiting service rather than an idealized goal. Wilson noted that by 1976, the "unauthorized absence rate" fell "27 percent since 1975; the average monthly desertion rate has gone down 29 percent; the confined population has gone down 24 percent. . . . the reenlistment rate has gone up 46 percent; and the deserter at large population has gone down 24 percent." Wilson also emphasized that the release of 272 recruiters in 1975 and early 1976 occurred for widely various reasons from general "indifference" to the recruiting service to a fundamental "inability to communicate" with potential recruits rather than not meeting their recruiting goals.[106] The Wilson–Barrow era of Marine Corps history initiated a serious transition in recruitment practice that continues today.

By the end of the 1970s, the Marine Corps had ostensibly turned a corner on racial issues, though these issues never seemed to go away fully. The new Unit Deployment Program, initiated in 1977, seemed to alleviate much of the racial unrest that had long pervaded III MAF units on Okinawa. Stabilizing unit personnel "churn" played "a major role in the

---

[106] Hearing, 26 May 1976.

reduction of racial tension and indiscipline rates." By 1979, "the situation had improved to the point that Headquarters Marine Corps' network of equal opportunity consultants rarely had anything to report." Nine years later, progress with race relations had reached a point in which "there were only three confirmed cases of racial discrimination that year and all were resolved to the satisfaction of the complainant."[107]

By the mid-1970s, it was evident that the Wilson-Barrow team would make significant changes for the future Marine Corps. The issue of recruitment and personnel in the AVF era had created several self-inflicted wounds for the Service that both generals were determined to overcome. During Wilson's congressional testimony in June 1976, an aide allegedly had informed Wilson that "Parris Island had not yet disbanded its motivational platoon, to which Wilson replied, 'By the time I go back in there to resume testimony, that platoon will be gone'." Allegedly, the Marines in South Carolina immediately brought bulldozers from "the Beaufort Marine Corps Air Station to fill in the Motivation Ditch, and the Motivation Platoon was indeed gone—*forever*."[108] It was a bold statement, but it was one that was highly illustrative of Wilson's determination to create a new high-quality Marine Corps that was trained at boot camp with dignity and without lowering professional standards. Starting with Wilson, the emphasis on quality personnel shifted into high gear.

---

[107] Packard, " 'The Marine Corps' Long March'," 93–95.
[108] *Making Marines in the All-Volunteer Era: Recruiting Core Values, and the Perpetuation of Our Ethos* (Quantico, VA: Lejeune Leadership Institute, Marine Corps University, 2018), 19. Although the story may have been apocryphal, it was highly illustrative of Wilson's determination to reform the recruit depots following the McClure and Hiscock incidents.

CHAPTER FOUR

The Cushman-Wilson-Barrow Era

On New Year's Day 1972, the office of the Commandant of the Marine Corps was about to change over. President Richard M. Nixon had nominated Lieutenant General Robert E. Cushman Jr. to become the 25th Commandant the previous November. Cushman was a familiar figure to Nixon. He had previously served in the late 1950s as then–Vice President Nixon's military national security advisor, and he later served as Nixon's deputy director of the Central Intelligence Agency (CIA) in 1969–71 before being nominated as the next Commandant. Cushman possessed a sterling combat record and had been awarded the Navy Cross after the Battle of Guam in 1944, during which Louis H. Wilson Jr. was one of his company commanders. Cushman's most significant combat action was at the Battle of Iwo Jima in 1945, during which he commanded the 2d Battalion, 9th Marines. In 1968, Cushman replaced Lieutenant General Lewis W. Walt as the commander of III Marine Amphibious Force (III MAF). When Cushman returned from Vietnam in 1969, Nixon appointed him to the position of CIA deputy director. While this post could easily move him into being the director of the CIA, Cushman personally felt that, if offered, "no Marine could ever possibly turn down the job of Commandant."[1]

---

[1] Col John Grider Miller, "Robert Everton Cushman Jr.," in *Commandants of the Marine Corps,* ed. Allan R. Millett and Jack Shulimson (Annapolis, MD: Naval Institute Press, 2004), 411–13.

Cushman's nomination surprised many Marine Corps insiders. He had recently retired as a general officer shortly before joining the CIA, and most in the Service expected either Medal of Honor recipient and then-Assistant Commandant of the Marine Corps General Raymond G. Davis Jr. or the highly regarded Headquarters Marine Corps chief of staff Lieutenant General John R. Chaisson to get the position. To make matters more difficult for Cushman, both the Chief of Staff of the U.S. Army, General William C. Westmoreland, and the Chief of Naval Operations (CNO), Admiral Elmo R. Zumwalt Jr., publicly stated that they wished Nixon had selected Chaisson. Cushman, however, proved he was a tough and dedicated bureaucratic infighter. He recognized that the end of the national draft would make recruitment more difficult than anyone had previously supposed 18 months prior to its conclusion. Cushman was adamant that the Marine Corps needed to return to its core competency of amphibious warfare and called for a "close partnership with the Navy."[2] The CNO was not as interested with this prospect. More concerned with rising Soviet naval power and its threat to the North Atlantic Treaty Organization (NATO) alliance, the Navy did not want to have to focus on amphibious warfare as well at that moment.

Early on, Cushman faced an immense challenge in resetting the force that had just emerged from Vietnam. Although personnel problems were significant, Cushman confronted an even more significant dilemma in articulating a new role and mission for the Marine Corps. Thanks to the War Powers Resolution, most post-Vietnam national security planners thought it was extremely doubtful that the United States would engage in any sort of overseas contingency operation, believing that the Services would only fight large-scale wars in the future. Referred to as the "Vietnam Syndrome," the concept suggested that the unpopular war had so damaged the social and national security fabric of the United States that it would go to great lengths to avoid any similar conflicts.[3]

---

[2] Miller, "Robert Everton Cushman Jr.," 412, 418.
[3] See Martin Kalb, "It's Called the Vietnam Syndrome, and It's Back," Brookings, 22 January 2013.

Figure 30. Gen Robert E. Cushman Jr.

Source: official U.S. Marine Corps photo.

During this same time frame, the Soviet Union had taken advantage of the U.S. preoccupation in Southeast Asia and greatly strengthened its military might. Under the leadership of Soviet Navy admiral Sergey G. Gorshkov, for instance, the Soviet Union had quietly created the second largest navy in the world, had acquired several naval infantry brigades (the Soviet equivalent of the Marine Corps), and possessed nearly 400 modern submarines, two small aircraft carriers, and hundreds of highly lethal surface combatants. By the mid-1970s, the Soviet Navy was a respected blue-water force. As early as 1968, Gorshkov warned that "the flag of the Soviet navy now proudly flies over the oceans of the world. Sooner or later, the U.S. will have to understand that it no longer has mastery of the seas."[4] The primary mission of Gorshkov's largest Soviet fleet, based at Polyarny, Murmansk Oblast, was to pry open NATO's Greenland-Iceland-United Kingdom (GIUK) gap to interdict the Atlantic sea lines of communication between Europe and the United States, primarily with their submarines. Gorshkov believed that a strong navy was absolutely necessary if the Soviet Union ever went to war with NATO. In this scenario, Gorshkov argued, being able to fight the U.S. Navy would allow the Soviet Union to interdict any potential reinforcement of NATO forces in Europe from the U.S. and Canadian mainland.[5]

Throughout the immediate post-Vietnam War era, Gorshkov's Red Fleet and the blue-water threat it represented attracted the primary focus of his counterparts in the U.S. Navy. Although wanting to center the Marine Corps on amphibious operations, Cushman recognized that his Service, for the first time since the Battle of Belleau Wood in 1918, may need to take part in a major ground war somewhere in Europe, most likely on NATO's Northern Flank to defend the GIUK gap. In addition, for the first time since 1945, the Marines would have to fight against a real

---

[4] Adm Sergey Gorshkov, quoted in "Power Play on the Oceans," *Time*, 23 February 1968, 23. An excellent article on the impact of Adm Gorshkov and Soviet maritime strategy of the 1980s is Donald Chipman, "Admiral Gorshkov and the Soviet Navy," in *The Legacy of American Naval Power: Reinvigorating Maritime Strategic Thought, an Anthology*, ed. Paul Westermeyer (Quantico, VA: Marine Corps History Division, 2019), 192–208.

[5] Donald Chipman, "Admiral Gorshkov and the Soviet Navy," *Air University Review* 33, no. 5 (July–August 1982): 28–47.

Figure 31. Soviet Cold War-era naval vessels

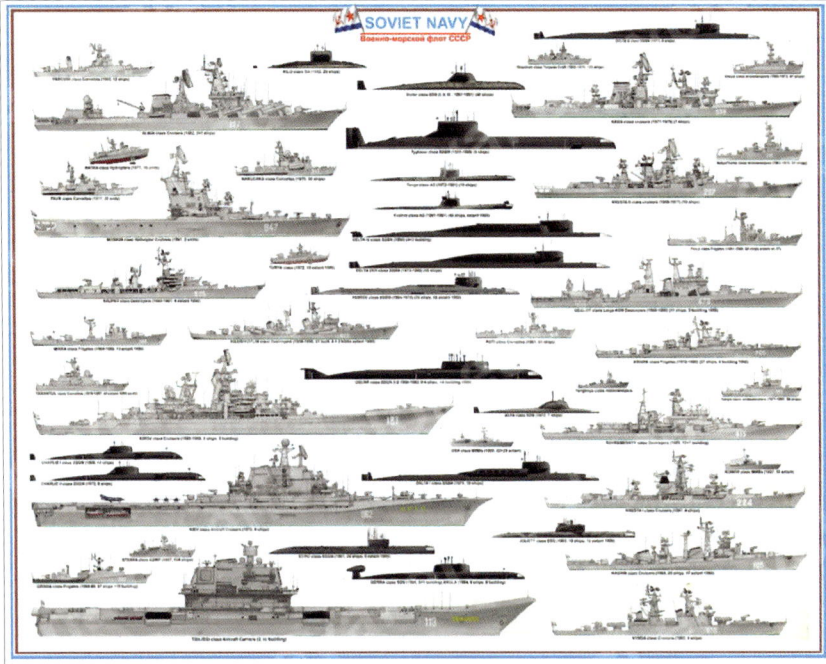

Source: official U.S. Navy photo.

military peer competitor to the United States. Furthermore, the Soviet military was mechanized, technologically advanced, and capable of rapid maneuver over great distances, whereas the Marine Corps had been fighting mainly as light infantry in the jungles of Vietnam for the previous eight years.

This meant that the Marine Corps was not organized or equipped at the time to fight an enemy with the military capabilities of the Soviet Union, which was something Cushman had to address. During the 1950s and 1960s, in anticipation of a more streamlined and lighter Fleet Marine Force (FMF), the Marine Corps acquired more capable helicopters. New helicopter technology enabled Marines to fly far inland over traditional beach obstacles, seawalls, and close-in sea and land mines and to get ashore in a fraction of the time compared to using conventional, slow-moving armored amphibious vehicles. Before and during the Vietnam War, however, the Marine Corps moved to divest itself of any organizational force structure that was incompatible with helicopters being the primary conveyor of Marine Corps tactical assault forces. Further, the Marine Corps in the Vietnam era reduced its strength by 2,000 personnel, created a fourth rifle company for its infantry battalions, transferred the command of division tank battalions to force troops, expanded the division reconnaissance company into a battalion, and added an anti-tank battalion with 45 M50 Ontos self-propelled tank destroyer weapons to each division.[6] The Cold War-era Marine aircraft wing incorporated more rotary-wing squadrons at the expense of fixed-wing elements, and the Service placed large segments of heavy engineers, bridging, and self-propelled artillery into force troops. In sum, the Vietnam War-era Marine division was lighter and more infantry-centric, giving it the ability to lift a significant amount of its combat power over long distances with new helicopter technology. By the 1970s, it was clear that this table of organization would not do if the division was required to deploy to Europe to face off against highly mechanized Soviet combat brigades.

From the late 1950s and into the early 1970s, the Marine Corps emphasized the key "principles of austerity and mobility." Yet, all this came at a cost. The Marine Corps was taking a "substantial calculated risk with fire support and logistical capability. It was assumed that Marine close air support and naval gunfire would offset the losses in artillery and tank fire in the early stages of an assault."[7] In 1957, the 21st Commandant of the Marine Corps, General Randolph M. Pate, accepted the majority of recommendations from a commission under Lieutenant General Robert E. Hogaboom, known as the Hogaboom Commission, that called for a lighter Marine division to take advantage of new helicopter technology. In 1972, Cushman inherited this Marine Corps, which was indeed

---

[6] LtCol Eugene W. Rawlins, *Marines and Helicopters, 1946–1962*, ed. Maj William J. Sambito (Washington, DC: Marine Corps History and Museums Division, 1976), 73–74. *Ontos* means "the thing" in Greek. The weapon consisted of a self-propelled platform for six 106-millimeter recoilless rifle rocket launchers. While the Marine Corps abandoned this weapon by the mid-1970s, single ground-mounted 106-millimeter recoilless rifles remained in the inventory throughout that decade.

[7] Allan R. Millett, *Semper Fidelis: The History of the United States Marine Corps* (New York: MacMillan, 1980), 527.

Figure 32. Marine Corps M-50 Ontos in Vietnam

Source: official U.S. Marine Corps photo.

lighter and faster but lacked heavier combat power at all levels. While suitable for small-unit infantry operations in Vietnam, it placed the Marine Corps at yet another role and mission crossroads in the mid-1970s.[8]

At this time, inflation in the United States caused the actual cost of personnel and equipment to skyrocket. Cushman noted at a general officers conference at Headquarters Marine Corps in July 1973 that the Service had "asked for more FY 74 money than last year, but it is buying us less." While the financial situation for the Marines was difficult, Cushman optimistically pointed out that "there is a growth in a general awareness of the utility of Marines and amphibious forces in a wide range of missions. This has come about, in part, through our heavy schedule

---

[8] Mark A. Olinger, *Conceptual Underpinnings of the Air Assault Concept: The Hogaboom, Rogers and Howze Board* (Arlington, VA: Institute of Land Warfare, Association of the United States Army, 2006).

Figure 33. LtGen Robert E. Hogaboom

Source: official U.S. Marine Corps photo.

of participation in amphibious exercises in the NATO arena." Cushman further reported that the Service was going to have to work hard to overcome its equipment modernization issue. He also cautioned that "we need to state our case effectively and to be able to *prove* it, if necessary."[9]

During his term as Commandant, Cushman handled another critical issue related to the status of Marine Corps aviation. The Corps possessed multiple assumptions linked to its aviation. The Service believed that its ground operations demanded abundant close air support, especially when not in range of naval gunfire and heavy artillery, and that any landing forces needed helicopters to carry troops and supplies as well as gunships and planes to provide protection.[10] The Marine Corps' tactical aviation had to remain organic to the Marine air-ground task force (MAGTF), as the Corps could no longer entirely rely on Navy aircraft carriers for support as had been provided during World War II and the Korean War. Moreover, due to the new Soviet blue-water fleet, the U.S. Navy's carriers would have to deal primarily with Soviet submarines and surface combatants. Admiral Elmo R. Zumwalt summed up the Navy's attitude about its carriers in testimony before Congress. Using the potential of supporting a Marine landing in Norway as an example, he stated that the Navy would "have to get our carriers out of there just as soon as the Marines were ashore and had an airstrip," similar to when they did so in World War II "under a far more permissive situation." In this case, the Navy would have to take these actions "in order to get back to the job of protecting the sea lines of communication against a tremendous threat."[11] Needing on-call tactical aircraft for its maneuver element ashore, and to not be reliant on a floating platform that could move away on a moment's notice, the Marine Corps heavily invested in short airfields for tactical support (SATS), which could be quickly con-

---

[9] Gen Robert E. Cushman Jr., "Corps Operations Facing Austerity," *Marine Corps Gazette* 57, no. 8 (August 1973): 2–3.

[10] Millett, *Semper Fidelis*, 611.

[11] *Hearings on Fiscal Year 1974 Authorization for Military Procurement, Research and Development, Construction Authorization for the Safeguard ABM, and Active Duty and Selected Reserve Strengths, before the Senate Committee on Armed Services*, pt. 2 (13 April 1973) (testimony of Adm Elmo R. Zumwalt Jr., USN), hereafter Hearing, 13 April 1973; and Martin Binkin and Jeffrey Record, *Where Does the Marine Corps Go from Here?* (Washington, DC: Brookings Institution, 1976), 44.

structed and operated like the deck of a stationary aircraft carrier. Depending on the terrain, SATS airstrips required approximately 10 days to build, although some experts argued it could take as few as 5 days.

This issue was the primary reason behind the Marine Corps becoming enamored with new vertical and/or short takeoff and landing aircraft (V/STOL), such as the British-built Hawker-Siddeley AV-8A Harrier, in the early 1970s. The Navy hated the Harrier and deemed it largely unsuitable for their large-deck amphibious ships, but the Marine Corps hoped to initially replace all its obsolete A-4 Skyhawk attack aircraft with V/STOL platforms. Further, the V/STOL concept was "uniquely suited to amphibious, force in readiness aviation." The aircraft was truly expeditionary and was available to deliver ordnance long before a more-conventional airstrip could be built. It could also employ existing improved roadways for short takeoff and landing operations. Although the aircraft was "less capable and more difficult to fly and maintain than originally anticipated," the Marines felt that the V/STOL option was "the ultimate solution to its fixed-wing basing problems."[12]

Even with this solution, Cushman's issues with fixed-wing aviation continued. This time, the issue was about which aviation missions the detachments should take on. During Cushman's tenure as Commandant, the Navy decided to go with the Grumman F-14 Tomcat air superiority fighter, which, like the Vietnam-era McDonnell Douglas F-4 Phantom II, could provide both close air support and attack capabilities if necessary, although the Navy did not envision it conducting attack missions often. The problem with the Tomcat from the onset was its extraordinary cost. During fiscal year 1974, Cushman told the members of the Senate Committee on Armed Services that he frankly did not need the Tomcat and much preferred the venerable F-4J Phantom II aircraft due to its superiority in delivering close air support, a demand required to follow Marine Corps amphibious doctrine. Zumwalt countered Cushman, stating

---

[12] Millett, *Semper Fidelis*, 612. Millett noted that by 1979 the Marine Corps had experienced 33 AV-8A crashes. For a short time, the decision was made that no Harriers were flown by lieutenants fresh out of flight school. The second-generation AV-8B produced by McDonnell-Douglas was a much more effective and safer aircraft.

Figure 34. Hawker Siddeley AV-8A Harrier attack aircraft

Source: official U.S. Marine Corps photo.

that he was worried about having "to maintain superiority over a beach-head" and that the Navy would "want to get away from the beachhead" as quickly as possible "in order to have sea room and the flexibility of being able to deal with the threat of distances." In sum, Zumwalt argued that the Navy would have to abandon the Marines to "a very great air threat under some circumstances if they did not have the F-14 ashore."[13] Consequently, Zumwalt urged the Marines to acquire the F-14, an excellent air superiority fighter that could simultaneously defend aircraft carriers and a beachhead but with far less attack capability, which was exactly the opposite type of aircraft the Marines needed.

In a rather incredible turnaround, Cushman changed his mind. Instead of purchasing the 138 F-4J Phantoms he had first planned on procuring for approximately $890 million (USD), he ordered 68 Tomcats,

---

[13] Hearings, 13 April 1973; and Binkin and Record, *Where Does the Marine Corps Go from Here?*, 45–46.

"enough to equip 4 squadrons," at an estimated $1 billion (USD).[14] Meeting with the CNO and the secretary of the Navy over a weekend, Cushman stated that he saw the F–14 purchase as ultimately benefitting the Marine Corps with an advanced fighter long before the McDonnell Douglas F–18 Hornet fighter aircraft would be available in the early 1980s. Some observers, however, saw this as a "Machiavellian" move that the Navy commonly employed since World War II, in which they saddled the Marines with equipment they did not want so that the "so-called Marine

Figure 35. Tomcat aviation patch for Marine Fighter Squadron 531

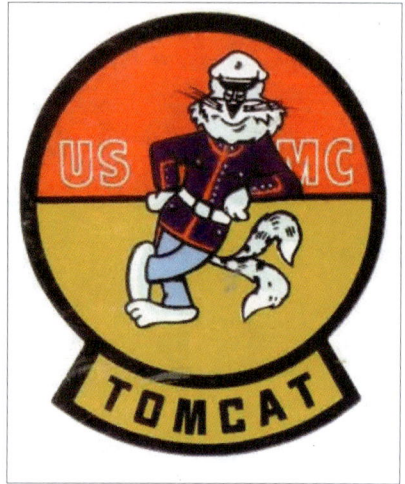

Source: official U.S. Marine Corps photo.

F–14s will later, if not sooner, end up on Navy flight decks while the Marines get lower quality F–18 Hornets in return."[15] According to Cushman, because the Navy Department buys all naval aviation platforms for both Services, it essentially forced the F–14 procurement on the Marine Corps by refusing to replace its aging fleet of F–4s and A–4s. Essentially, he claimed, the Navy had the Marine Corps "over a barrel." Cushman later admitted that it was "embarrassing" to explain to Congress why he changed his mind on the F–14.[16]

Nevertheless, soon after taking over as Commandant from Cushman in 1975, General Wilson moved to cancel the purchase of F–14s. He announced that the Marine Corps would retain the F–4J until the F–18 became available. Wilson argued that in doing so, "we are reaffirming a basic concept that air defenses in the initial stages of an amphibious

---

[14] Binkin and Record, *Where Does the Marine Corps Go from Here?*, 46.

[15] Binkin and Record, *Where Does the Marine Corps Go from Here?*, 46; and Robert D. Heinl Jr., "Marine Corps in the Middle on Fighter Plane Choice," *Detroit (MI) News*, 22 May 1975.

[16] Gen Robert E. Cushman Jr., USMC (Ret), Oral History Transcript, 9 November 1982 session, Benis M. Frank interviewer, Marine Corps History Division, Quantico, VA, 1984, 354.

operation will be provided by carrier based Navy aircraft."[17] Wilson was basically demanding that the Navy remain true to its role of protecting Marine beachheads until the Marines established their organic aviation ashore. In Wilson's mind, five days was not an excessive amount of time for the Navy to wait, and the interval could be reduced with the purchase of more Harriers.

Before Wilson became intertwined in this controversy about aviation platforms, Cushman was involved in selecting Wilson as his replacement before permanently retiring from active duty. Cushman had announced early in 1975 that he intended to retire before the end of the fiscal year (30 September 1975) due to a pay inversion issue that was going to affect general officer retired pay for those who left after 1 October 1975. Cushman strongly desired that his Assistant Commandant, General Earl E. Anderson, succeed him. While Cushman's choice for his successor might not seem controversial on the surface, he made it so by requesting that all active-duty lieutenant generals announce their plans to retire, making them ineligible for the position of Commandant. Wilson, now a lieutenant general and the commanding general of Fleet Marine Force, Pacific (FMFPAC), and highly dissatisfied with the state of the Service, refused Cushman's request. After a frank conversation with Cushman, Wilson made it clear that even if Cushman demoted him to the rank of major general, he would not retire. Cushman was not inclined to force the issue, telling Wilson that he was fine with him remaining at FMFPAC and that if he were "taking a run at the job, okay."[18]

General Anderson's selection as Commandant would have introduced additional issues. At the start of his Marine Corps career in 1940, Anderson primarily served as an infantry officer on sea duty with the Navy, including being aboard USS *Yorktown* (CV 5) when the Japanese sank it at the Battle of Midway. He survived the ordeal and eventually transferred into naval aviation, earning his gold wings as a Marine pilot before the

---

[17] Binkin and Record, *Where Does the Marine Corps Go from Here?*, 46–47.
[18] Cushman, Oral History Transcript, 404–5.

war ended.[19] In the entire history of the Marine Corps to that point, no officer connected with naval aviation had ever been allowed to rise higher than the position of Assistant Commandant, which lasted until the historic appointment of General James F. Amos as the 35th Commandant of the Marine Corps in 2010. Not having a naval aviator as Commandant had become a sort of tradition for the Marines.[20]

To make matters worse, Cushman sent out a supposedly anonymous survey to the Marine Corps general officer corps before his retirement that asked them to name who they believed should be the next Commandant. Despite the promise of anonymity, it had allegedly been coded so that Cushman could specifically tell who supported the selection of Anderson and who did not. In his oral history, Cushman admitted that he had "geographically coded" the letters and, when asked about it, stated that he thought that all the previous Commandants had coded the questionnaires by region vice knowing who actually wrote them.[21] The entire issue quickly mushroomed. In a report in *The Washington Post*, staff writer Michael Getler wrote that "roughly half of the 70 Marine generals expressed some private concern that forms they received from the commandant's office may have been typed in a fashion that would identify them and their views to other officers."[22] Both Cushman and Anderson vehemently denied these charges, but the report had damaged trust for them among the Marine Corps general officer corps.

In hindsight, it is hard to see what the alleged coding tempest was about other than creating a gigantic rift in the general officer corps. For those officers, the real issue seemed to be that the secret coding concept implied that Cushman was worried about his generals possibly "stuffing the ballot box" against his preferred successor in Anderson.[23] Despite

---

[19] "General Earl Edward Anderson, USMC (Deceased)," Marine Corps History Division, accessed 20 January 2024.

[20] G. H. Dodenhoff, "Why Not an Aviator Commandant?," U.S. Naval Institute *Proceedings* 125, no. 8 (August 1999).

[21] Cushman, Oral History Transcript, 395–98.

[22] Michael Getler, "Dispute Embroils Marines: Possible Coding of Letters Alarms Generals," *Washington Post*, 10 April 1975, A1; and Cushman, Oral History Transcript, 395–98.

[23] Gen Wallace M. Greene, quoted in Michael Getler, "Marine Chief Admits 'Mistake' on Idea to Code Officer Poll," *Washington Post*, 12 April 1975, A4.

this perception, the analogy did not fit, as the selection of the next Commandant is not done by an internal vote. In modern times, the decision has been predicated on the combined recommendation of the outgoing Commandant, the secretary of the Navy, and, since 1947, the secretary of defense. The president would then make the final selection. Sometimes the president agreed with the picking of the recommended nominee. When there was no consensus reached, however, the president would either receive a list of officers to interview or directly select from, or the president made their own choice, such as when Nixon selected Cushman in late 1971.[24] Cushman believed that the early selection of his successor would allow these general officers to "gauge their prospects for advancement under his successor" and consider their plans for retirement, including early retirement, which could "ultimately save them money."[25] Instead, both Cushman and Anderson retired on 1 July 1975.

Wilson received his historic appointment soon after this controversy. Like his predecessor, he needed to focus on the continued actions related to personnel issues, recruiting, recruit training, and growing Marine Corps-wide indiscipline. At the same time, he had to articulate an amphibiously oriented future role and mission for the Service in the new War Powers Resolution era. Wilson did not intend for the Marine Corps to become akin to a second land army. Rather, he believed in preparing Marines to fight engagements across the entire spectrum of combat, but at no time should the Service lose sight of its expeditionary nature. Wilson wanted to create a disciplined Marine Corps that was light enough to quickly get to the fight but heavy enough to win. At his change of command ceremony, Wilson stated that he was going to "call upon all Marines to get in step, and to do it smartly."[26]

---

[24] Edgar F. Puryear Jr., *Marine Corps Generalship* (Washington, DC: National Defense University Press, 2009), 28–42. Puryear had extraordinary access to many still-living Commandants when he wrote this book. A significant part of his second chapter deals with how an officer is selected for Commandant. As Puryear demonstrated, every successful candidate for the office of the Commandant blazed a uniquely individualistic path on their way to the top.

[25] Miller, "Robert Everton Cushman Jr.," 426.

[26] Col David H. White Jr., USMCR, "Louis H. Wilson, Jr.," in *Commandants of the Marine Corps*, 429.

Figure 36. Gen Louis H. Wilson Jr.

Source: official U.S. Marine Corps photo.

Figure 37. Gen Robert H. Barrow

Source: official U.S. Marine Corps photo.

Just two months later, Wilson provided remarks at a District of Columbia Navy League "welcome aboard" luncheon. He emphatically stated that when he "called upon all Marines to get in step" after taking over in July, it was "a direct invitation to those on the fringes—marginal performers in one or more areas—to make the extra personal effort to catch up with the rest of the Corps." He emphasized that this targeted both "the visible indicators of self-discipline, such as physical fitness and military appearance—to include weight control" and the "less tangible indicators as well—the ones that show up in performance on the job." Wilson vividly illustrated for his audience the new direction that he was giving to the recruiting service about finding quality future Marines at all costs. Consequently, he told them that from now on, "recruiters are to report directly to the Commanding Generals at the Parris Island and San Diego recruit depots," an arrangement that would "provide positive command attention to the recruiting effort by the same commander" who held the responsibility to train them. "Our goal," Wilson summarized, "is to have a Corps where three out of four Marines have a high school diploma when they come in, and the remainder have completed 10th grade or higher." While Wilson insisted on ensuring quality recruits, he also stressed "taking aggressive action to remove from our ranks those Marines who have demonstrated their inability or unwillingness to get with the program, noting that they had discharged "about 1500 of these substandard performers" already.[27] Wilson was convinced that a smaller, more disciplined Marine Corps was better than the one he inherited from his predecessor in the long run.

Wilson faced daunting challenges in just about every conceivable area of the Service. Captain Arthur S. Weber Jr. described his experience with the 4th Marines on Okinawa in 1975 in extremely dismal terms. He asserted that "the Battalion Landing Teams that go afloat are not adequately prepared for combat or amphibious assault." Furthermore, in his own company, Weber said that three out of his four noncommissioned officers had "departed before we began our lock-on training." Weber

---

[27] Gen Louis H. Wilson Jr., "New Directions Remarks," District of Columbia Navy League 1975 Welcome Aboard Luncheon, 10 September 1975.

lamented that most of his Marines lacked even basic combat skills, made more difficult by their lack of ability to train due to a shortage of adequate on-island training ranges and money for fuel and ammunition. He was especially concerned that "known and identified trouble-makers with lengthy disciplinary problems could not be separated because of a higher headquarters 'quota' policy." Considering these problems and the lack of training, Weber openly questioned the wisdom of III MAF remaining on Okinawa, arguing that "the Marine Corps may be overstating its current capabilities and leaving many very serious problems unsolved."[28]

Starting with his first year as Commandant, Wilson decided to do what he could to improve discipline and morale in the ranks. He instinctively knew that the chronic troublemakers and discipline cases were holding everyone back. Consequently, he created a tool for unit commanders that enabled them to remove the worst offenders. It was common in the mid-1970s for some Marines to have dozens of nonjudicial punishment offenses recorded in their service record books. Most of the offenses, such as unauthorized absences, disrespect, and missing from their appointed place of duty, were minor. Some received summary courts-martial for drug and alcohol offenses or barracks theft. Others who committed more serious crimes were usually given a special court-martial and were often, but not always, expelled via a bad conduct discharge.[29]

By late 1975, Wilson believed that he had to "implement a radical personnel proposal" to address the situation. With the consent of Secretary of Defense James R. Schlesinger, Wilson reformed the structure

---

[28] Capt Arthur S. Weber Jr., "Unsolved Problem Areas," *Marine Corps Gazette* 59, no. 6 (June 1975): 41–42. The author reported to the 3d Marine Division in January 1980, and the situation did not seem to have changed much from Weber's experience in 1975. The single-year rotational tour was largely to blame for the difficulties of III MAF, but this situation vastly improved once the unit rotational program (UDP) became the norm by the mid-1980s. The author returned to Okinawa in 1986–87 in command of Delta Battery, 2d Battalion, 11th Marines, based out of Camp Pendleton, CA. Delta Battery was the first Marine Corps artillery unit to participate in the UDP. Thanks to the program, the situation on Okinawa had changed for the better, although the training areas for the 3d Marine Division were still inadequate and remain so today.

[29] Author's personal recollections; and Phillip Carter and Owen West, "Dismissed!," *Slate*, 2 June 2005.

around "the authority to discharge unworthy Marines," moving it "down to the battalion level." Under this new arrangement, known as the "expeditious discharge program," battalion commanders "quickly cut 6,000 undesirables." All the U.S. Services felt the reverberations of Wilson's system, which paved the "way for the subsequent military performance surge" under the administration of President Ronald W. Reagan.[30] In this pursuit of keeping the ranks filled with good, high-performing Marines, Wilson recognized that the process would likely cause the Marine Corps to temporarily fall below its congressionally prescribed personnel levels. He trusted that it was worth the risk. He believed that the renewed vigor and emphasis that his reformations provided toward recruiting would bring high-quality young people into the Marine Corps, resulting in better behaved and higher-performing Marines.

Wilson's reforms had a clear effect within a year. One reporter from *The Boston Globe* reflected on this transformation when commenting in June 1976 on the conclusion of what he believed had been a highly successful exercise at sea conducted by the 6th Marine Regiment. "Last year the same could not be said," he wrote. In the previous year's exercise, Operation Solid Shield 1975, the reporter recorded that "although 2500 men of the 3000 man 6th Regiment were slated to participate, only 1200 of them waded ashore." The reasons that the other 1,300 Marines remained in the barracks varied. According to *The Boston Globe*, 800 of the Marines were "awaiting undesirable discharges," another 200 had gone absent without leave or were counted as deserters, and another 267 "had been administratively reassigned to the brig where they were imprisoned for a variety of offenses."[31] When Colonel Harold G. Glasgow took command of the 6th Marine Regiment in May 1975, he discovered a shockingly disheveled situation. Glasgow "found that 294 of his Marines were carried in an unauthorized leave status and 231 more were either confined or under restraint." Reflecting the lack of institutional discipline at the moment, he mentioned being "lucky if one in every five Marines" gave

---

[30] Carter and West, "Dismissed!"

[31] Walter V. Robinson, "The Marines' Toughest Fight: Long Battle for Respectability," *Boston (MA) Globe*, 6 June 1976.

him a salute. Additionally, he believed that "between 10 and 15 percent" of the regiment's personnel were "intentionally trying to fail their physical fitness tests." Even more alarming to Glasgow, the 2d Marine Division "discharged as undesirable 2400 men," 1,027 of whom came from the 6th Marines' 3,000-man force, in 1975. Between December 1975 and June 1976, the 2d Marine Division lost an additional 600 Marines under Wilson's expeditious discharge program.[32]

Although increasing the quality of the Marines in the ranks, weeding out the troublemakers came with pitfalls outside of the Service. In November 1975, the city council of Oceanside, California, asked Wilson to revise the longstanding Marine Corps policy of discharging personnel at their final duty station. They asked that these discharges take place in the Marine's hometown of record instead. The council argued that the troublemakers discharged from the 1st Marine Division stationed nearby "tended to settle in Oceanside, adding to the city's problem of Marine-related crime."[33] Wilson rejected the request, contending that it would place too large an administrative burden on his hard-pressed recruiting stations. The council had good reason for this concern. Two years earlier, 10 murders occurred in Oceanside, and "Marines were involved in seven of them." Similarly, between October 1973 and May 1974, "more than 800 Marines were arrested in the city and were responsible for 40 percent of all crimes," including "murders, rapes and assaults."[34]

Wilson went to work on education, training, and ensuring that Marines met the Service's standards. He established a "combined-arms-exercise-college" at the Marine Corps Base Twentynine Palms, California. Wilson loved the broken desert terrain, the "112-degree temperatures, sand in the faces, fire and movement skirmishes that tore up the deck, the sky and confidence. Twentynine Palms had everything but comfort, and the general wanted it that way." He also formed an "avia-

---

[32] LtGen William K. Jones, *A Brief History of the 6th Marines* (Washington, DC: Marine Corps History and Museums Division, 1987), 146.

[33] Dave Polis, "Marines Reject Discharge Change," *San Diego (CA) Union*, 1 November 1975.

[34] Nathan Packard, "The Marine Corps' Long March: Modernizing the Nation's Expeditionary Forces in the Aftermath of Vietnam, 1970–1991" (PhD diss., Georgetown University, 2014), 120.

tion weapons and tactics squadron" that would encourage "totally integrated, top-drawer air-ground operations" at Marine Corps Air Station Yuma, Arizona. After arriving at Headquarters Marine Corps in Washington, DC, Wilson noted numerous out-of-shape and improperly groomed Marines, making them a target for reform from day one. In one memorable moment following his first staff meeting there, he stated that the building looked like a "rat's nest" and that the personnel were "unkempt and out of shape." Wilson bluntly pointed out that "if I see a fat Marine, he's in trouble, and so is his commanding officer." Wilson was so serious about this issue that Secretary of Defense Schlesinger frequently jested that "an old Gunny" could "lose 13 pounds just by keeping the Commandant's picture on the fridge."[35]

In a 31 December 1975 letter to Senator Samuel A. Nunn Jr. (D-GA), Wilson responded to the Senate Committee on Armed Services' inquiry on the status of Marine Corps personnel quality, mission, and force structure. Nunn had long been a fierce critic of the all-volunteer force, and he sensed that the program was floundering. Wilson countered Nunn's instincts. He reported that he believed the expeditious discharge program resulted in the reduction of "unauthorized absences by 28 percent" and desertions by "24 percent . . . during the first five months of the current fiscal year as compared to the same period" the previous year. Wilson noted further efforts to upgrade the quality of Marine Corps enlistees. He happily reported that during December 1975, one division had "a group of some 700 Marines" join out of a recruit depot that consisted of 87 percent "high school graduates possessing an average GCT (general classification test) score of 100." The previous year, he recorded, 55 percent of the "accessions in that division were graduates and the average GCT score was 85."[36] Wilson took the improvement of the overall quality of the entire Marine Corps intensely serious, believing that the future institutional existence of the Service depended on this exact issue. To ad-

---

[35] Cyril J. O'Brien, "General Louis H. Wilson Jr.," *Leatherneck*, April 2003, 32–33.
[36] Gen Louis H. Wilson Jr. to Senator Sam Nunn, 31 December 1975, Gen Louis H. Wilson Historical Reference Files, Marine Corps History Division, Quantico, VA.

dress it, he was more than willing to accept a smaller Marine Corps if it meant getting a higher quality, better-disciplined Marine in the ranks.

While Wilson had made great strides toward correcting the personnel issue, in January 1976, two Brookings Institution scholars, Martin Binkin and Jeffrey Record, published a study titled *Where Does the Marine Corps Go From Here?*, which was a blistering critique of the Service. The publication stirred up intense backlash, but many experts believed that a reasoned debate about the Service's future role had been long overdue. In sum, Binkin and Record argued that "the need for the Corps' principal mission—amphibious warfare—is less apparent than in the past." They contended that since the Vietnam War, the Marine Corps had slowly devolved into an anachronistic organization "increasingly haunted by its [own] limitations." They believed that forced-entry amphibious assaults required the Corps to maintain robust tactical aircraft wings and that the high cost of possessing a "third air force" for a dubious future contingency could only come at the "expense of the cross-country mobility and fire power Marine ground forces need to meet contingencies they are most likely to face."[37]

As Commandant, Wilson remained extremely serious about training and readiness. In his posture statement to the House Committee on Armed Services (HASC) for fiscal year 1979, he stated that "operational readiness is at the apex of our efforts, for it is the cornerstone of our existence as a fighting organization."[38] Wilson proudly highlighted the success of the newly established Marine Corps Air Ground Combat Training Center at Twentynine Palms, which also hosted "approximately ten combined arms exercises." He added that the Marine Corps "participates annually in over forty major exercises around the globe," including "cold weather training in Europe and Korea, jungle training in Panama and the Philippines, and numerous amphibious exercises in other parts

---

[37] "What's Next for the U.S. Marine Corps?," *Brookings Bulletin* 13, no. 1 (Winter 1976): 15–17.
[38] Gen Louis H. Wilson Jr., *Statement of General Louis H. Wilson, Commandant of the Marine Corps before the Committee on Armed Services, U.S. House of Representatives on Marine Corps Posture, Plans, and Programs for FY 1979 through 1983* (Quantico, VA: Marine Corps University Library, 1978), 2, hereafter *Statement for FY 1979 through 1983.*

of the world."[39] While Wilson admitted to the HASC membership that he had spent much of his first two years as Commandant on "restoring the quality" of enlistees and on "training and readiness of the Fleet Marine Forces," he emphasized that the Marine Corps was now fully capable of successfully fighting and operating across the entire spectrum of potential combat operations.[40] Wilson was slightly premature in his assessment, especially if the Service was required to operate in a cold-weather environment.

Finally, Wilson pointed out that his personnel campaign seemed to be paying off dividends. He cited numerous "selected quality indicators" to reflect its success. Between fiscal years 1977 and 1978, the Service's "unauthorized absence rate" fell by 22 percent, the desertion rate dropped 33 percent, and the "confined population" declined by 13 percent. Wilson noted that a "30 percent decrease in major command special courts-martial convictions" continued a downward trend. "More significantly," he wrote, "losses for reasons other than normal expiration of active service are down by 46 percent." He also argued that "quality improvement has had a positive effect on the Corps' retention rate." According to Wilson, this rate rose "from 12.3 percent in fiscal year 1976 to 17.3 percent in fiscal year 1977, and 20.1 percent in the first three months of this fiscal year." All of these changes led to a "high state of unit morale" throughout the Service, which he saw personally while visiting multiple field commands.[41]

Wilson also fundamentally disagreed with the amphibious warfare naysayers at the Brookings Institution. During an interview, CBS News correspondent Ike Pappas asked Wilson if he thought that amphibious warfare was out of date. Wilson promptly replied that "critics have said that before. They were wrong then and just as wrong now." Wilson's strategic point of view for the future, however, was not wedded to amphibious warfare alone. Instead, he reported to Congress that the Marine Corps was "a ready, mobile, general purpose force with amphibious

[39] Wilson, *Statement for FY 1979 through 1983*, 6–7.
[40] Wilson, *Statement for FY 1979 through 1983*, 1–2.
[41] Wilson, *Statement for FY 1979 through 1983*, 12.

expertise . . . a global force in readiness," and that he opposed "special-ized restructuring for combat in Europe only."[42] Wilson believed that the growing push to reorient the Marine Corps to exclusively participate in NATO missions was too restrictive for a predominately seaborne Service designed to project power globally. He pointed to other potential areas of concern where the Marine Corps could be more useful if not ideal for the specific situation. While the Marines could greatly assist NATO's mission, the Service could not become solely focused on it.

Binkin and Record were not the only people who saw amphibious warfare as an outdated concept. In his annual report to Congress, Secretary of Defense Schlesinger believed that the price tag for amphibious shipping alone made it wise to consider whether the Marine Corps should maintain its focus on this core competency. In a classic understatement, Schlesinger wrote that "amphibious forces are not cheap."[43] Representative Leslie Aspin Jr. (D-WI), a long-serving member of the HASC, was even more blunt when noting that the Marines "are in trouble."[44]

To diminish the perspective that the Marines were too light and slow to assist during an assault on NATO, Marine Corps leadership argued that its amphibious forces were ideally suited to protect NATO's flanks. In 1976, during Operation Teamwork 76, a significant exercise of strategic mobility, the Marine Corps successfully landed about "8,000 Marines in Norway on NATO's northern flank." The following year, the commanding general of Fleet Marine Force, Atlantic (FMFLANT), Lieutenant General Robert H. Barrow, was tasked with conducting a similar exercise called Operation Nifty Nugget on NATO's southern flank in Turkey. When asked about how his Marines might handle the vaunted threat of Soviet armor, Barrow shot back that "we have at least reached the threshold and maybe crossed it in making the tank obsolete, or near obsolete on the battlefield." In Barrow's mind, he believed that the U.S.

---

[42] White, "Louis H. Wilson Jr.," 429–30.

[43] Nathan Packard, "Giving Teeth to the Carter Doctrine: The Marine Corps Makes the Case for Its Strategic Relevance, 1977–1981," *International Journal of Naval History* 12, no. 2 (Summer 2015): 6.

[44] George C. Wilson, "Marines to 'Invade' Turkey to Stress Value in European War," *Washington Post*, 10 August 1977.

military had reached a point in the late 1970s "where some guy tucked away in the woods over there with a precision-guided weapon is going to knock the bejesus out of that 60-ton tank. . . . Ten years from now the guy who brings a lot of tanks to the battlefield may be bringing liabilities rather than assets."[45]

Because the Marine Corps was now considering operating in snowy Norway, Wilson ordered the reinvigoration of cold-weather training for all Marine Corps operating units. Since the height of the Korean War in 1951, the Service had maintained the Mountain Warfare Training Center (MWTC) in Pickel Meadow near Bridgeport, California. Located deep within the Sierra Nevada mountains, this camp's establishment was related to the experience of Marines in Korea. During the Chosin Reservoir campaign (26 October–15 December 1950), for example, the 1st Marine Division reported "7,313 non-battle casualties," most of which "had been attributed to frostbite and other types of cold injuries suffered in the 0 to -35 degree weather." After taking part in this fight, 1st Marine Division commander Major General Oliver P. Smith tasked his brightest staff officer, Brigadier General Merrill B. Twining, with creating a camp and training syllabus that would allow all Marine Corps replacements to first experience some cold-weather training before arriving in Korea. Twining's selection of Pickel Meadow, which sat on the same latitude as the Korean Peninsula, as a cold-weather training site most closely approximated the terrain and weather Marines would experience.[46] With the Corps once again looking at operating in a frigid environment approximately 25 years later, Pickel Meadow suddenly took on renewed importance.

No longer fighting in the heat of Vietnam, the Marine Corps in the mid- to late 1970s needed to review its cold-weather and mountain warfare operational doctrine. Brigadier General Alfred M. Gray Jr., the commanding general of the 4th Marine Amphibious Brigade (4th MAB), was

---

[45] Wilson, "Marines to 'Invade' Turkey to Stress Value in European War."
[46] MajGen Orlo K. Steele, USMC (Ret), and LtCol Michael I. Moffett, USMCR (Ret), *U.S. Marine Corps Mountain Warfare Training Center, 1951–2001* (Washington, DC: Marine Corps History Division, 2011), 19–22.

Figure 38. Official logo for the Marine Corps Mountain Warfare Training Center

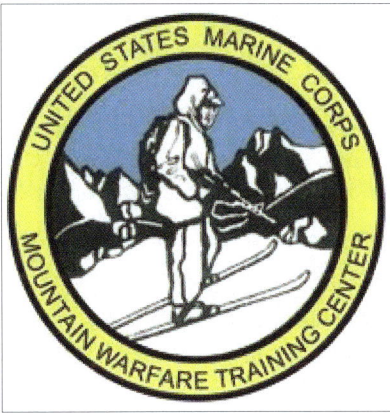

Source: official U.S. Marine Corps photo.

one of the leading proponents for improved cold-weather doctrine. During Teamwork 76, Gray proved the efficacy of landing Marines in Norway. In a further demonstration of flexibility, the 4th MAB then "backloaded" from Norway and went to Denmark, where "it participated in a second and much larger exercise known as Bonded Item." Moreover, the "flexibility and air-ground combat capabilities of the 4th MAB pleasantly surprised many NATO observers."[47] As a result of these early successes, the Marine Corps planned for more exercises in northern Europe during that decade.

Yet, most of the post-exercise reports from the late 1970s observed that the Marine Corps seemed far less mobile and efficient in cold environments than their allies from the British Royal Marines and the Norwegian Army. Primarily, this problem was related to familiarity and training. The 4th MAB, stationed on the East Coast of the United States, lacked an appropriate cold-weather training site where it "could conduct pre-deployment training under terrain and climate conditions" similar to those found in a Norwegian winter. Camp Drum, a former World War I-era National Guard center located near Plattsburgh, New York, was available. Although it could get quite cold and snowy there, the ground was relatively flat and did not resemble the mountainous terrain of Norway. Even with more predeployment cold-weather training, the Marines still required better equipment for operating in deep snow, something that they had not had to do since the Chosin Reservoir campaign.[48]

Consequently, in 1979, the Marine Corps commissioned Northrop Services to "identify landing force deficiencies" in areas related to "a

---

[47] Steele and Moffett, *U.S. Marine Corps Mountain Warfare Training Center*, 98.
[48] Steele and Moffett, *U.S. Marine Corps Mountain Warfare Training Center*, 98–99.

cold-weather amphibious assault, including naval support systems directly involved with ship-to-shore movement" as well as to study how the Marines needed to improve its cold-weather training syllabus at the MWTC. The Northrop study uncovered that the Corps was definitely capable of operating in dry-snow cold-weather conditions but "did not have the capability to conduct winter warfare operations under extreme-cold conditions (temperature range -25 to -65 degrees F)."[49] The report also stressed that the Marines needed to improve their dilapidated cold-weather training facility at Pickel Meadow and address the "excessive weight of cold-weather equipment, clothing, and supplies" that the Service had held over from the Korean War, which "added to the normal load of weapons and ammunition that the infantryman is required to carry and man-haul on sleds." This added weight, it found, reduced an individual Marine's "effectiveness in combat and greatly increases the possibility of cold weather injury and defeat."[50]

Finally, the study recommended that the Marine Corps adopt the practice of dedicating certain units specifically to winter warfare, something that their European allies had already done. It emphasized that these units assigned solely to the cold-weather mission would be most effective in accomplishing its objectives. The Marine Corps, however, had the habit of rotating any number of battalions through its cold-weather operations training syllabus. Each exercise in Norway usually saw a different Marine ground unit undergoing winter warfare training for the only time in a Marine's standard four-year enlistment. However, the 4th MAB headquarters staff was routinely given command of these training missions, allowing them to pass along plenty of lessons to the supporting force structure. The more the 4th MAB headquarters staff took on these operations, the better they got at them. Their subordinate attached units, however, were usually a different story.

---

[49] Steele and Moffett, *U.S. Marine Corps Mountain Warfare Training Center*, 99. By admitting that it was not capable of operating in extreme-cold conditions, the Marine Corps "not only narrowed the scope of training programs, but literally saved millions of dollars when it came time to procure new sets of cold weather clothing and equipment." See, Steele and Moffett, *U.S. Marine Corps Mountain Warfare Training Center*, 99n.

[50] Steele and Moffett, *U.S. Marine Corps Mountain Warfare Training Center*, 99–100.

During the late 1970s, there was also a belief within the U.S. Department of Defense (DOD) that in any overseas regional contingency, seaborne Marines would be "too slow to give the White House the responsiveness it desired." If a seaborne unit was required in Asia or the Middle East, for instance, the "transit times via amphibious shipping from the United States" would take "anywhere from one to two months." To make matters more complicated, the new administration of President James E. "Jimmy" Carter Jr. entered the White House in 1977 having promised to drastically cut the nation's defense budget. The low-priority amphibious mission of the Marine Corps soon found itself in serious financial straits. The Service also saw its top-priority McDonnell Douglas AV–8B Harrier II attack aircraft program drastically reduced by more than half, from $173 million to $85 million (USD). Secretary of Defense Harold Brown further recommended pushing back or slowing down procurement of the new Landing Ship, Dock 41 (LSD 41) destined to replace many of the Navy's slower-moving Vietnam-era amphibious ships.[51] Wilson became frustrated with the CNO, Admiral James L. Holloway III, who he believed was paying "lip service" to the Harrier program while "working behind the scenes" to weaken the secretary of the Navy's "support of the aircraft." Wilson was so upset with the CNO that he "confronted him personally" about the matter.[52]

With these struggles emerging, Binkin and Record argued that it was the right time for the Marines to implement significant changes. They suggested that the Service "shift its principal focus from seaborne assault to a more appropriate mission, such as garrisoning America's remaining outposts in Asia or defending central Europe." They also thought that Marine Corps tactical aviation, a necessity for amphibious assaults, consumed too much of its operating budget. If the Marines eschewed am-

---

[51] Packard, "Giving Teeth to the Carter Doctrine," 6. This excellent article provides the best and most concise information on how the Carter Doctrine and the ongoing argument emanating from the Brookings Institution against the amphibious warfare mission spelled trouble for the Marine Corps in the late 1970s and early 1980s.

[52] Gen Louis H. Wilson Jr., USMC (Ret), Oral History Transcript, 25 June 1979 session, BGen Edwin H. Simmons interviewer, Marine Corps History Division, Quantico, VA, 187–88, hereafter Wilson Oral History Transcript.

phibious warfare, Binkin and Record contended, they would no longer need their own organic and costly aviation assets, especially because the Air Force or the Navy could provide close air support in the future. The two analysts estimated that the Marine Corps could save approximately $2 billion (USD) in overall procurement while also saving $300 million per year. Binkin and Record proposed four options for Congress and the DOD to consider. All of them envisioned a vastly smaller Marine Corps that no longer centered on amphibious warfare. Binkin and Record concluded that the Marines are "well suited for amphibious operations in the Third World, where U.S. intervention now seems increasingly unlikely, and less well-suited for combat in key areas—Europe, Northeast Asia, and the Middle East—to whose security U.S. policy now assigns highest priority."[53]

Just as Commandant Randolph Pate did in 1956 when he found himself at a role and mission crossroads, Wilson formed a board of senior officers to consider what the Service should look like in the post–Vietnam War era. Named after its senior member, Iwo Jima veteran and Marine Corps major general Fred E. Haynes Jr., the Haynes Board received the task of providing recommendations for a restructuring of the 1960s-era FMF. At that time, Haynes was Wilson's director of operations. He had carte blanche to recommend any sweeping changes that the board saw fit, except the recently "reorganized combat support echelon in the FMF, which Haynes could study but not offer organizational changes."[54]

The Haynes Board submitted its report in March 1976.[55] Its recommendations made clear that the board members were worried about the possibility of a NATO mission and Marine Corps divisions being underweight in armor, personnel carriers, self-propelled artillery, and anti-armor assets. The board endorsed adding more artillery to the division, including heavy self-propelled batteries and antiaircraft missile batter-

[53] "Amphibious, Air Units Criticized by 2 Analysts," *San Diego (CA) Union*, 2 February 1976.
[54] Kenneth W. Estes, USMC (Ret), *Marines under Armor: The Marine Corps and the Armored Fighting Vehicle, 1916–2000* (Annapolis, MD: Naval Institute Press, 2000), 178. Estes is a retired tank officer who later became a Marine Corps historian.
[55] See Allan R. Millett, "The U.S. Marine Corps: Adaptation in the Post–Vietnam Era," *Armed Forces & Society* 9, no. 3 (Spring 1983): 363–92, https://doi.org/10.1177/0095327X8300900301.

ies, and activating "the fourth company in the assault amphibian battalions (to increase the tactical mobility of the infantry)." It also advocated improved combined arms training for all Marine Corps combat units. Most importantly, the board suggested creating two mobile assault regiments, with one placed on each coast of the United States. These regiments would consist of "two combined infantry battalions, two tank battalions, one assault amphibian battalion, and a self-propelled artillery battalion." This would increase the number of tank battalions from three to four, but it was suggested that the battalions would be made smaller, "using the same total number of tanks in the current inventory."[56] These regiments needed to be ready to deploy to Northern Europe or the Middle East on a moment's notice. However, the report did not address how these proposed forces would get to the fight.

As Marine Corps historian Kenneth W. Estes notes, the "tentative doctrine for fighting with combined arms using the existing FMF organizations" came out of two publications—*Education Center Publication 9-3 (1978)* and *Operational Handbook 9-3B (1980)*. When the Service attempted to apply such mechanized warfare concepts during various 1980s-era combined arms exercises at Twentynine Palms, the Marines never achieved the "doctrinal cohesion" they so fervently sought. According to Estes, this struggle to achieve cohesion resulted from Marine Corps leadership being uneasy "with mechanization," doubting the costs of moving heavy equipment, and overemphasizing "an almost mythical state of 'lightness' " among the infantry units.[57]

During a Pentagon news briefing on 24 March 1976, Haynes told the assembled gathering that the board's report was intended to provide the Commandant with some force restructuring options that accounted for the intensity of combat that the Marines would likely face in the coming 10–20 years. The board heavily leaned on the lessons from the Yom Kippur War of 1973 to create their suggestions. Haynes pointed out to the amphibious warfare naysayers in the audience that everyone tended to "confuse the World War II amphibious assault with today's reality."

---

[56] Estes, *Marines under Armor*, 178–79.
[57] Estes, *Marines under Armor*, 179–80.

He saw the Marine Corps as part of a more significant Joint effort. New equipment scheduled to come online in the 1980s, he believed, opened up far more beaches to future amphibious force commanders than had been available to World War II–era Marines, which was something he had personally experienced during the Battle of Iwo Jima. Haynes was also adamant that the Marines needed to remain a global force-in-readiness rather than focus solely on European missions. As a result, the Haynes Board pointedly did not recommend that the Marine Corps add any more tanks to its inventory. Instead, it advised reshuffling those it already had for better combat distribution. Haynes reiterated again and again to the gathered reporters that the board purposely "avoided optimizing for a particular theater."[58] Wilson was generally pleased with the various alternatives that the board provided, but he favored options that did not transform the Marines into an U.S. Army-like mechanized force. He also concurred with the Haynes Board that the future of Marine Corps tactical aviation resided in the rapid acquisition of the Harrier aircraft.

The 1970s was a time of tremendous national security fluidity. The decade began with the conclusion of U.S. military involvement in Southeast Asia and yet another high-intensity conflict between Israel and its Arab neighbors. It ended with an Islamic and highly anti-American revolution in the formerly allied nation of Iran and a full-fledged Soviet invasion of Afghanistan. Interspersed between these concerning events were at least two major economy-disrupting oil embargoes that the Organization of the Petroleum Exporting Countries levied against the United States. Suddenly, U.S. national security planners became increasingly concerned with continued access to Middle Eastern energy. Such activity began what defense analyst Michael T. Klare called "resource wars" in the Middle East and the Persian Gulf. Consequently, in addition to remaining watchful over the rising Soviet Bloc threat to NATO, national security planners needed to consider the impact of anti-Western powers blocking or controlling the flow of Middle Eastern oil to the United States and its allies. President Carter officially addressed this issue in

---

[58] MajGen Fred Haynes, Pentagon News Briefing Transcript, 24 March 1976, MajGen Fred Haynes Biography Files, Marine Corps History Division Reference Branch, Quantico, VA, 1–3.

1980 when he created what became known as the Carter Doctrine: "the United States will not permit a hostile state to acquire the ability to obstruct the free flow of oil from the Gulf to major markets in the West."[59]

It was one thing for the United States to say it would defend the Persian Gulf, but quite another to possess the ability to do so. As Record noted, "No area of the world is more distant from the United States than the Persian Gulf." According to Record, the region is 7,000 nautical miles by air; 8,500 nautical miles by sea when going through the Suez Canal; and a whopping 12,000 nautical miles by sea around the Cape of Good Hope. Moreover, with Iran no longer a Western ally, the United States did not have any friends in the region that would allow U.S. forces to permanently establish any forward-bases there. If the U.S. military needed to go to the Gulf, it would have had to do so mainly via the sea.[60]

Reflecting on the growing importance of the Persian Gulf to U.S. national security affairs, U.S. forces became more present in the region. The U.S. Navy initiated a considerable increase in its aircraft carrier battle groups operating in the Arabian Sea and Indian Ocean. In 1976, the Navy deployed only 3 percent of its carrier force there for a total of "19 carrier ship days." By 1980, it deployed 51 percent of its carrier force for a total of "836 carrier ship days."[61] Similarly, small Marine expeditionary units (MEUs) made more frequent visits to the Gulf region. To make matters even more confusing for U.S. defense planners, the nations of Iraq and Iran went to war in September 1980. Heavily armed Iraq had always assumed that it would win an easy victory over its Persian neighbor, which still reeling from the Iranian Revolution of 1979. Iranian resistance stiffened after a few weeks of intense fighting, however, and the stalemated war dragged on for a full eight years. While U.S. war planners had always worried about an outside entity such as the Soviet

---

[59] Michael T. Klare, *Resource Wars: The New Landscape of Global Conflict* (New York: Metropolitan Books, 2001), 53.

[60] Jeffrey Record, *The Rapid Deployment Force and U.S. Military Intervention in the Persian Gulf*, 2d ed. (Cambridge, MA: Institute for Foreign Policy Analysis, 1983), 19.

[61] Adam Siegel, Karen Domabyl, and Barbara Lingberg, *Deployments of U.S. Navy Aircraft Carriers and Other Surface Ships, 1976–1988* (Alexandria, VA: Center for Naval Analyses, 1989), 13, 15, 21, 26–27; and Michael A. Palmer, *Guardians of the Gulf: A History of America's Expanding Role in the Persian Gulf, 1833–1992* (New York: Free Press, 1992), 107.

Union threatening access to Middle Eastern oil, perhaps it would come "from a Middle Eastern state" instead.[62] This possibility forced the U.S. defense planners to consider whether the United States would intervene militarily against an internal Middle Eastern threat to oil.

President Carter's national security advisor, Zbigniew K. Brzezinski, was unequivocal about the matter. He sincerely believed that with the exit of the British from the region in the 1950s and the overthrow of the Shah of Iran in 1979, the United States "could no longer avoid primary responsibility" for defending the Middle East under the new political arrangement and security conditions. "The only other option," Brezinski argued, "was to withdraw, a policy that was politically, economically, and militarily unthinkable."[63] The new Carter Doctrine clearly signaled to the world that the United States was committed to preventing either internal or external aggressors from hindering the Gulf oil flow.

In addition to addressing the concern of securing access to oil, U.S. forces in the Middle East would have to face some of the most heavily armed states in the world, including ones that were antithetical to its interests. Consequently, Brzezinski, in conjunction with the Joint Chiefs of Staff and Secretary of Defense Brown, floated the idea of creating a Rapid Deployment Joint Task Force (RDJTF) for potential use at global flashpoints, especially in the Middle East. Early on in his administration, Carter released Presidential Directive 18, which he signed on 24 August 1977. In the directive, the president commanded that the United States must "maintain a strategic posture of 'essential equivalence'," reaffirm its forward defense strategy for NATO in Europe, and sustain a "'deployment force of light divisions with strategic mobility' for global contingencies, particularly in the Persian Gulf region and Korea."[64]

In March 1980, the U.S. military established the RDJTF at MacDill Air Force Base in Tampa, Florida. By the following spring, the RDJTF consisted of "four Army divisions (about 100,000 troops), the equiv-

---

[62] Palmer, *Guardians of the Gulf*, 109.
[63] Palmer, *Guardians of the Gulf*, 110.
[64] Zbigniew Brzezinski, *Power and Principle: Memoirs of the National Security Adviser, 1977–1981* (New York: Farrar, Straus, and Giroux, 1983), 177.

alent of two Marine divisions, and associated airlift, sealift, and logistical capabilities."[65] In a major coup for the Marine Corps, Brown named future Commandant Lieutenant General Paul X. Kelley as the first RDJTF commander. In late 1979, General Barrow, now Commandant of the Marine Corps, provided a potential reason for Kelley's appointment despite the Army strongly advocating for the command. He believed that seaborne forces had a greater capability to deliver "combat power" than "the linear kinds of delivery that you get out of airlift . . . I think, clearly, the sea is the way to go." The Marine Corps could play a central role in this capability:

> The big problem—I've said this publicly and I've said it many times, not as criticism but just an observation—relates to *when* those forces move forward in a crisis situation. I haven't seen one of these crises yet that didn't have a lot of indicators saying that it was getting worse. If we could somehow act before the period of extremis—the eleventh-hour kind of thing, when perhaps it's too late for anyone to do too much about it—then naval forces, meaning this great Navy and Marine Corps force projection capability, could be moved forward to these areas in a timely manner, and perhaps not even have to be employed. The mere fact that they have done so—moved forward—would serve as a deterrent. This is one of the values of our capability— to reassure friends that might need to be reassured in the area. Should all of that fail, and you do need them, *there they are*, with the whole capability ready to go ashore in a matter of hours and days as opposed to being strung out physically and being dependent upon a permissive environment on the receiving end.[66]

The most prominent issue with the RDJTF revolved around strategic mobility and late-notice crisis response. The distance between the Middle East and the United States became the true enemy for U.S. security forces. To resolve this issue, Deputy Secretary of Defense W. Graham

---

[65] Brzezinski, *Power and Principle*, 456.
[66] "A Conversation with General Barrow," *Sea Power*, November 1979, 32, emphasis in original.

Claytor Jr. announced on 5 March 1980 that the Pentagon would position seven cargo ships that carried "enough equipment and supplies for a 10,000-man Marine Amphibious Brigade plus several squadrons of Air Force fighters" in the Indian Ocean.[67] Claytor stated that the military planned "to start loading these ships in May, and to have them loaded and on their way to the selected anchorage before the end of June."[68] Kelley and others associated with the RDJTF recognized that the primary weakness of the entire program resided in the amount of strategic lift available to the force. In this case, it decidedly lacked both available airlift and sealift. President Carter's 1979 defense budget included projected improvements toward strategic mobility, but they would not come online until sometime in 1983. Consequently, the RDJTF created a stopgap strategic lift measure that secured the immediately needed sealift. Labeled the Near-Term Prepositioned Ships (NTPS) program, it consisted of "three roll-on-roll-off (RO/RO) ships, two break-bulk ships, and two tankers."[69] Claytor and Kelley hoped to have these assets in place at the Diego Garcia naval anchorage in the Indian Ocean no later than mid-1980. This design was meant to serve as a preemptive strategy for getting U.S. forces into a crisis region before the other side—the Soviet Union, for instance—could build up any preponderant military power.

Before implementing the NTPS concept, Carter authorized a highly classified operation to free U.S. hostages, including Marine embassy guards, held by Iran since the overthrow of the Shah in 1979. A series of unfortunate events ultimately doomed the rescue attempt, code named Operation Eagle Claw. First and foremost, the Joint Staff insisted that the operation remain highly classified and compartmentalized until the absolute last moment. Consequently, many servicemembers involved in the mission did not see or understand how to execute the exceptionally complex plan. Next, the rescue mission command decided to use eight U.S. Navy Sikorsky RH-53D Sea Stallion minesweeping helicopters for

---

[67] Quinlan, *The Role of the Marine Corps in Rapid Deployment Forces*, 11; and Vernon A. Guidry Jr., "Rapid Deployment Force to Get First Components," *Washington (DC) Star*, 6 March 1980, 12.

[68] Quinlan, *The Role of the Marine Corps in Rapid Deployment Forces*, 11.

[69] Quinlan, *The Role of the Marine Corps in Rapid Deployment Forces*, 11; and Guidry, "Rapid Deployment Force to Get First Components."

Figure 39. USNS *Sisler* (T-AKR 311)

Source: official U.S. Navy photo.

the operation. It seems that they chose this aircraft for the mission because they were to be launched from the deck of USS *Nimitz* (CVN 68),
and another Service's aircraft might arouse suspicion. Moreover, at that
time the Sea Stallion did not have in-flight refueling capability and had
a notoriously poor operational readiness rating. A post-operation inquiry headed by retired U.S. Navy Admiral James L. Holloway III found that
"operational security was enforced too zealously" by the two mission
commanders, Joint Task Force commander Army major general James
B. Vaught and Army Delta Force antiterrorism group founder Colonel
Charles A. Beckwith. This accounted for much of the lack of coordination between the various armed Services providing equipment and personnel for this mission.[70]

---

[70] Paul B. Ryan, *The Iranian Rescue Mission: Why It Failed* (Annapolis, MD: Naval Institute Press,
1985), 53–59, 75–76, 115–16; and Maj William C. Flynt III, USA, *Broken Stiletto: Command and
Control of the Joint Task Force during Operation Eagle Claw at Desert One* (Fort Leavenworth, KS:
School of Advanced Military Studies, 1995), 27–29.

Almost from the start of the operation, things went wrong. The eight Sea Stallions, callsign "Bluebeard," took off from *Nimitz* shortly after 1900 on 24 April 1980. Two hours into the mission, one of the helicopter crews suddenly reported a serious mechanical problem. The crew landed, and, after inspecting the craft, decided that it could no longer fly. A following Sea Stallion, Bluebeard 8, picked up the downed crew and proceeded with the mission. After crossing into Iran, the remaining helicopters ran into a massive sandstorm, locally known as a "haboob." The storm caused Bluebeard 5 to have an electrical problem, forcing it to return to *Nimitz*. Now down to just six helicopters, they all made it to the selected rendezvous site inside Iran, called Desert One, where six Lockheed Martin C-130 Hercules aircraft, including three Lockheed Martin EC-130 refueling planes, waited for them. After landing, Bluebeard 2 indicated a major hydraulic failure, making it unflyable as well. Believing that six functioning helicopters were the bare minimum necessary for mission success, both Vaught and Beckwith now recommended aborting the mission. Vaught sent this request to the White House, which Carter quickly approved. As the Sea Stallions and the C-130s prepared to depart, Bluebeard 3 accidentally turned into a nearby EC-130 refueler. Both aircraft exploded into a fireball, killing eight U.S. servicemembers, three of whom were Marine Corps noncommissioned officers. The conflagration soon spread to some of the other aircraft. The surviving servicemembers and helicopter crews boarded the remaining C-130s to evacuate. In the confusion, "some of the helicopters could not be reached for 'sanitizing' and their classified material . . . fell into the hands of the revolutionary government."[71]

The news of the disaster at Desert One shocked the United States in its aftermath. Both the Senate and House of Representatives convened investigations into the matter immediately. Carter's approval rating plummeted during a critical election year. The calamity proved to be one of the many reasons he lost his reelection bid to Ronald W. Reagan that fall. The DOD convened a board of inquiry under the leadership of

---

[71] Charles Tustin Kamps, "Operation Eagle Claw: The Iran Hostage Rescue Mission," *Air & Space Power Journal en Español* 18, no. 3 (2006).

Figure 40. Desert One crash site

Source: official U.S. Department of Defense photo.

recently retired CNO Admiral Holloway that included Major General Alfred M. Gray Jr. as the Marine Corps representative. The *Rescue Mission
Report*, also known as the Holloway Report, extensively examined the
operation in detail and came away with two major conclusions. First, it
determined that an overriding concern about operational security caused
much of what went wrong during the mission. Second, instead of using
an established Joint task force (JTF) organization, the Joint Chiefs of Staff
had to create a JTF from scratch, which included steps to "find a commander, create an organization, provide a staff, develop a plan, select
the units, and train the forces." This issue left the Joint Chiefs of Staff
without "an organizational framework of professional expertise around
which a larger tailored force organization could quickly coalesce."[72]

---

[72] *Rescue Mission Report* (Washington, DC: U.S. Joint Chiefs of Staff, 1980), 60.

217

One of the most important recommendations to come out of the Holloway Report was the need for the DOD to establish a new permanent Joint task force. The Counterterrorist Joint Task Force (CTJTF) would act as "a field agency of the Joint Chiefs of Staff with permanently assigned staff personnel and certain assigned forces."[73] The committee suggested that the CTJTF, under the direction of the National Command Authority, "would plan, train for, and conduct operations to counter terrorist activities" that threatened U.S. "interests, citizens, and property" abroad. Most importantly, the committee members believed the CTJTF should report directly to the Joint Chiefs of Staff and maintain a staff with representatives from the four main Services who were chosen for "their specialized capabilities" in various forms of special operations.[74]

Based on this recommendation, Congress created a separate and distinct U.S. Special Operations Command (USSOCOM) in 1987. Since the disestablishment of Marine raider battalions in 1944, Commandants of the Marine Corps traditionally resisted creating special operations forces beyond a few internally focused force reconnaissance companies. In the 1980s and 1990s, the Marine Corps also argued that their MEUs could be trained to have a special operations capability before deploying with forward operating naval forces. At the time, most Commandants believed that the detachment of Marine Corps special forces from their traditional roots diminished the overall combat power of the MAGTF. All of this changed following the terrorist attacks of 11 September 2001. Commandants General James L. Jones Jr. and General Michael W. Hagee saw that the growth of transnational terrorism required Marine Corps participation within USSOCOM, and they took immediate steps to provide a permanent Marine regiment of highly trained special operators for the command. In 2014, Commandant General James F. Amos redesignated

---

[73] *Rescue Mission Report*, vi.
[74] *Rescue Mission Report*, 61.

these Marine Corps special operations units as Marine raider battalions in honor of their World War II ancestors.[75]

Toward the end of Wilson's tenure as Commandant, then-Brigadier General Gray led the 4th MAB—increasingly known in the Service as the "Carolina MAGTF" due to its predominate home base being Camp Lejeune, North Carolina—in one more major NATO exercise. In 1978, the 4th MAB, built primarily around Colonel Gerry H. Turley's Regiment Landing Team 2 (RLT 2) and Marine Aircraft Group 20, was invited to participate in Operation Northern Wedding/Bold Guard 78. The planning for this prestigious event in northern Europe took more than 10 months, and the 13-day exercise "resulted in the execution of three extremely complex and large-scale NATO operations."[76] The Carolina MAGTF played a central role in the successful campaign.

Using the sea as a major avenue of approach, the 4th MAB conducted a series of exercises. The 4th MAB, along with 40 Royal Marine commandos who embarked in Royal Navy equipment, demonstrated their ability to rapidly reinforce NATO's northern flank with landings that took place in the United Kingdom's Shetland Islands, which approximated the weather and terrain of nearby Norway. In the next phase—officially Northern Wedding—RLT 2 and its attachments backloaded its forces, sending them to Denmark. The final portion—Bold Guard—took place in Schleswig-Holstein, Germany, and included participation from an early incarnation of the NTPS concept. The 4th MAB and its companion forces overcame significant difficulties and challenges to make the entire exercise a success according to its Navy-Marine Corps planners.[77]

Northern Wedding/Bold Guard 78 was significant for several reasons. First and foremost, it proved that the amphibious warfare mission could still support NATO operations. Turley noted that it demonstrated that

---

[75] Col John R. Piedmont, USMCR, *Det One: U.S. Marine Corps Special Operations Command Detachment, 2003–2006* (Washington, DC: Marine Corps History Division, 2010), 1–17; and Jon Harper, "Marine Corps Special Operators Renamed 'Marine Raiders'," *Stars and Stripes*, 6 August 2014.

[76] Col Gerald H. Turley, USMCR, to BGen Alfred M. Gray Jr., Commanding General, 4th Marine Amphibious Brigade, 2 November 1978, in *Exercise Northern Wedding/Bold Guard 78: Post Deployment Report*, vol. 1 (Camp Lejeune, NC: 2d Marine Division, Fleet Marine Force, 1978), 1.

[77] "Executive Summary," *Exercise Northern Wedding/Bold Guard 78*, vol. 1.

opponents would not have the ability to stop a combined Navy-Marine Corps team that works "with a little mutual give and take, interspersed with some serious arguments." Second, the exercise finally put to rest the rumblings of critics that the Marines were too light for the NATO mission. During the operation, the 4th MAB possessed a "balanced task organization" as well as a sufficient amount of "armor mobility and anti-tank assets . . . to accomplish that mission." While the brigade admittedly could have used more mechanized equipment, the commanders noted that it possessed enough "armor, mechanized, or tank-killing assets" to accomplish the entire mission.[78]

The operation provided volumes of lessons from the aspects of command and control, intelligence, operations and training, logistics, and communications. Most importantly, the Service's armored amphibious vehicle, the United Defense Industry Landing Vehicle, Tracked, Personnel 7 (LVTP 7), had performed effectively as an armored personnel carrier in land operations.[79] After adding some recommended weapons improvements to the LVTP 7, the ability of this ubiquitous amphibious vehicle to act as a dual sea-land armored personnel carrier increased the combat value of the Corps toward the overall NATO mission. Robust Marine Corps tactical aviation was also significant to the success of Northern Wedding/Bold Guard 78.

Shortly after this triumph, Wilson prepared to retire from active duty in 1979, after having served the Marine Corps for more than 38 years. Since that time, he has been credited for quite literally saving the Service from self-destruction. He was also the first Commandant to serve as a permanent standing member of the Joint Chiefs of Staff, which came about due to an inter-Service insult to his rank and office. Prior to Wilson's tenure, the Commandant was not considered a full-time member of the Joint Chiefs of Staff, only being invited into meetings when the group discussed issues that directly impacted the Marine Corps. In the decades since the Korean War, the Commandant usually attended most of the meetings because few matters in that time did not in some way

---

[78] "Executive Summary," 1-1.
[79] *Exercise Northern Wedding/Bold Guard 78*, vol. 1, 7-38.

affect the Marine Corps. In another tradition that existed before the permanent establishment of the vice chairman of the Joint Chiefs of Staff in 1986, the chairman of the Joint Chiefs of Staff would appoint an acting chairman, usually based on seniority, when they were not available for scheduled meetings.

The insult that changed the structure of the Joint Chiefs of Staff struck Wilson the year before he retired. Wilson's legislative affairs assistant, Brigadier General Albert E. Brewster, later recounted that, at some point in early or mid-1978, "a most unusual situation had arisen" when all the members of the Joint Chiefs of Staff except for Wilson were "scheduled to be absent." In that circumstance, Brewster recalled, Wilson presumed that he had "co-equal status with the members of the Joint Chiefs of Staff" under Public Law 416 and that he would become acting chairman. The other members of the Joint Chiefs of Staff, however, denied this setup because they considered the Commandant as a "part time" member only. Instead, the other chairs "designated the Vice Chief of Staff of the Air Force as the Acting Chairman."[80]

Wilson was livid about this turn of events. He immediately sent for Brewster. "When I walked into his office," Brewster noted, "he was in the most agitated state I had ever witnessed in him." The Commandant immediately declared, "I will not allow the Marine Corps to be insulted like this. You will not believe what the Joint Chiefs have just done to insult the Corps!" Wilson directed Brewster to take speedy action on this issue. He demanded that Brewster do what he could to get the law changed so that the Commandant became "a regular, full-time member" of the Joint Chiefs of Staff. Brewster replied that doing so would "require an Amendment to the National Security Act of 1947, and Title 10 of the [U.S.] Code," an obstacle that did not seem to faze Wilson. Fortunately for Brewster, Senator John C. Stennis (D-MS), the powerful chairman of the Senate Committee on Armed Services and a strong supporter of Wilson, seemed inclined to assist with legislation to make the Commandant a permanent member of the Joint Chiefs of Staff. Qui-

---

[80] BGen Albert E. Brewster, USMC (Ret), "The Commandant of the Marine Corps and the JCS," *Marine Corps Gazette* 92, no. 3 (March 2008): 63.

etly working behind the scenes, Stennis noted in committee testimony that "I never have seen the Marines fail to arouse some interest around matters that concern them. They usually have good reason for their position, and they give more for the military dollar, in my opinion, than anybody else. I am supporting it on those principles."[81]

Consequently, Wilson became a permanent standing member of the Joint Chiefs of Staff on 20 October 1978. Due to the shrewdness of Wilson and Stennis, the legislation making Wilson a full member of the Joint Chiefs of Staff caught Secretary of Defense Brown; the chairman of the Joint Chiefs of Staff, Air Force general David C. Jones; and the other Service chiefs completely unaware. In Wilson's own words, "No one knew what I was doing." He added that "I did tell—not ask—tell the Secretary of the Navy the day before what I was doing and requested his confidence." When Jones called Wilson to complain about what happened, Wilson told the chairman to "stand up and be counted. If you don't want the Commandant as a member of the JCS, I suggest you call Senator Stennis to get this through." Jones allegedly responded, "You know I can't do that," and dropped the issue altogether.[82] Wilson provided his successor with more influence than any previous Commandant on the Joint Chiefs of Staff.

General Barrow was closely associated with Wilson during their years of service and held similar attitudes toward maintaining a highly disciplined Marine Corps that could act as a globally oriented force-in-readiness. Barrow, like Wilson, approached his service in the Marine Corps with a subtle steely resolve. Like his two most immediate predecessors, Cushman and Wilson, Barrow possessed a superb combat record, having fought with distinction in three wars. He successfully commanded the 9th Regiment, 3d Marine Division, during the Vietnam War, including leading it through one of the war's most successful combat missions, Operation Dewey Canyon, in 1969. At the time, the 3d Division fought as part of the U.S. Army's XXIV Corps, commanded by Lieutenant General Richard G. Stilwell, which was under the command

---

[81] Brewster, "The Commandant of the Marine Corps and the JCS," 63–64.
[82] Wilson Oral History Transcript, 324–25.

of III MAF. After Dewey Canyon, Stilwell, who rarely lavished praise on others, called Barrow "the finest regimental commander in Vietnam."[83]

Barrow returned to the United States in 1972 and took over Marine Corps Recruit Depot, Parris Island, South Carolina. He worked hard to eradicate problems that he found there. For example, soon after arriving, Barrow performed an inspection at the Parris Island medical facilities and learned, to his dismay, that the doctors had treated 23 recruits for broken jaws in 1972 alone. Although the injuries were reported to have occurred because they had fallen in the shower or on the obstacle course, Barrow was convinced that they resulted from recruit mistreatment. He believed this issue had become part of the system over time, but knew it was "very bad" and "very wrong."[84]

At the same time, Barrow constructed reforms that addressed physical abuse. They were "designed to end physical abuse and harassment of recruit trainees by drill instructors." These actions included ensuring "closer supervision by officers," all of which had "worked well." While Barrow was adamant that the Corps keep up its high physical training standards, he also "demanded that there be no more 'excess stress' on recruits, including 'nose-to-nose yelling' by drill sergeants."[85]

Barrow also worked to increase the quality of recruits who went through Parris Island. Frankly, the number of unqualified recruits coming through the recruiting depot in the early 1970s shocked the new base commander. He believed that fewer than half of his recruits had graduated high school before reaching Parris Island, which "didn't seem to bother anyone at Headquarters."[86] Barrow complained to the recruiting command so frequently that the Assistant Commandant at the time

---

[83] BGen Edwin H. Simmons, USMC (Ret), "Robert Hilliard Barrow," in *Commandants of the Marine Corps*, 446; and LtGen Willard Pearson, *Vietnam Studies: The War in the Northern Provinces, 1966–1968* (Washington, DC: Department of the Army, 1975), 67–70.

[84] Gen Robert H. Barrow, USMC (Ret), Oral History Transcript, 26 January 1989 session, BGen Edwin H. Simmons interviewer, Marine Corps History Division, Quantico, VA, 2015, 345–51; and Nathan Packard, "Congress and the Marine Corps: An Enduring Partnership," *MCU Journal* 8, no. 2 (Fall 2017): 17, https://doi.org/10.21140/mcuj.2017080201.

[85] Douglas Martin, "Robert Barrow, a Marine Corps Reformer Who Became Commandant, Dies at 86," *New York Times*, 31 October 2008.

[86] Barrow Oral History Transcript, 337–38; and Simmons, "Robert Hilliard Barrow," 446–47.

referred to him as the "troublemaker in Parris Island."[87] Still, Barrow remained adamant that he would not confine Parris Island to a long-standing policy that no more than 10 percent of recruits could receive a discharge due to unsuitability prior to graduation. Through his previously mentioned reforms, he also fought against the general attitude of Marine Corps leadership that the Service's boot camp experience could fix any negative attributes of the new recruits. Consequently, the Parris Island boot camp attrition rate under Barrow rose to nearly 25 percent of recruits being discharged before graduation. He stayed resolute in ensuring that only those recruits who truly demonstrated the best qualities needed to become a Marine would wear the cherished eagle, globe, and anchor after graduation. He later stated that he "was afraid" that his unhappiness with recruit quality at the time caused him to take it out "on the recruiting service."[88]

When Wilson became Commandant, he immediately brought Barrow to Headquarters Marine Corps and made him his deputy chief of staff for manpower. Barrow actually volunteered for this position due to his interest in and experience with Marine Corps personnel issues. The Commandant immediately tasked Barrow with making improvements in recruit training and within the recruiting service.[89] Wilson strongly believed that a high school diploma could act as a strong indicator of a recruit's success. He felt that new enlistees who met the challenge of graduating from high school would most likely succeed at boot camp over any who quit school before joining the Marine Corps.[90]

After a short stint as the commanding general of FMFLANT between 1976 and 1978, Barrow received a promotion to full general and became Assistant Commandant of the Marine Corps. Few doubted that Barrow would succeed Wilson as Commandant in 1979, which came true. The highly respected Marine Corps chief of staff, Iwo Jima veteran Lieutenant

---

[87] Barrow Oral History Transcript, 337.
[88] Barrow Oral History Transcript, 337–38; and Simmons, "Robert Hilliard Barrow," 446–47.
[89] Barrow Oral History Transcript, 13 December 1989 session, 369.
[90] Millett, *Semper Fidelis*, 618–22.

General Lawrence F. Snowden, was likely Barrow's "closest contender," but both men went out of their way to "not campaign for the job."[91]

In his last posture statement as Commandant of the Marine Corps (fiscal year 1980), Wilson was pleased to report the progress the Service had made in its recruitment changes to Congress. According to Wilson, "A comparison of selected quality indicators between fiscal year 1976, the first year of our quality improvement program, and fiscal year 1978 documents the effectiveness of those initiatives." He noted a sharp reduction in the "rates of unauthorized absence and desertion" as well as a "27 percent reduction in the confined population" that marked a "continued downward trend in discipline rates," which also reflected "the 42 percent decrease in major command special court-martial convictions."[92]

As defense writer L. Edgar Prima wrote in the June 1979 issue of *Sea Power* magazine, Barrow was inheriting Wilson's legacy of focusing on the "appearance, conduct, and performance" of individual Marines. Yet, he also took on "serious hardware and personnel problems."[93] Barrow also got a budget wracked by inflation and one that was slightly less than the previous fiscal year. Consequently, the Service did not have enough money in the budget to make the necessary improvements in vehicles, weapons, and especially cold-weather gear for the potential Norway mission.

The poor quality of the Marine Corps' cold-weather gear and its underwhelming operational performance came under scrutiny in the early 1980s. *The New York Times* openly criticized these elements after the newspaper's defense reporters observed a NATO training operation that involved U.S. Marines in March 1979. Like Cushman and Wilson, Barrow needed to maintain a critically underfunded Service that also sustained its strategic mobility to conduct combat missions across various

---

[91] Barrow Oral History Transcript, 17 December 1991 session, 424–25; and Simmons, "Robert Hilliard Barrow," 448–49.

[92] Gen Louis H. Wilson Jr., *Statement of General Louis H. Wilson, Commandant of the Marine Corps on Marine Corps Posture, Plans, and Programs for FY 1980 through 1984* (Quantico, VA: Marine Corps University Research Library, 1979), 11.

[93] L. Edgar Prina, "Wilson's Legacy, Barrow's Inheritance: A Combat-Ready Corps," *Sea Power* 21, no. 6 (June 1979): 38–40.

environments—from the desert sands of the Middle East to the Arctic Circle of Norway. This challenge emerged just as the Navy decreased the number of purpose-built amphibious ships and the DON started making these vessels a low priority. Barrow needed to convince the DOD and the secretary of the Navy to "restore the proposed new LSD (Landing Ship Dock) 41's ship program to the Five-Year Defense Plan." Significantly for the Marine Corps, although President Carter had not requested more funds for the Service's top aviation program, the Harrier AV-8B, the House and Senate Committees on Armed Services "authorized $180 million in the FY 1980 budget" for developing the highly valued second-generation Harrier II.[94]

The Marine Corps' budget struggles lasted until President Reagan's administration came to power in 1981. The Service had been so underfunded throughout the 1970s that Reagan's secretary of the Navy, John H. Lehman Jr., believed that the department had funded nearly everything that Barrow had requested. These wishes included "a series of light-armored vehicles, suitable for airlifting ashore; the Mark-19 grenade launcher; the M-198 155mm howitzer; and a host of field equipment, including shelters, containers, motor transport, and material handling and service support equipment."[95] Lehman considered the equipment improvements long overdue despite still having concerns that the new acquisitions may have made the Marines a bit too heavy.

Shortly before Reagan's election, the Congressional Budget Office (CBO) published a paper in May 1980 that offered several alternative approaches to the MPS program. The organization believed that the program was too costly and that even the ground-based set would need to vary significantly depending on the environment where operations took place. This study mirrored some of the same criticisms found in Binkin and Record's 1976 publication. The CBO study noted that the majority of the investment for the Marine Corps budget went to "modernize and maintain the Marine aviation component." In the nine years between 1970 and 1979, the CBO reported, "nearly two-thirds of all pro-

---

[94] Prina, "Wilson's Legacy, Barrow's Inheritance," 38–40.

[95] John F. Lehman Jr., *Command of the Seas* (New York: Charles Scribner's Sons, 1988), 160.

curement funding was spent on Marine air wings."[96] The CBO believed these "financial constraints" caused a curtailment in the Marine Corps' "research and development (R&D) efforts," which also instigated the Service's consideration for substantial troop reduction.[97] To make matters worse, since 1971, no new amphibious ship construction had been authorized, and the Carter administration had delayed production of the long-awaited LSD 41 program. Meanwhile, the Navy's amphibious ship inventory had diminished to just 64 ships—barely enough to lift a single MAF. Finally, the CBO believed that the Marine Corps could not "simultaneously reorient" itself to missions in both Norway and the Middle East "while maintaining its general-purpose" of forward deployments and amphibious operations unless it received an enlarged budget and took on significantly extended deployments. Yet, longer deployments, the CBO argued, "would aggravate one of the Marines' most persistent manpower concerns" and have a negative effect on morale and reenlistments.[98]

Barrow and Lehman fundamentally disagreed with that CBO assessment. Lehman was in the process of creating his much-touted "600 ship Navy," including making more MPS shipping available in active service. Furthermore, Lehman's proposal to resurrect four World War II-era *Iowa*-class battleships, all equipped with long-range 16-inch naval guns, offered ideal naval gunfire support platforms for Marine Corps landing operations. Although the battleships' antiquated oil-fired engines made them expensive to deploy, USS *New Jersey* (BB 62) returned to active duty in 1982. Soon after, USS *Iowa* (BB 61), USS *Missouri* (BB 63), and USS *Wisconsin* (BB 64) joined the fleet as well. Each battleship carried a traditional Marine detachment of approximately 60–80 troops, usually under the command of a captain, primarily for shipboard security duties. Additionally, these units usually crewed one of the ship's numerous 5-inch gun

---

[96] Alice M. Rivlin, *The Marine Corps in the 1980s: Prestocking Proposals, the Rapid Deployment Force, and Other Issues* (Washington, DC: Congressional Budget Office, 1980), 6. Today's U.S. Navy amphibious shipping inventory stands at more than 30 vessels. Although more capable than 1980s, today's purpose-built amphibious fleet can provide lift for only two Marine expeditionary brigades.

[97] Rivlin, *The Marine Corps in the 1980s*, 7.

[98] Rivlin, *The Marine Corps in the 1980s*, 29–30.

Figure 41. McDonnell Douglas F/A-18 Hornet fighter aircraft

Source: official U.S. Marine Corps photo.

mounts, which was typically emblazoned with an eagle, globe, and anchor on its armored turret. The four battleships had their Korean War–era antiaircraft batteries replaced with 14 McDonnell Douglas Harpoon antiship missiles and 32 McDonnell Douglas Tomahawk Land Attack Missiles but retained all three of their massive 16-inch gun turrets. In addition, the Navy added the new General Dynamics Phalanx close-in weapons system, which was designed to defend against threats from sea-skimming cruise missiles, to the recommissioned vessels. Battleship duty was considered a choice assignment for any Marine.[99]

Lehman also pushed to get the long-delayed LSD 41 program back on track, which resulted in the production of USS *Whidbey Island* (LSD 41). Barrow attended this vessel's keel laying "at the Lockheed shipyard

---

[99] Lehman, *Command of the Seas*, 115, 120, 158–60; and Bill Keller, "The Navy's Brash Leader," *New York Times Magazine*, 15 December 1985.

Figure 42. Landing Craft, Air Cushion (LCAC)

Source: official U.S. Navy photo.

in Seattle, Washington, on 4 August 1981."[100] *Whidbey Island* was the first ship to use its well deck to launch and recover at least four new landing craft air cushion vehicles that skimmed across the water at nearly 40 knots and could deliver Marines and heavy equipment ashore much farther and faster than the old World War II–era landing craft utility boats.[101]

Meanwhile, Barrow continued Wilson's policies related to personnel. High school graduates now represented more than 90 percent of enlistees every year. Barrow continued to stress physical fitness and personal appearance. He significantly increased the number of women serving in the Marine Corps, although he remained adamantly opposed to them serving in the infantry. He went straight after the long–simmering drug abuse problem that plagued all the U.S. military Services throughout the 1970s. Barrow was fortunate that, for the first time, the military could

---

[100] Simmons, "Robert Hilliard Barrow," 453–54.
[101] Lehman, *Command of the Seas*, 181.

now randomly drug test everyone using a standardized urinalysis test that was accurate and easy to administer. Barrow declared a war on Marines' drug use in a February 1982 All Marines Message. He expressed a zero-tolerance policy for all commissioned and noncommissioned officers who were caught using any form of illicit drugs.[102] Junior enlisted members faced disciplinary punishment that could potentially end up as an "other than honorable discharge." In just a few years, random tests of large Marine Corps units, such as an airwing or infantry regiment, revealed less than 1 percent drug use in the ranks. After retiring, Barrow remarked that this score was "better than any institution in America. Hands down. Maybe the Girl Scouts can do better."[103]

Around this time, there was growing dissatisfaction with the "cumbersome" arrangement of the RDJTF command. The chairman of the Joint Chiefs of Staff, General Jones, and Army leadership wished for the RDJTF to fall under U.S. European Command (USEUCOM), which typically came under the command of an Army four-star general. Navy and Marine Corps leadership favored it falling under U.S. Pacific Command (USPACOM), which the Navy long dominated. Both commands already had extensive far-flung responsibilities. Consequently, the Joint Chiefs of Staff made a compromise arrangement that created a new unified headquarters, U.S. Central Command (USCENTCOM), which would encompass the region around the Middle East. Activated at MacDill Air Force Base on 1 January 1983, Army lieutenant general Robert C. Kingston became its first commander. As part of the creation of this new command, the Marine Corps and Army, with approval from the Joint Chiefs of Staff, established an informal agreement that the two Services would alternate having command over USCENTCOM, at least until the 1990s. Marine Corps General George B. Crist became the second USCENTCOM commander between 1985 and 1988, making him the first Marine general to serve as a unified force commander. The biggest problem for the new command was that while USEUCOM and USPACOM had substantial standing forces already in place, USCENTCOM did not. To remedy this

---

[102] "Chronologies—1982," Marine Corps University, accessed 27 April 2023.
[103] Lehman, *Command of the Seas*, 164.

issue, the United States made "three access agreements" with the regional states of Oman, Kenya, and Somalia for the use of their facilities in "emergency situations." During Operation Bright Star the following year, however, Oman, even after receiving "100 million dollars in military assistance," only gave a Marine landing force permission to proceed inland "just 4 miles" from the Arabian Sea and they could stay ashore for "just 30 hours."[104] In sum, if USCENTCOM needed to "send a combat force" to a trouble spot, it would have to "start from almost zero in terms of combat power and support structure in the region."[105]

This situation made the MPS program even more critical for USCENTCOM, but its leadership would have other concerns as well. The Carter administration created the RDJTF without substantially increasing the military's size or budget. Any forces intended for the Persian Gulf would naturally come at the expense of other unified commanders and their missions. For example, USCENTCOM included the 82d Airborne, 101st Airborne, 9th Infantry, and 24th Mechanized Infantry Divisions from the Army. All these units, however, were based in the United States and were mainly reserved for NATO contingencies.[106] Similarly, the Marine brigade originally assigned to USCENTCOM was assigned to U.S. Marine Corps Forces, Pacific, with its main role being to support USPACOM contingencies.

The lack of friendly ports available to the United States in the Persian Gulf region was a primary driving factor in the final decision to preposition military equipment there. The Army had long prepositioned land-based critical resources in Central Europe using a program called Prepositioning of Material Configured to Unit Sets (POMCUS).[107] This program's implementation was not surprising, given the Army's preoccupation during the Cold War with the defense of Central Europe, the

---

[104] David Isenberg, *The Rapid Deployment Force: The Few, the Futile, the Expendable* (Washington, DC: Cato Institute, 1984), 4.

[105] Isenberg, *The Rapid Deployment Force*, 453; Cynthia Watson, *Combatant Commands: Origins, Structure, and Engagements* (Santa Barbara, CA: Praeger, 2011), 126; and Palmer, *Guardians of the Gulf*, 115–17.

[106] Record, *The Rapid Deployment Force and U.S. Military Intervention in the Persian Gulf*, 52–53.

[107] Douglas I. Bell, *Just Add Soldiers: Army Prepositioned Stocks and Agile Force Projection* (Carlisle, PA: U.S. Army Heritage and Education Center, U.S. Army War College, 2021).

Fulda Gap in West Germany, and the overall NATO mission. The Army practiced reinforcing its robust standing forces in Europe from the continental United States during a yearly exercise called Reforger. This program came with the advantage that the United States knew where the threat to NATO would come from. In the Persian Gulf or the Horn of Africa, threats could emanate from a variety of sources.[108]

Lieutenant General Paul X. Kelley argued that forces sent to the Middle East would naturally have to be task-organized for the actual mission. As the RDJTF commander, Kelley had the responsibility of tailoring the force package to the mission. When called on, he would deploy the proper units, sending forces as small as a single Marine amphibious unit (MAU) to one as large as several Army divisions and a division-size MAF. In sum, Kelley believed that the RDJTF's mission was to provide a "central reservoir of forces" based in the continental United States that could be drawn on to "cope with a specific contingency."[109] Consequently, according to Barrow, the Marine Corps would offer scenario specific contributions to USCENTCOM. This approach allowed for more cost-effective contributions from the Marine Corps, as sealifting heavy equipment was cheaper than airlifting it. This method also enabled the MPS program to quickly shift to other locations around the globe in an emergency. The MPS and USCENTCOM seemed to be a match made in heaven. Kelley believed MPS was the ideal solution to the strategic mobility problem that could emerge during potential operations in and around the Persian Gulf.[110] All the Marine Corps needed was a nearby friendly port where it could offload combat gear.

Yet, the Commandant recognized the program's limitations. In congressional testimony, Barrow cautioned that the MPS was a "means of enhancing our strategic mobility only. It is not a substitute for United States ability to project power into a hostile environment." Barrow and the other Joint Chiefs were convinced that the U.S. military could only

---

[108] Col Matthew Morton, USA, "We Were There: Reforger Exercises Designed to Counter Soviet Threat," Association of the United States Army, 24 March 2022.

[109] Record, *The Rapid Deployment Force and U.S. Military Intervention in the Persian Gulf*, 53.

[110] Gen Paul X. Kelley, "One Telephone Call Gets It All: Maritime Prepositioning for Crisis Response Enhancement," in *The Legacy of American Naval Power*, 209–15.

deploy MPS in a "non-hostile scenario" and that the Navy still needed to retain some amphibious assault capacity if necessary. He cautioned that "we must not be lulled into the perception that commercially designed and crewed ships are substitutes for war ships."[111] The decreasing number of purpose-built amphibious ships, which the Marine Corps could use in a national emergency, concerned Barrow even more. Approximately 64 amphibious ships, always a low priority with the Navy, existed in 1981.[112] Hence, it might take too much time for the Navy to gather enough amphibious ships to lift just a single MAF, making it problematic for the force to contribute to future combat contingency operations related to either NATO or USCENTCOM.

During this time, Secretary Lehman proved to be a significant contributor to the DON's trajectory. He opposed the formation of the RDJTF, referring to it as just another "layered bureaucracy."[113] During his extensive tenure, Lehman wished to couple recent revolutionary improvements in technology with a new maritime strategy that would enable the United States to retain total maritime superiority, despite the secretary of defense allegedly not approving of his use of this phrase. He also sensed that the biggest adversary to his vision was not necessarily going to be the rising Soviet blue-water threat—although this was indeed a considerable concern—or any other outside entity. As he wrote in *Command of the Seas*, the Pentagon bureaucracy worried him the most. Lehman believed that these defense insiders were fundamentally "allergic to different thinking and jealous of its prerogatives." Even if they could surmount this issue, he asked, "Would the navy system deliver? Could the Pentagon and the contractors actually build the ships and planes to the plans and budget?"[114]

Lehman especially had concerns about the apparent threat of the Soviet Navy. He warned that the Soviets and their Warsaw Pact allies

---

[111] Record, *The Rapid Deployment Force and U.S. Military Intervention in the Persian Gulf*, 65.

[112] Norman Polmar, "The U.S. Navy: Amphibious Lift," U.S. Naval Institute *Proceedings* 107, no. 11 (November 1981).

[113] James Kitfield, *Prodigal Soldiers: How the Generation of Officers Born of Vietnam Revolutionized the American Style of War* (New York: Simon and Schuster, 1995), 237.

[114] Lehman, *Command of the Seas*, 116.

had outpaced the United States in shipbuilding by 1980. By that time, he claimed, those nations "were outproducing us by two to one in major combatants and by five to one in submarines."[115] At this rate, U.S. maritime superiority, which previous DOD and DON leadership long assumed would be a given, was in jeopardy if the United States was forced to fight in more than one major regional contingency simultaneously. President Reagan emphasized a dedication to naval superiority during the recommissioning ceremony of the venerable battleship *New Jersey* on 28 December 1982. In his remarks, Reagan, who Lehman introduced, announced in part:

> Maritime superiority for us is a necessity. We must be able in time of emergency to venture in harm's way, controlling air, surface, and subsurface areas to assure access to all the oceans of the world. Failure to do so will leave the credibility of our conventional defense forces in doubt. . . . This 58,000-ton ship, whose armor alone weighs more than our largest cruiser, is being recommissioned at no more than the cost of a new 4,000-ton frigate. The "Big J" is being reactivated with the latest in missile electronic warfare and communications technology. She's more than the best means of quickly adding real firepower to our Navy; she's a shining example of how this administration will rebuild America's Armed Forces on budget and on schedule and with the maximum cost-effective application of high technology to existing assets.[116]

Unlike many of his Navy counterparts, Barrow maintained a reasonably good working relationship with Lehman throughout his time as Commandant. Further, unlike many parsimonious defense officials in the Carter administration, Reagan's defense bureaucrats, including Lehman and Secretary of Defense Caspar W. Weinberger, staunchly backed the

---

[115] Lehman, *Command of the Seas*, 129, 132–33. Today, the People's Republic of China far surpasses the United States in the number of warships built in a single fiscal year.

[116] Ronald W. Reagan, "Remarks at the Recommissioning Ceremony for the U.S.S. *New Jersey* in Long Beach, California" (speech, Recommissioning Ceremony of USS *New Jersey*, Long Beach, CA, 28 December 1982).

Figure 43. USS *New Jersey* (BB 62)

Source: official U.S. Navy photo.

Marine Corps. All three men strongly supported the Carter Doctrine that initially called for the use of at least a brigade of Marines as a principal element of the RDJTF for any contingency in the Persian Gulf. Lehman, however, favored an even stronger approach. He allowed the Marine Corps to increase the extent of amphibious shipping prepositioned near the Gulf. These forces could sustain an entire MAF with adequate provisions for at least 30 days of fighting ashore while also maintaining enough capacity for at least an additional brigade deployed to another regional contingency at the same time. After receiving the reinforced MPS plan that Barrow prepared, Lehman presented it to the Defense Review Board, which approved and adopted it. To fulfill this plan, Lehman and Barrow intended to increase the capability of "the force we inherited" so that it could "deploy a single MAF in one theater and, independently, a single MAB in another." To deal with the problems related to multiple

235

tasking of its forces, Lehman and Barrow proposed raising the Marine Corps personnel ceiling from 188,000 to 200,000 Marines.[117]

While the Marine Corps containing approximately 200,000 personnel could be helpful, it would still be of limited use if Marines could not rapidly deploy into two climatically different environments. In 1981, Barrow believed that the best and most cost-effective place for the Marines to conduct cold-weather training was at MWTC. However, at that time, the facilities at the MWTC were decrepit at best. Many of the center's staff members and their families were housed in temporary doublewide trailers.[118]

In 1980, Brigadier General Americo A. Sardo, the director of training at the MWTC, with the support of Colonel John W. Guy, the commanding officer of the MWTC, commissioned a study that rejected the idea of creating a permanent Marine staffing presence at Camp Drum, New York, due to it not being cost-effective. Moreover, the concept paper argued that the amount of snow and terrain at Camp Drum did not provide enough overlap with Norway's environment. Pickel Meadow, however, was already a Marine Corps facility that could be turned into a first-rate cold-weather doctrine and training site with some money dedicated to facility improvement and staffing levels. At a briefing on this issue, the Marine Corps deputy chief of staff for manpower, among others, strongly opposed the proposal because he felt that the MWTC was unnecessary and cost too much.[119] Others in the room seemed to agree. Barrow then asked his Assistant Commandant, General Kenneth McClennan, his thoughts. McClennan argued that "the Marine Corps could ill afford to give up this valuable training capability, especially with the

---

[117] Lehman, *Command of the Seas*, 158–59.

[118] Steele and Moffett, *U.S. Marine Corps Mountain Warfare Training Center*, 99–106.

[119] Steele and Moffett, *U.S. Marine Corps Mountain Warfare Training Center*, 100–2. This briefing to the Commandant of the Marine Corps had been delayed for half a year due to BGen Sardo having suffered a major heart attack. Sardo was still recovering at the time of the presentation in March 1981. In the early 1980s, Steele, then a colonel, commanded the prestigious Marine Barracks, Washington, DC, during part of the time that Barrow was Commandant. Both Steele and Barrow were present for an evening parade conducted for Reagan by the barracks Marines just a few months after he had survived an assassination attempt by John Hinckley Jr. The author also attended this event. The evening parade was Reagan's first public event after being seriously wounded.

Corps' growing commitment to the reinforcement of northern Norway." McClennan's firm, declarative statement appeared to undermine the arguments of the naysayers.[120]

Throughout the briefing, Barrow never indicated his feelings on the MWTC issue. He did, however, declare to the entire gathering that "whatever we may decide gentlemen, we shall never again allow our people to be housed in trailers." As Barrow often did, he traveled to MWTC to see the situation for himself. Colonel William H. Osgood had just taken command of the facility when Barrow traveled there. After receiving the news that the Commandant and his wife, Patricia Ann Barrow, as well as the sergeant major of the Marine Corps, Leland D. Crawford, would soon arrive to tour the center, Osgood and his subordinate, Major Edward J. Robeson IV, arranged for the Commandant to travel around the base and meet with the Marines there in an informal setting, believing that it would allow for Barrow to "gain a better perspective of the base and its activities." Although Barrow clearly enjoyed chatting with the instructors and their students, according to Osgood, he was "noticeably upset by the run-down condition of the base and its facilities." During the tour, Barrow did not let on his thoughts concerning the future of the MWTC. As he was preparing to leave, however, he discovered that many Marine Corps families lived almost 153 kilometers away at the naval ammunition depot in Hawthorne, Nevada. As he got into his staff vehicle, Barrow told Osgood, "I want you to start moving people out of that housing area and have them relocate to the Highway 395 corridor." Additionally, Osgood was to call Barrow once a week to report how many families remained in Nevada.[121]

Despite its dilapidated condition, Barrow ultimately decided to invest in the MWTC. Major General Orlo K. Steele later asked Barrow why he decided to save the facility despite his hatred for its "substandard living conditions," which he likened to a "squatter's camp." Barrow responded that he feared that the Marine Corps would possibly be required to give the property back to the U.S. Forest Service if it did not use the facili-

---

[120] Steele and Moffett, *U.S. Marine Corps Mountain Warfare Training Center*, 101–2.
[121] Steele and Moffett, *U.S. Marine Corps Mountain Warfare Training Center*, 102–3.

ty. In addition, as the former commanding general of FMFLANT, he had "gained some appreciation for the difficulties and complexities associated with our commitment to north Norway." Most importantly, having had "first-hand experience in Korea," he believed that "mountain training, summer, and winter, requires skills that just cannot be taught or learned by the seat of the pants."[122] Clearly, the Chosin Reservoir campaign had come full circle for Barrow.

Barrow did not cease his reforms with cold-weather training. In the late spring of 1982, approximately 10,000 Marines of the 1st Marine Division from nearby Camp Pendleton, along with significant elements of the Army, including the 82d Airborne Division, conducted an eight-day Joint exercise across arid southeastern California near Fort Irwin and Twentynine Palms. Called Gallant Eagle 82, the Joint exercise—one of the most extensive that the U.S. military conducted during the Cold War—tested the ability of units assigned to the rapid deployment force to operate in the harsh desert conditions they would potentially face in the Persian Gulf. Most notably, the 82d Airborne conducted its largest airdrop since World War II during Gallant Eagle 82, although it also came with a black mark. Photographs of the air operation, with dozens of Air Force airplanes filling the skies, harkened back to the unit's legendary airdrop made in Normandy, France, in 1944. Yet, on the first day of the exercise, 1 April 1982, the leaders running the exercise decided to drop 2,300 paratroopers in high-gusting winds. Wind readings immediately prior to the drop indicated acceptable wind speeds for the exercise, causing the planners to go ahead with the exercise, but the decision just as the wind increased resulted in the deaths of 6 soldiers and the injury of another 158. Although this regrettable occurrence put a negative spot on the exercise, the Joint Staff scheduled subsequent Gallant Eagle exercises every two years throughout the 1980s, but with an increased focus on safety.[123] These exercises were critical toward enabling the Services

---

[122] Steele and Moffett, *U.S. Marine Corps Mountain Warfare Training Center*, 103.
[123] "Massive Joint-Service 'Gallant Eagle' Exercise Concluding," AP News, 2 August 1986; "Chronologies—1982"; and "36 Years Ago the Military Had Its Biggest Airborne Operation Blunder in History," PopularMilitary.com, 30 March 2018.

to get ready for an actual Persian Gulf contingency—Operation Desert Shield and Operation Desert Storm (1990–91).

Throughout much of late 1982 and up through his retirement at the end of June 1983, Barrow increasingly became focused on the events taking place in the volatile setting of Beirut, Lebanon. This area of the Middle East did not fall within the RDJTF commander's area of responsibility. Instead, it was under the purview of USEUCOM and the Navy's Sixth Fleet. In 1982, with little warning for the United States, the IDF, led by General Ariel Sharon, invaded southern Lebanon. Long frustrated by rocket, artillery, and terrorist attacks in northern Israel, the state of Israel was determined to rid this region of its Palestine Liberation Organization (PLO) fighters permanently. In an unexpected turn of events, Sharon's armored columns advanced all the way to the Lebanese capital and surrounded large pockets of PLO fighters in West Beirut. This action brought them into direct contact with a wide variety of Muslim militia bands as well as the Syrian Army, which had been occupying Lebanon's nearby Beqaa Valley for some years. Due to Syria's position as a client state of the Soviet Union, the Israeli incursion into West Beirut brought the possibility of Soviet intervention in the region ever closer.[124]

Thanks to the timely involvement of U.S. assistance, especially of special envoy Philip C. Habib, the Israelis and the PLO agreed to a convention that resulted in the establishment of a multinational peacekeeping force, including soldiers from France and Italy, that would provide safe passage of PLO fighters to the island of Cyprus and Tripoli, Tunisia. Reagan, against Weinberger's advice, authorized the landing of the 32d MAU, the Sixth Fleet's designated landing force commanded by Colonel James Mead, in Beirut as part of the multinational force. At a press conference, Reagan informed reporters that he did not envision the Marines staying ashore for more than 30 days. In a pleasant surprise, the Marines pulled off their part of the mission mainly without a hitch and returned to their ships approximately 17 days later. The PLO fighters had

---

[124] LCdr Bradley M. Jacobs, USCG, *Operation Peace for Galilee: Operational Brilliance-Strategic Failure* (Newport, RI: Naval War College, 1995).

been peacefully removed, and the threat of Soviet intervention greatly diminished.[125]

As was often the case in the volatile world of the Lebanese Civil War
and its politics during the 1980s, events beyond the control of most of
the major actors involved with the removal of PLO fighters served to de-
rail any prospects for a lasting peace. Since at least the 1920s, Lebanon's
constitution only allowed for a Maronite Christian to hold the office of
president. By the 1980s, however, its population was now mostly majority Muslim, which created tensions over this aspect of the government
that eventually erupted. On 14 September 1982, Lebanon's newly elected president, Bachir Gemayel, along with 26 other major Christian Phalange party leaders, were killed in a bomb attack on their headquarters.
The Federal Bureau of Investigation concluded that a terrorist organization called the Syrian Social Nationalist Party, which had direct links to
the state of Syria, was responsible for the attack.[126] In revenge, Christian
militia fighters slipped into two predominately Lebanese and Palestinian
Shiite refugee camps—Sabra and Shatila—and massacred approximately
700 people, although the actual figure remains in dispute to this day. The
horrific images of the death and destruction in the refugee camps created an immediate international outcry for an end to the senseless violence. Soon after, Reagan ordered the 32d MAU ashore for a second time.
The Marines landed without fully understanding the cultural or political
details behind the various factions struggling for control over Lebanon.

---

[125] Jacobs, *Operation Peace for Galilee*; and Benis M. Frank, *U.S. Marines in Lebanon, 1982–1984*
(Washington, DC: Marine Corps History and Museums Division, 1987), 11–21.

[126] Frank, *U.S. Marines in Lebanon*, 11–21; and Neil A. Lewis, "U.S. Links Men in Bomb Case to
Lebanon Terror Group," *New York Times*, 18 May 1988, A6. In 1982 and 1983, Lebanese national
politics was exceptionally complex due to its ongoing civil war that seemed both politically
and religiously motivated. The international community also criticized Ariel Sharon for the
indirect role he may have played in allowing the Christian militias to carry out their attacks
on the camps and then doing nothing about it once it was underway. Syria was implicated in
their training and support of highly violent Shia-inspired militias in and around Beirut. Few
members of the Reagan administration, except for special envoy Philip C. Habib, were well-
versed on what was going on inside Lebanon. Three members of the Syrian Social Nationalist
Party were arrested in 1987 for illegally crossing into Vermont through the U.S.-Canadian
border with the intent to conduct a car bomb attack. Fortunately, they did not make it far.

During the initial 17-day stint in Beirut, Mead remained tremendously concerned about the lack of intelligence he received on the political factions and militias that roamed the city. He believed that only sheer luck and the military professionalism of his Marines kept the initial operation ashore in volatile Beirut from devolving into something more tragic.[127] In their return to Beirut, Mead and his Marines once again received little guidance as to their new role as "peacekeepers" other than the Reagan administration's general explanation that the Marines were there to "provide a presence in Beirut that would in turn help establish the stability necessary for the Lebanese government to regain control of their capital."[128]

To avoid initiating the application of the War Powers Resolution clock, which would allow the Marines to remain ashore for as long as the president desired without having to consult Congress, the Reagan administration went to great lengths to demonstrate that the role of the Marines ashore was not a combat role. This placed the Marines in West Beirut in the unenviable position of managing an undefined, open-ended presence mission with little ability to discern or affect the fast-moving events taking place behind the scenes there. To make matters worse, the administration later authorized ships from the Sixth Fleet, such as the battleship *New Jersey*, to provide naval gunfire support for the Lebanese national forces associated with the Amine Gemayel government against Shia militia groups in their deepening civil war. To the militias, this activity meant the United States was no longer a neutral peacekeeper but was now an active combatant in support of the Lebanese national forces. This unfortunate decision initiated a tremendous change on the ground for the Marines. These same militias began targeting the Marines, both at the airport and while on patrol, with sniper and indirect weapons fire. In March 1983, for the first time since the Vietnam War and Koh Tang Island, the Marines suffered combat casualties. Nevertheless, the vul-

---

[127] Robert Fisk, *Pity the Nation: The Abduction of Lebanon* (New York: Atheneum, 1990), 359–75; Seth Anziska, "A Preventable Massacre," *New York Times*, 16 September 2012; and Frank, *U.S. Marines in Lebanon*, 19, 22–30.
[128] Frank, *U.S. Marines in Lebanon*, 22–23.

nerability of the peacekeeping Marines in Beirut appeared to be part of the administration's overall strategy to avoid risking Congress invoking the War Powers Resolution. While the Marines were authorized to harden their positions and to respond with force if attacked, the administration was adamant that the Beirut International Airport must remain open to the general Lebanese public when possible.[129]

In February 1983, the 32d MAU, now redesignated as the 22d MAU, relieved the 24th MAU, which had earlier replaced the 32d MAU in peacekeeper positions in and around the Beirut International Airport. Once on the ground, Mead's Marines settled into conducting routine patrols in West Beirut and ensuring that the international airport positions were adequately manned. Mead received numerous high-level visitors, including U.S. Navy admiral William J. Crowe Jr., commander in chief of U.S. Naval Forces Europe. For many of the 22d MAU Marines, this was their second deployment to Beirut in less than five months, which gave them knowledge of the combustible environment. Amazingly, it had been one of the worst winters in Lebanon's history, trapping many snowbound civilians in the mountains and hills east of the city. Fortunately, the cold-weather training, tracked vehicles, and helicopters of the 22d MAU were instrumental in evacuating dozens of those civilians. Only a few weeks later, on 15 March 1983, an Italian patrol from the multinational force was ambushed, resulting in the death of one soldier and the wounding of several others. The next day, a grenade thrown from an upper-story window slightly wounded five U.S. Marines on foot patrol. These incidents clearly indicated that things were not improving in West Beirut.[130]

A month later, the United States suffered its most significant casualties of the peacekeeping effort. On 18 April 1983, a terrorist drove a stolen van filled with explosives into the wall of the U.S. embassy, blasting a portion of the front side of the embassy wide open. The explosion de-

---

[129] Peter Huchthausen, *America's Splendid Little Wars: A Short History of U.S. Military Engagements, 1975–2000* (New York: Viking, 2003), 45, 48–63; and Lehman, *Command of the Seas*, 315–18.

[130] Frank, *U.S. Marines in Lebanon*, 49–53, 57–58.

stroyed entire sections of the embassy's outer walls, exposing the offices inside. The blast killed 63 people, including Marine security guard Corporal Robert V. McMaugh and 16 other Americans. The U.S. ambassador, Robert S. Dillon, was in his top-floor office when the explosion took place and emerged unscathed. With the U.S. embassy now totally unusable, the British ambassador, Sir David Roberts, invited Dillon and his staff to set up temporary shop in his embassy. The British ambassador also took the extraordinary step of asking that the 22d MAU to provide security for the now combined embassy and its grounds, which was perhaps the "first time in history that you have U.S. Marines guarding a British embassy."[131]

May 1983 was a watershed moment for the multinational force. After weeks of negotiation, the United States brokered the 17 May Agreement between Israel and the Lebanese national government. Notably, the Syrians and the antigovernment Shia militias operating in West Beirut and the nearby Chouf Mountains refused to participate in the proceedings, having had no intention of going anywhere. Once the details of the agreement had been finalized, the IDF withdrew from Beirut completely by September 1983. Although this was an apparently positive step on the surface, the de facto buffer that the IDF provided between the Shia militias, the Maronite Christians (with whom the militias were also fighting with at that moment), and the multinational force was now gone. The Marines soon found themselves face-to-face with heavily armed militias, Syrians, and others antithetical to the interests of the United States. Throughout the long hot summer of 1983, the Beirut International Airport and other multinational force positions began receiving random mortar and artillery fire, ostensibly fired by either the Shia militias or Lebanese national forces in response.[132] The situation for the Marines around the airport, however, was about to get much worse. Unless the perpetrator of such attacks could be positively identified, the rules of engagement did not allow the Marines to return fire.[133]

[131] Frank, *U.S. Marines in Lebanon*, 60–63.
[132] "Retreat from Beirut," *PBS Frontline*, 26 February 1985.
[133] Frank, *U.S. Marines in Lebanon*, 64.

Figure 44. Comdt Paul X. Kelley at a news briefing, August 1983

Source: official U.S. Marine Corps photo.

As Barrow prepared to retire from the Marine Corps after 41 years of continuous and dedicated service, he received accolades from various political and military leaders from around the globe. Reagan, who always seemed to have a soft spot in his heart for the Marines, told the Commandant at his change of command ceremony at the Marine Barracks in Washington, DC, that "under your stewardship, the Marines have never been better prepared or led."[134] During his time as a Marine, Barrow was only the second leatherneck to have received the U.S. Army's Distinguished Service Cross during the Vietnam War. In addition to that medal and his Navy Cross awarded for gallantry in action at Funchilin Pass during the Chosin Reservoir campaign in 1950, Barrow had been awarded two Defense Distinguished Service Medals, the Navy Distinguished Ser-

---

[134] Cyril O'Brien, "Giants of the Corps: General Robert H. Barrow," *Leatherneck*, January 2009, 39.

vice Medal, the Silver Star, three Legions of Merit, and the Bronze Star. Throughout his career, he had fought with distinction in three major conflicts and was instrumental in carrying on the legacy and policies of his predecessor Louis Wilson. It can be fairly said that both Wilson and Barrow truly remade the Marine Corps and created a more professional, better equipped, more lethal fighting force than had been seen since the Battle of Belleau Wood.

CHAPTER FIVE

Beirut, Grenada, and the Reagan Era

On 1 July 1983, General Paul X. Kelley succeeded General Robert H. Barrow, becoming the 28th Commandant of the Marine Corps. Known throughout the Service as P. X. Kelley, he was a natural choice for the position. Since 1981, Kelley had operated as Barrow's Assistant Commandant and had previously been the first commanding general of the Rapid Deployment Joint Task Force (RDJTF). During the early 1960s, Kelley had served as an exchange officer with the British Royal Marines' 45 Commando in Aden, Yemen, and 42 Commando in Southeast Asia. He possessed a sterling combat record as commander of the 2d Battalion, 4th Marines during the Vietnam War and had been awarded the Silver Star, two Bronze Stars, and the Legion of Merit with valor devices. During 1970–71, he commanded the 1st Marine Regiment, the last major Marine Corps ground unit in Vietnam, earning a second Legion of Merit award.[1]

Kelley assumed the position of Commandant intent on carrying on with organizational reforms started under Generals Wilson and Barrow. He was also a strong advocate of the newly developed Maritime Prepositioned Shipping (MPS) program. Like his immediate predecessor, Kelley benefitted from generous defense budgets that enabled him to continue

---

[1] *Hearing before the Committee on Armed Services United States Senate, on Nominations of General Paul X. Kelley to Be Commandant of the Marine Corps, Richard L. Armitage to Be Assistant Secretary of Defense (International Security Affairs), Chapman B. Cox to Be Assistant Secretary of the Navy (Manpower and Reserve Affairs)*, 98th Cong., 1st Sess. (24 May 1983) (biographical sketch of Gen Paul X. Kelley, USMC), hereafter Hearing, 14 May 1983.

the vital recapitalizing of the Marine Corps during the administration of President Ronald W. Reagan. As Commandant, Kelley also continued to emphasize high-quality recruitment, training, and combined arms exercises conducted primarily at Marine Corps Air Ground Combat Center Twentynine Palms, California. At the same time, he actively supported the Marine Corps contribution to the North Atlantic Treaty Organization (NATO) mission in Norway and the Mediterranean. Still, Kelley faced a significant challenge, as he inherited a confusing situation with the Marine amphibious units (MAUs) assigned to peacekeeping duties in Beirut, Lebanon, since 1982.[2]

The 17 May Agreement of 1983, signed by Israel, the United States, and the Lebanese national government, required the Israel Defense Forces (IDF) to withdraw all units then in and around the vicinity of Beirut to positions south of the Awali River in southern Lebanon at some point. The Lebanese Armed Forces (LAF) would take over the Israeli positions once they left. The Lebanese were also committed to the establishment of a security zone along the northern Israeli border to provide a safeguard against Palestine Liberation Organization (PLO) forces trying to return to the region. However, the Syrian government, which also occupied significant portions of Lebanon, did not sign the 17 May Agreement. Further, the agreement did not have the backing of the increasingly violent Muslim militias then engaged in a virtual civil war. This did not bode well for the Marines of the 24th MAU, commanded by Colonel Timothy J. Geraghty. Moreover, the entire U.S. military chain of command seemed totally unaware of the potential consequences of a political agreement that all the Muslim forces still fighting in Lebanon had condemned.[3]

From early in its six-month deployment, the 24th MAU faced increasing hostility from the Muslim militias in and around West Beirut. As the IDF prepared for its withdrawal, the militias grew emboldened and planned to fill the coming power vacuum. Most of the Marines of the 24th MAU manned checkpoints around a perimeter that encompassed the Beirut International Airport. About 1.5 kilometers away from that pe-

---

[2] Hearing, 24 May 1983.
[3] "Retreat from Beirut," PBS *Frontline*, 26 February 1985.

Figure 45. Distribution of religious groups in Lebanon

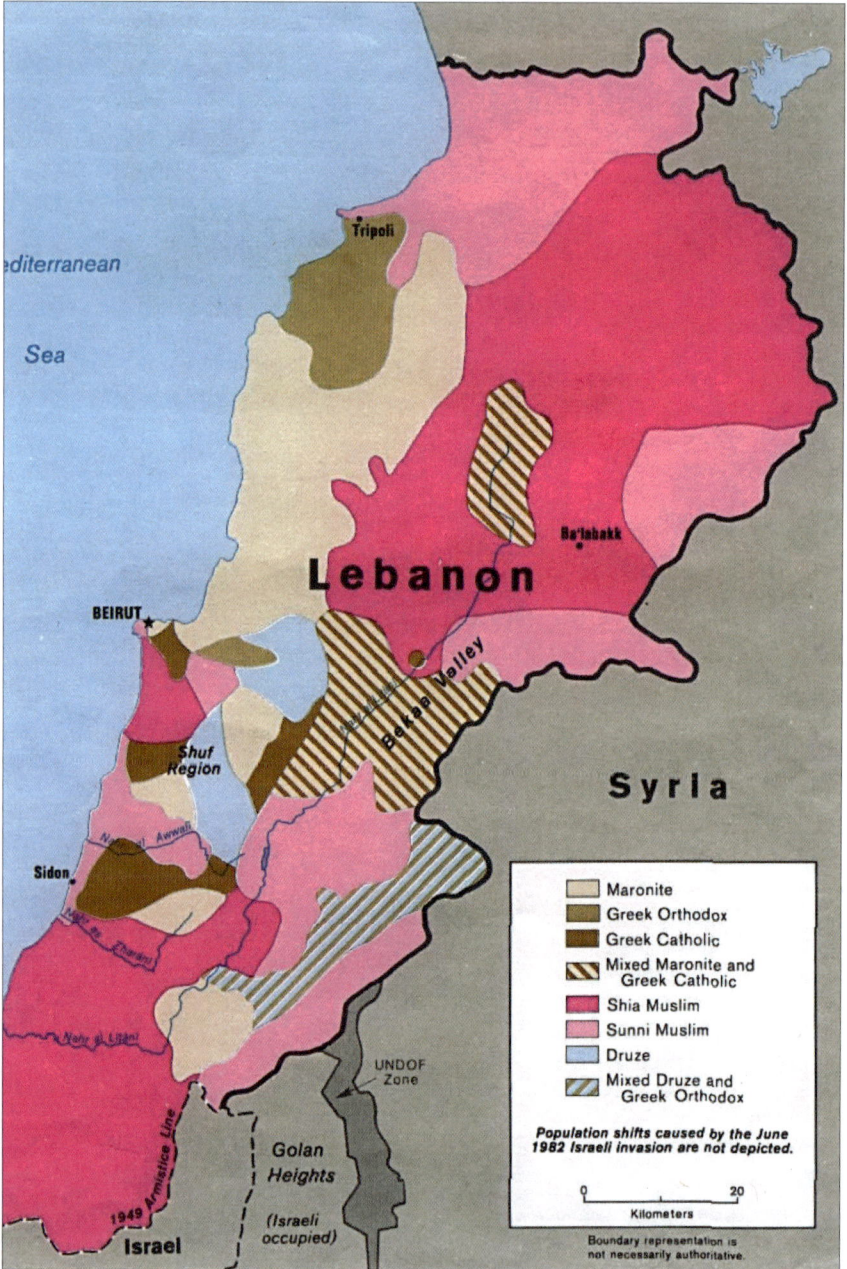

Source: Perry-Castadeña Library Map Collection, University of Texas.

Figure 46. The zones of the mulitnational forces in Beirut

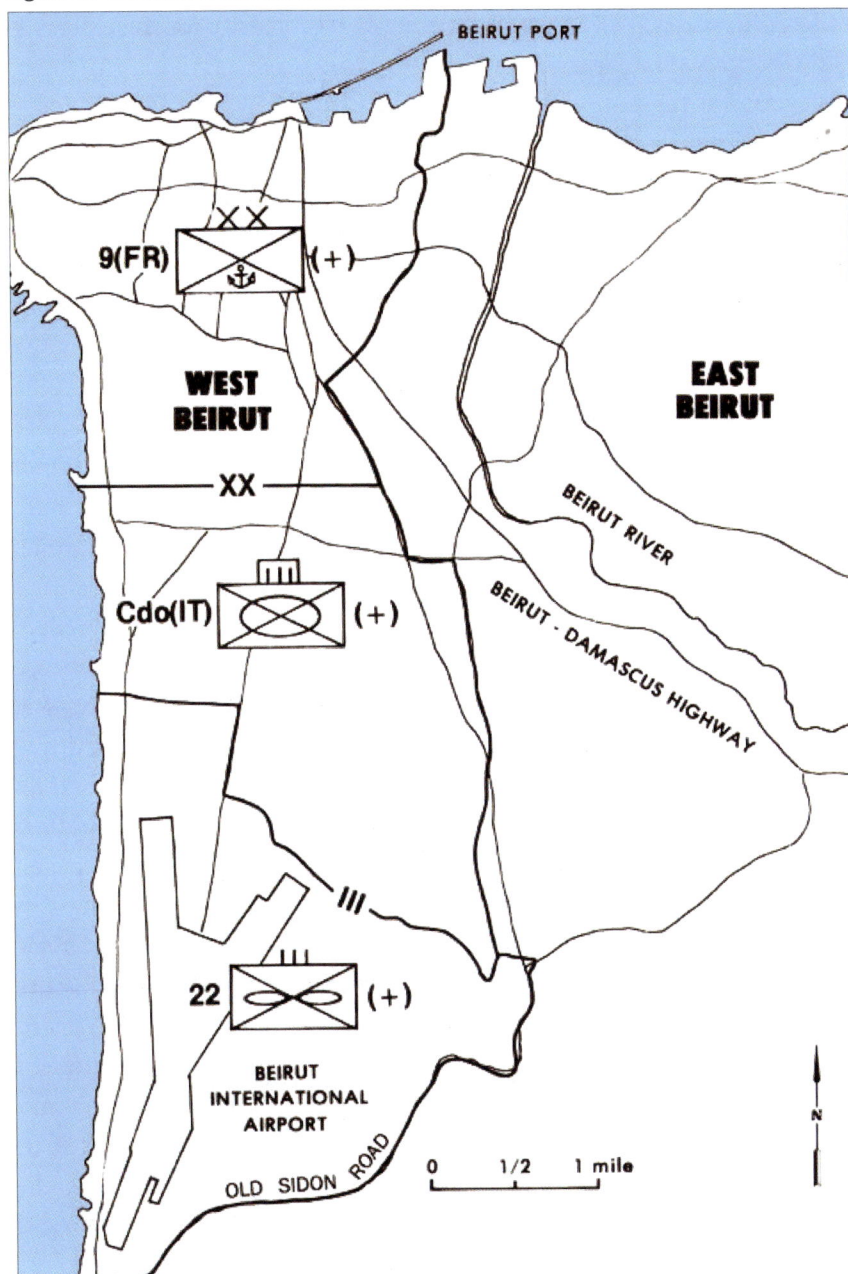

BEIRUT PORT

9(FR)

XX

(+)

WEST
BEIRUT

EAST
BEIRUT

XX

BEIRUT RIVER

Cdo(IT)

(+)

BEIRUT - DAMASCUS HIGHWAY

III

22

(+)

BEIRUT
INTERNATIONAL
AIRPORT

OLD SIDON ROAD

N

0   1/2   1 mile

Source: courtesy of the *Marine Corps Gazette*.

rimeter, one rotating rifle company of Battalion Landing Team 1st Battalion, 8th Marines (BLT 1/8), was assigned as security for the Lebanese University. A small detachment of 24th MAU Marines assisted the Marine guard at the temporary U.S. embassy, which was collocated within Great Britain's diplomatic compound. Starting in July, the Marines at the university and the airport endured random rocket-propelled grenade attacks, while larger Soviet-made 122-millimeter Katyusha rockets began landing squarely within the airport perimeter. The militias claimed that these strikes occurred unintentionally, resulting from errant rounds fired in their increasingly violent civil war.[4]

By August 1983, the local forces increased their direct attacks on the Marines and Lebanese officials. On 10 August 1983, while special envoy Robert C. McFarlane, a retired Marine lieutenant colonel and President Reagan's deputy national security advisor, visited the 24th MAU headquarters, Muslim gunners briefly fired on the airport perimeter, which resulted in one injury. Most likely, the attack was meant to get the attention of the Reagan administration through McFarlane. Around this time, a formerly obscure Muslim faction, the Druze, kidnapped three Lebanese national government cabinet ministers, holding them hostage briefly. Walid K. Jumblatt, the Druze leader, claimed responsibility for both the airport shelling and the kidnapping. As a result, McFarlane broke off all discussion with the Druze. Before this attack, Jumblatt had openly allied with the Syrians, who gave the Druze military advice and assistance.[5]

The Marines experienced increased tension with militia groups in the local community as well. The Marines of the 24th MAU, like their predecessors in the 22d MAU, conducted vehicular patrols in the "Shiite Muslim quarter of Hay-es-Salaam," a large refugee camp. The main routes to and from the Beirut International Airport went right through the middle of Hay-es-Salaam. Some sardonic Marines dubbed the ramshackle settlement "Hooterville," which was notable for the lack of

---

[4] "Retreat from Beirut," PBS Frontline, 26 February 1985; and Benis M. Frank, U.S. Marines in Lebanon, 1982–1984 (Washington, DC: Marine Corps History and Museums Division, 1987), 57, 63, 74–76.
[5] Eric Hammel, The Root: The Marines in Beirut, August 1982–February 1984 (Pacifica, CA: Pacifica Military History, 1985), 112–14.

young men in and around the neighborhood.[6] Initially, the local population in Hay-es-Salaam greeted the Marines with friendliness. Colonel
James M. Mead and his staff with the 22d MAU had even roamed the city
with relative ease. The situation drastically changed for Colonel Geraghty
and the 24th MAU by August 1983. Marines on patrol started to notice
anti-Americanisms literally written on the walls alongside portraits of
the Iranian leader Ayatollah Ruhollah Khomeini that were plastered in
numerous locations where they had once been absent. in other places,
young Lebanese boys taunted the Marines as they drove by chanting:
"Khomeini good, America no good."[7]

In the aftermath of the bombing at the U.S. embassy in Beirut in April
1983, the 24th MAU also received assignments to provide security for the
recently created combined diplomatic compound of the United States and
United Kingdom or the Lebanese University campus, located about 1.5 kilometers outside the perimeter around the Beirut International Airport,
during which they experienced an increase in violence. First Lieutenant
Peter J. Ferraro commanded the 3d Platoon of Company A, 1st Battalion,
8th Marines, which was sent to provide additional security for the combined embassy compound soon after arriving in Beirut. Ferraro believed
that the security arrangements in and around the compound was much
more robust in response to the bombing of the original U.S. embassy,
serving as a deterrent against further attacks. Thanks to concrete physical barriers, Ferraro rerouted traffic away from the compound altogether.
Later, a single car bomb attack occurred in a parking lot near the compound, but it caused minimal damage and no one was killed or wounded. Ferraro's platoon, along with the rest of Company A under Captain
Paul Roy, was later reassigned to Check Point 76, located near the center
of the airport perimeter. For weeks in August and September 1983, Roy's
company, among others in the MAU, endured "snipers, mortars, rock-

---

[6] Hammel, *The Root*, 43, 46, 49. The Marines may have likely taken the name "Hooterville"
from a popular American TV sitcom, *Petticoat Junction*, which ran 1963–70. In the show, the
town was decidedly agrarian and somewhat backward in a comedic way.
[7] Hammel, *The Root*, 110.

ets, artillery, and machinegun fire," which caused multiple casualties.[8] During this time, Ferraro noted, it seemed that the unit was in the middle of a firefight "nearly every night."[9] By the end of September, Company A was sent to secure the Lebanese University. The change of venue did not bring much respite from the snipers or shelling.

Throughout August and into September 1983, Colonel Geraghty entertained many high-ranking visitors, including senior members of Congress; U.S. Army general John W. Vessey Jr., the chairman of the Joint Chiefs of Staff; Secretary of the Navy John H. Lehman Jr.; and Commandant Kelley. During this period, Geraghty's defensive position at the airport became increasingly dangerous and the 24th MAU faced significant difficulties. Heavily armed Muslim militias and Syrian military forces based in the rugged Chouf Mountains surrounded the city, and locals in the nearby neighborhoods of West Beirut frequently confronted the patrols. Yet, the 24th MAU had to respond carefully to sniper and indirect fire attacks due to it receiving restrictive rules of engagement. Geraghty's Marines could return fire with mortars, artillery, and even on-call naval gunfire to a limited extent, but only after scout-sniper teams found the militias and confirmed that they were purposely targeting the MAU. Even then, due to the fiction that all the rounds from the militias landing within the airport perimeter were not deliberately intended for the Marines, the 24th MAU was initially required to fire nonlethal illumination rounds before using more lethal means. Fortunately, Geraghty had a highly accurate 155-millimeter artillery battery, Battery C, 1st Battalion, 10th Marines, supporting the MAU. Commanded by Captain Robert C. Funk, Battery C fired precise fire missions that suppressed or destroyed

---

[8] Col Peter J. Ferraro, USMC (Ret), interview with the author, 12 June 2019, hereafter Ferraro interview. Ferraro had an extraordinary 34-year career as both an enlisted Marine and an officer. He is still currently working for the Marine Corps as a civilian director in Manpower and Reserve Affairs, Quantico, VA. Ferraro also recalled being visited while in Beirut by then McNeil-Lehrer news correspondent James H. Webb Jr.—a man who was a Marine captain, a Navy Cross recipient, secretary of the Navy, and a U.S. senator during his life—who observed the excellent fire discipline of all the Marines he encountered while there. Webb was astounded that several times each day in checking on his Marines at their posts, Ferraro had to dash across an open roadway that was subject to sniper fire. Fortunately for Ferraro, their aim was not good, and he was never hit.

[9] Ferraro interview.

numerous Muslim indirect fire weapons systems. The battery saw extensive action throughout its entire time ashore. Later, naval gunfire, especially that provided by the U.S. Navy task force just offshore, was also extremely effective against the militia gunners.[10]

The Lebanese Civil War continued to escalate throughout August 1983. On 28 August, heavy fighting broke out between the LAF and Muslim militias just beyond the airport perimeter. One study described the exchange between the two sides "as great as that on a 200-yard rapid-fire string of the Marine Corps qualification course." During two subsequent days, "over 100 rounds of 82mm mortar and 122mm rocket fire landed in the airport area, with the shells landing as close as one kilometer in front of Marine positions." Although difficult to prove, mortar fire from Walid Jumblatt's Druze militia seemed to occasionally directly target the Marine lines. One Druze mortar team decided to ignore 5-inch illumination warning rounds that USS *Belknap* (CG 26) fired at them. Geraghty ordered his artillery battery to return fire, and a round landed right in the middle of the mortar team, which likely resulted in the deaths of several Druze members.[11] Despite this success, Company A still suffered two casualties, Staff Sergeant Alexander M. Ortega Jr. and Second Lieutenant Donald G. Losey, in a different mortar attack. They were the first Marines killed in action since the *Mayaguez* affair in 1975.

As September 1983 began, the Marines at the Beirut International Airport and the Lebanese University were regularly taking incoming rounds of all calibers. On 4 September, allegedly without notifying the Lebanese national government, the Israel Defense Forces "began redeploying its troops from the Chouf and Alayh districts to the Awali River in southern Lebanon." By this point, everyone recognized that the Israelis were anxious to depart Beirut as soon as possible, but the LAF was not prepared to "fill the vacuum left by the Israeli withdrawal" at that time any more "than it had been on 17 May, when the Israeli-Lebanese

---

[10] Lawrence Pintak, *Beirut Outtakes: A TV Correspondent's Portrait of America's Encounter with Terror* (Lexington MA: Lexington Books, 1988), 125–26; and Col Timothy J. Geraghty, USMC (Ret), *Peacekeepers at War: Beirut 1983—The Marine Commander Tells His Story* (Washington, DC: Potomac Books, 2009), 45, 50–53.

[11] Frank, *U.S. Marines in Lebanon*, 77–78.

Agreement was signed."[12] Consequently, the Israeli withdrawal precipitated increased fighting between the Druze, Christian Phalange militias, and the LAF in the mountains, as each party violently jostled for power and position. Meanwhile, Geraghty and U.S. Navy captain Morgan R. France, the commodore of the Sixth Fleet task force, which provided direct support of the operation ashore, were required to provide the LAF with significant amounts of small arms and artillery ammunition. In one day alone, the LAF expended several thousand rounds of artillery ammunition. In a poignant 10 September situation report, Geraghty informed U.S. Navy vice admiral Edward H. Martin at Sixth Fleet headquarters in Naples, Italy, that he was gravely concerned that the requirements levied on the MAU were quickly becoming overwhelming. Furthermore, he noted that the United States had "changed the rules" by turning into an "active participant" in the fighting. Having taken part in numerous forms of support for both the LAF and the U.S. Office of Military Cooperation (OMC), including providing equipment, "training, intelligence, [and] security," the Marines had become active allies in Beirut. This left Geraghty in a precarious situation: "With each bombardment of the airport and increase in the number of casualties sustained, my ability to influence those factions who desire to involve us militarily has declined. . . . In effect, I have reached my limit of response given the capabilities of the weapons within my force and the constraints of the current rules of engagement."[13]

At the Beirut International Airport, the intermingling of the Marine and Lebanese Army positions created problems for the U.S. presence as well. Because the two forces "were sometimes interspersed," including at certain checkpoints, the opposing militias claimed they could not tell the difference between them. Jumblatt loudly complained that LAF helicopters and jets that attacked his forces in the Chouf mountains were parked only yards from the Marine Corps' airport perimeter. Jumblatt later told reporters that he had a message for the U.S. Marines: "Stay away from Lebanese army positions. It's better for them and better for

---

[12] Frank, *U.S. Marines in Lebanon*, 81.
[13] Geraghty, *Peacekeepers at War*, 68–69.

me."[14] Jumblatt, of course, used this situation as an excuse whenever a Druze round landed—purposefully or not—within the Marines' perimeter. Geraghty dismissed Jumblatt's warning and noted that "our mission is to support the Lebanese government and the Lebanese Armed Forces in reestablishing sovereignty and control within their own country," which the Marines planned on fulfilling.[15]

Even more disconcerting was the late August appearance of approximately 300 Lebanese Shiite militants, who immediately made their presence known to the Marines stationed at the airport. Trained by the violently anti-American Iranian Islamic Revolutionary Guard Corps (IRGC), these fighters, at first, conducted sniper and grenade attacks on both the Marines and other foreign peacekeepers. Infiltrating the slums in and around Hay-es-Saleem, they sometimes wore the "rust-mottled camouflage similar to that of the Syrian army," but some also "could be identified by their red armbands."[16]

These militiamen, calling themselves the Islamic Amal, were a breakaway group from the larger Amal militia that had been fighting in the Lebanese Civil War for years. Their leader, Hussein al-Musawi, was a former military commander in the Amal militia who had broken from the group over its failure to adequately oppose the Israeli invasion of Lebanon in 1982. Considered "staunchly pro-Khomeini," the Islamic Amal most likely received advisement, "sophisticated Soviet Dragonoff sniper rifles, and a plan to provoke the Marines" from the IRGC.[17]

Because these militias targeted U.S. forces at a wide variety of locations, the U.S. units in the multinational force received expanded rules of engagement on 31 August 1983. These changes allowed the Marines to "fire artillery in defense" of the OMC, "the U.S. Embassy, the Special Negotiator's team, and other U.S. government organizations in Lebanon," which included ones housed in the Lebanese Ministry of Defense. Furthermore, the Joint Chiefs of Staff directed the U.S. forces to provide U.S.

---

[14] Pintak, *Beirut Outtakes*, 120. Pintak was an award-winning journalist with considerable experience in the Middle East.

[15] Pintak, *Beirut Outtakes*, 123.

[16] Pintak, *Beirut Outtakes*, 129.

[17] Pintak, *Beirut Outtakes*, 129.

Army target acquisition data and intelligence directly to the LAF, but with a 20-minute delay built in so as not to appear that the United States was directly providing the information to the Lebanese. While the U.S. European Command (USEUCOM) opposed this decision, it aligned with Special Envoy McFarlane's preferred strategy of greater involvement by the United States in offering direct support of the LAF.[18] As a reinforcement insurance policy, The Chief of Naval Operations (CNO), Admiral James D. Watkins, had previously ordered the Okinawa-based 31st MAU under Colonel James H. R. Curd to stand by offshore as a floating reserve. The 31st MAU arrived off the coast of Lebanon on 12 September 1983.[19]

The Islamic Amal, among other Muslim militias, were serious trouble for the Marines at the Beirut International Airport. On 6 September 1983, two more Marines, Lance Corporal Randy W. Clark and Corporal Pedro J. Valle, fell victim to Muslim shellfire. The following day, aircraft from USS *Eisenhower* (CVN 69) and the French aircraft carrier FS *Foch* (R 99) made a show of force when they streaked over the city but did not drop any ordnance. Lebanese president Amine Gemayel demanded more troops from the multinational force to support the government. The widely respected LAF brigadier general Ibrahim Tannous asked that the United States provide direct support for his army units at Suq al Gharb, an inconsequential mountain town over which his government forces had just taken control at high cost to his vaunted, U.S.-trained 8th Brigade. At risk of being overrun by the Druze and others, Tannous and Gemayel hinted that Syria, Iran, and the PLO were involved. PLO leader Yasser Arafat, "anxious not to look as if he was being left out, announced that some of his men were indeed involved in the battle as 'volunteers'."[20]

This turn of events made it appear that the Gemayel government was up against foreign powers, which McFarlane accepted as true. McFar-

---

[18] Ralph A. Hallenbeck, *Military Force as an Instrument of U.S. Foreign Policy: Intervention in Lebanon, August 1982–February 1984* (New York: Praeger, 1991), 78.

[19] Frank, *U.S. Marines in Lebanon*, 87–88.

[20] Pintak, *Beirut Outtakes*, 160–62. There is some indication that PLO participation at Suq al Gharb was never as great as the Gemayel government had insinuated. The alleged presence of "foreign fighters" assisting the Druze certainly gave the Lebanese national government significant leverage with Washington, DC.

lane wished to shell Muslim positions at Suq al Gharb to send a message to those forces opposed to the Lebanese national government. Geraghty balked at this suggestion, rightly thinking that escalating the fight made his forces especially vulnerable to retaliation. McFarlane used Army brigadier general Carl W. Stiner to place pressure on Geraghty to use the naval gunfire at his disposal. Although Stiner was not in Geraghty's or Captain France's chain of command, he was serving as McFarlane's liaison officer with the Lebanese government and reported directly to the chairman of the Joint Chiefs of Staff, General Vessey. Due to continued attacks on the airport, Geraghty partially gave in to Stiner's entreaties. On 16 September 1983, he requested the 5-inch guns of USS *Bowen* (FF 1029) and USS *John Rodgers* (DD 983) fire on Muslim artillery batteries inside Syrian-occupied territory. It was the "first time that U.S. Navy ships had struck [targets] behind Syrian lines."[21]

Most surprisingly about the situation was that the U.S. National Command Authority (NCA) gave Geraghty, a Marine colonel, full responsibility over most of the available military assets in the region. Additionally, Geraghty noted, he also "had the authorization of the president" to use such force if the situation met three specific conditions. First, he could employ this power if he concluded that Suq al Gharb, an important ridgeline held by Lebanese forces, "was in imminent danger of falling." Second, he could use it if "the attacking force was non-Lebanese." Finally, he had the authority to employ these assets if the Lebanese government "requested assistance." The Joint Chiefs of Staff order emphasized that the message was not meant to change the multinational force's mission.[22]

Both Geraghty and France complained to Vice Admiral Martin about Stiner's role in the existing command structure. They believed that Stiner's ability to directly contact them about the need to support LAF operations in the Chouf mountains created confusion among the forces. In a heated exchange with Stiner about his urging that the Marines more actively support the Lebanese Army then desperately fighting in the Chouf

---

[21] Pintak, *Beirut Outtakes*, 165–67.

[22] Geraghty, *Peacekeepers at War*, 64–65.

mountains, Geraghty shouted, "General, don't you realize we'll pay the price down here? We'll get slaughtered! We're totally vulnerable!" Soon after, Geraghty received a phone call from McFarlane, responding with what he said to Stiner. He wondered after getting "continuous calls to me to unleash our massive firepower against the Muslim factions," if the NCA had any concept about "where this fucking train was headed." Geraghty freely admitted that he disagreed with McFarlane and his staff over "the doom-and-gloom reports coming out of Suq al Gharb."[23]

Geraghty had his own troubles with the situation in the Chouf mountains. He had little to no confidence in the LAF on-scene commander, Colonel Michel N. Aoun, describing the Maronite Christian officer as "indecisive and prone to panic."[24] Moreover, Geraghty respected the firepower available to the Druze and Syrian forces in the nearby Chouf mountains. He believed that those elements had "upwards of 600 tubes" that "could be brought to bear" against the perimeter around the Beirut International Airport.[25]

A real tipping point in the fighting between the Muslim militias and the LAF took place on 19 September 1983 around a mountain village known as Suq al Gharb. It was here that things started to disintegrate for the Lebanese national forces. During the Muslim militia assault on the village, which included at least two battalions of infantry, their forces were also supported by heavy artillery and possibly some armor assets. As the day progressed, Geraghty noted that the LAF 8th Brigade was in danger of being overrun. He also confirmed intelligence that the Syrians—and possibly even the Iranians—were supporting the assault in some fashion. During the attack, Brigadier General Tannous contacted Geraghty and pleaded with him to provide fire support. His request met all the preconditions of the new rules of engagement. Now believing

---

[23] Geraghty, *Peacekeepers at War*, 65–66.

[24] Geraghty, *Peacekeepers at War*, 65–66. Col Aoun was, if anything, a survivor. He fought with the 8th Brigade, LAF, throughout the September 1983 "mountain war" in the Chouf against Walid Jumblatt's Druze militia and, with U.S. support, hung onto Suq al Gharb. He was a longtime opponent of the Syrian occupation of his country. Driven into exile in France in 1990, he returned to Lebanon in 2005 and was elected president in October 2016.

[25] Geraghty, *Peacekeepers at War*, 66.

that the Lebanese army only had a tenuous hold on the area, mainly due to a severe shortage of ammunition, Geraghty felt that he had no choice but to request help from the powerful Sixth Fleet or risk the failure of the entire U.S. effort in Lebanon. In addition to *John Rodgers* and *Bowen*, the Navy employed the weapons of USS *Arthur W. Radford* (DD 968) and of the large nuclear-powered cruiser USS *Virginia* (CGN 38). From these ships, the Sixth Fleet fired several hundred 5-inch rounds, as compared to just 72 total rounds that the *John Rodgers* and *Bowen* had expended three days earlier. Observers noted that the naval gunfire was tremendously effective, as the highly accurate gunfire pulverized Druze and other militia positions, forcing the attacking militiamen to flee in disarray.[26]

Geraghty saw the use of naval gunfire at Suq al Gharb as a significant turning point in the Lebanese Civil War, but he believed that he had no choice in requesting its assistance. As he later stated, the U.S. support of LAF operations in the Chouf mountains represented "a milestone." Being critical to the success of the overall LAF war effort, the action moved the United States away from "a previous, very careful, razor edge line of neutrality that we were walking, and treating all the Lebanese communities alike." The U.S. fire support, according to Geraghty, pushed its role "to a different category." Geraghty concluded that "it would have been unconscionable" for the United States to ignore the request for support at "a very critical time" due to the LAF running extremely low on ammunition.[27] Yet, Geraghty was left wondering if the LAF had purposely overblown the emergency. Nevertheless, after Suq al Gharb, the die was cast and "the rules of the game had changed forever with that decision, with its consequences unknown."[28]

Despite the potential negative consequences, the naval fire support for the LAF definitively changed the status of the U.S. units in the multinational force from peacekeepers to combatants. As early as 25 August 1983, Druze leader Walid Jumblatt publicly stated that "the Marines have bluntly and directly threatened us," which he considered "proof of the

---

[26] Geraghty, *Peacekeepers at War*, 70–72.
[27] Frank, *U.S. Marines in Lebanon*, 89.
[28] Geraghty, *Peacekeepers at War*, 72–73.

U.S. alliance" with the Christian-dominated Lebanese national government and its armed forces, which was primarily led by Christians.[29] To the Muslim militias, the connection between the LAF and the U.S. forces was unmistakable. Consequently, Muslim factional leaders saw any U.S. support as tilting the scales in favor of their long-standing enemies in the Lebanese Civil War.

Following the 19 September fighting around Suq al Gharb, Gemayel had forced the hand of the Western powers. Moreover, the nations in the multinational force had "staked their prestige on the creation of a strong Lebanese government," placing their strength fully behind Gemayel. "By failing to distance themselves from Gemayel," who had made it clear that he had little interest in "pursuing a path that would unite Christians and Muslims," the Western powers were drawn deeper "into a war that should not have been theirs."[30] In fact, the entire multinational force had a fundamental misunderstanding of the actual political situation on the ground, only basing their perception on what information they received from the Gemayel regime. As long as the Gemayel's government could claim it was on the verge of collapse, the United States and its allies in Lebanon felt obligated to respond to any perceived military or political setback. This situation was a primary reason why U.S. diplomats begged the IDF—to no avail—to postpone their departure from Beirut in September 1983. Everyone involved in the U.S. mission in Lebanon, including McFarlane, believed that the LAF was unprepared to assume the military responsibility that the Israelis would leave behind. Later events proved this suspicion correct.

After 19 September, "naval gunfire became the weapon of choice" for Geraghty and the 24th MAU. Geraghty was relatively fine with this decision because it provided his Marines at the airport an amount of physical separation from any kinetic response coming from offshore. It is hard to tell whether the Muslim militias picked up on this subtle distinction.

---

[29] *Report of the DOD Commission on Beirut International Airport Terrorist Act, October 23, 1983* (Washington, DC: Department of Defense, 1983), 59–60. Reporting the findings of what was known as the Long Commission, the report addressed the issues that led to the deadly terrorist attack on the Marine barracks on 23 October 1983.

[30] Pintak, *Beirut Outtakes*, 162–63.

In either case, the weeks after Suq al Gharb marked a moment of significant escalation. On 20 September, Muslim gunners using equipment acquired from the Soviet Union and Syria fired a KBM Kolomna surface-to-air missile at U.S. Navy aircraft. While the missile missed its target, it was a clear escalation in response to U.S. reconnaissance overflights. The following day, *Radford*, *John Rodgers*, and *Virginia* "fired 90 more 5-inch rounds on two targets."[31] Lawrence Pintak, an American journalist with a keen grasp of the situation in Lebanon, astutely noted that the U.S. naval gunfire support for the government forces clearly indicated to the Muslim forces that the United States was now fully committed to the Gemayel regime. He wrote, "In Ronald Reagan's world, the good guys wore white hats, and the bad guys wore black. In Lebanon, all the hats were gray."[32]

Two days later, fighting erupted around the airport yet again. Simultaneously, partisans attacked the French and Italian compounds. The French responded with airstrikes launched from the aircraft carrier *Clemenceau*. The LAF and Muslim militias engaged in exceptionally heavy fighting in the area around Hay-es-Saleem. Forward-based Marine checkpoints came under such intense fire that no reinforcements could reach them, causing Geraghty to order their dismantling. Even the MAU command post came under indirect fire. While this fighting occurred, Nabih Berri's Amal militia briefly captured and detained two U.S. Army personnel who took a wrong turn while driving through Beirut. The Amal members, being less militant than the breakoff Islamic Amal militia, quickly released the U.S. soldiers, minus a pistol, and their jeep to the French. Geraghty was especially concerned about this incident because he felt that it illustrated the "relative inability of the [U.S. Multinational Force in Lebanon] to respond to incidents of this nature and demonstrated the variety of threats to the [multinational force] and their possible consequences."[33]

---

[31] Frank, *U.S. Marines in Lebanon*, 89.
[32] Lawrence Pintak, *Seeds of Hate: How America's Flawed Middle East Policy Ignited Jihad* (Sterling, VA: Pluto Press, 2003), 150.
[33] Frank, *U.S. Marines in Lebanon*, 90.

Geraghty's Marines at the Beirut International Airport got a welcome respite from the nearly continuous shelling and sniper fire after diplomats brokered another tenuous ceasefire on 26 September. Geraghty used this break to rotate small groups from the 24th MAU to Navy ships offshore for a hot meal, a shower, and relief from the tension of standing guard. The Beirut International Airport, which was closed during the more severe fighting, reopened on 30 September. Even so, minor skirmishes continued taking place throughout West Beirut. By 9 October, things had seemingly quieted enough that the Joint Chiefs of Staff allowed the 31st MAU, "which had been kept afloat and at the ready throughout the September fighting," to steam for their Western Pacific home base. *Eisenhower* also received authorization to pull back to Naples, Italy, for an essential port visit.[34]

Despite this seemingly peaceful break, violence continued during the ceasefire. On 5 October, a Marine helicopter carrying McFarlane was shot at as it flew over Beirut. The round went through the cockpit glass, but fortunately no one was hurt. The following day, a second Marine helicopter took small arms fire that hit its rotor blade, but this story was missed due to the news that a car bomb attack in the city of Tyre killed the Israeli national guard chief. Just two weeks later, terrorist forces tried to do the same to Geraghty as he made his way back to the airport from the U.S. embassy. Further, between 9 and 14 October, snipers began firing at BIA security company personnel and inbound helicopters again. A sniper mortally wounded Marine Corps sergeant Allen H. Soifert, then on a perimeter patrol in a jeep, on 14 October. His death caused the suspension of all foot and vehicle traffic along the perimeter road as well as at the helicopter landing zone. Geraghty had intelligence that correlated the rising number of car bomb attacks with an increase in violent activity by "pro-Iranian Islamic fundamentalists."[35]

That same day, Geraghty reported to Vice Admiral Martin that this surge in sniping, car bombs, and command-detonated mine attacks against the multinational force units signaled yet another change in

---

[34] Hallenbeck, *Military Force as an Instrument of Foreign Policy*, 107.
[35] Geraghty, *Peacekeepers at War*, 80–84, 87.

tactics by the extremist partisans. While Geraghty took sensible force protection measures, such as putting parts of West Beirut off-limits to U.S. patrols, he assured Martin that he had no intention of retreating "behind an earthen berm and 'show the flag' only from the top of my flagpole." That Saturday, 22 October, things had calmed down enough for the United Service Organizations (USO) to sponsor a concert by the folk group MEGA, which was attended by all hands who could be spared from their security duties. According to Geraghty, it would be "the only USO performance during the 24th MAU's six-month tour in Beirut." He called it was a reasonably quiet night "by Beirut standards."[36]

While the 24th MAU enjoyed the USO performance, however, Islamic terrorists prepared for a suicide attack against both U.S. and French forces the following morning.[37] Two days before the USO concert, according to Pintak, the 24th MAU headquarters received an advisory message from the Central Intelligence Agency (CIA) that "two men with Mediterranean skin tones have been overheard in a Paris café saying that a major complex will be hit in Beirut." Even if true, this intelligence was extremely thin and contained no actionable information. As part of a daily CIA intelligence dump that included an inundation of raw and unanalyzed information, it only provided an educated guess that Islamic Amal was the most likely organization to launch the attack. The report also did not indicate what complex would be the target, what weapons would be used, nor when it would possibly take place.[38]

Geraghty arose early on the morning of 23 October to review the various messages that the MAU operations center had received during the night. Being a Sunday, most of the Marines in the 24th MAU worked a modified schedule during the slower-than-typical workday at the Beirut International Airport, including reveille taking place at 0630, an hour later than usual due to it being a Sunday morning. At 0622, a suicide bomber driving a 19-ton Mercedes Benz truck laden with approximately 2,000 pounds of high explosives picked up speed in an adja-

---

[36] Geraghty, *Peacekeepers at War*, 87–89.
[37] Geraghty, *Peacekeepers at War*, 89.
[38] Pintak, *Beirut Outtakes*, 191.

cent airport lot. The driver crashed through the triple-strand concertina wire and drove past the Marine sentry on duty who was armed but, as per procedure, did not have rounds in the chamber of his weapon. The truck then crashed into the four-story concrete BLT headquarters building, breaching its interior open atrium. The suicide bomber detonated the explosives packed into his truck that now sat below the mostly sleeping Marines. The blast devastated the concrete structure, causing its complete collapse.[39]

Figure 47. The band MEGA plays a USO show on 22 October 1983

Source: official U.S. Marine Corps photo.

The detonation's violent shock wave was tremendous, to an almost unbelievable level. Captain Robert Funk, the artillery battery commander, was sleeping in a heavily sandbagged bunker more than a kilometer away from the BLT building when the explosion occurred. He thought that his immediate position had just taken a direct hit from a rocket-propelled grenade because the shock wave heaved his cot with him in it to the other side of the bunker.[40] Closer to the scene, the blast had thrown the MAU operations center into disarray. Geraghty, who was in the operations center at the time, observed that the explosion seemed to have shaken the Marines on duty, but none had been injured. Eventually, he and his logistics staff officer, Major Robert S. Melton, went outside to see what had occurred and Melton exclaimed, "My God, the BLT building is gone!"[41] Once he recovered from his initial shock, Ger-

---

[39] Geraghty, *Peacekeepers at War*, 91–94.
[40] Capt Robert C. Funk, interview with the author, June 1988.
[41] Geraghty, *Peacekeepers at War*, 92. Geraghty stated that the requested BLT headquarters replacement staff, led by LtCol Edward Kelley, commanding officer of BLT 2/6, arrived within 36 hours. However, he noted that for unknown reasons, the additional security company he had asked for "had been scratched."

Figure 48. A cloud of smoke rises from the Marine barracks shortly after the bombing on the morning of 23 October 1983

Source: official U.S. Marine Corps photo.

aghty choked down his anger and, with the help of his MAU executive officer Lieutenant Colonel Harold W. Slacum, sent out an urgent flash message to Captain France to be prepared to receive mass casualties. At the same time, he notified the National Military Command Center in the Pentagon of the situation and requested immediate replacements for the

BLT 1/8 headquarters element along with an additional rifle company from Camp Lejeune. Geraghty wisely kept his MAU headquarters staff intact, and they assumed control of all the remaining security forces within and on the airport perimeter and at the university. Additionally, he assigned Major Douglas C. Redlich, the commander of MAU Service Support Group 24, to organize the rescue effort on the ground, including dispatching personnel, engineer platoons, and heavy equipment to help the site. Soon after Geraghty left for the scene of the explosion.[42]

Once Geraghty arrived, his first order of business was to take care of the wounded immediately. He described the scene as "surreal." Everything in the vicinity of the destroyed building was covered in thick gray ash. Geraghty later recalled observing "mangled, dismembered bodies" strewn about in "grotesque fashion," including seeing "one Marine still in his sleeping bag impaled on a large tree branch." He remembered thinking "here lie the fucking unintended consequences of getting sucked into an eight-sided civil war while trying to carry out a peacekeeping mission."[43] A later report from a Department of Defense (DOD) commission on the attack, known as the Long Commission for its chairman, retired Navy admiral Robert L. J. Long, seemed to find general agreement with Geraghty's assessment. While it did not find "a direct cause and effect linkage between Suq al Gharb and the terrorist bombing," it noted that the "prevalent view" among the staff of the commander of USEUCOM was that "there was some linkage between the two events."[44]

Tannous and Stiner reached the site soon after Geraghty. Tannous promised the delivery of heavy construction equipment to assist with the possible recovery of injured Marines still trapped in the rubble. Rafik Hariri, a wealthy Lebanese-Saudi construction philanthropist, quickly

---

[42] Geraghty, *Peacekeepers at War*, 94–97.

[43] Geraghty, *Peacekeepers at War*, 99–100.

[44] *Report of the DOD Commission on Beirut International Airport Terrorist Act*, 42, 59–60. The senior Marine Corps representative on the Long Commission was LtGen Lawrence F. Snowden. Snowden was a highly decorated veteran of the Battle of Iwo Jima and regarded throughout the Corps as one of its best officers.

Figure 49. Sketch of route taken by the Marine barracks suicide bomber

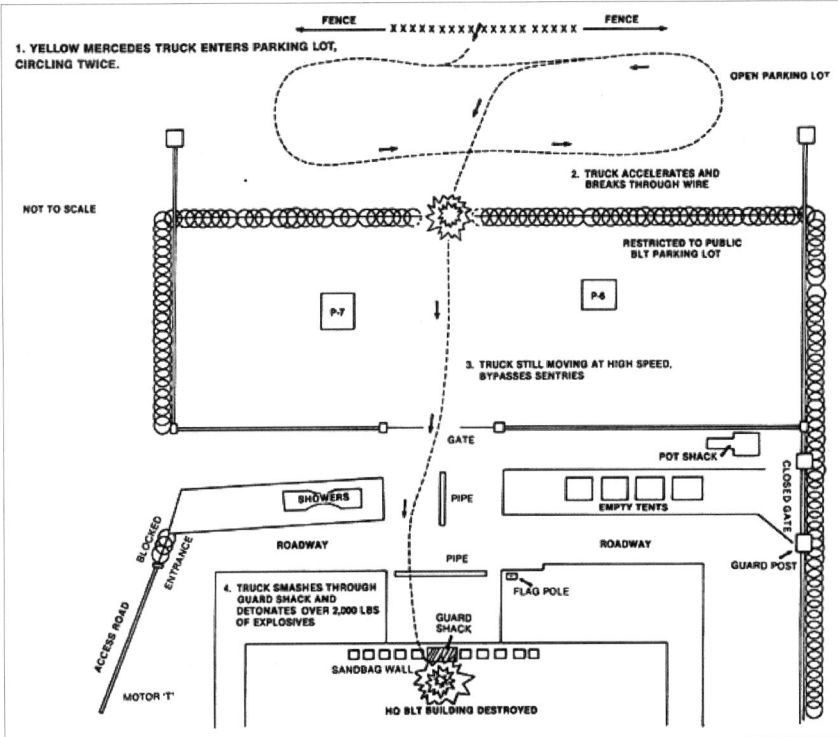

Source: *Report of the DOD Commission on Beirut International Airport Terrorist Act, October 23, 1983* (Washington, DC: Department of Defense, 1983).

supplied the equipment, which prevented the recovery operations from being delayed considerably.

After receiving the flash message from Geraghty, France sent in select teams of doctors, nurses, and corpsmen from his ships located just offshore to assist with the immediate medical recovery effort. The 24th MAU's sole physician ashore, U.S. Navy Reserve lieutenant John R. Hudson, perished in the blast, leaving the unit without medical care. Most of the gravely injured needed immediate care in a sophisticated trauma center. Although shipboard sickbay facilities were better than the faculties available to the Marines ashore, these survivors required the assistance of specialists, such as neurologists who could treat severe head trauma and orthopedic surgeons who could treat other critical injuries. Marine Medium Helicopter Squadron 162 (HMM-162), the helicopter squadron connected to the 24th MAU, immediately swung into action from the deck of USS *Iwo Jima* (LPH 2), bringing the squadron's flight surgeon, U.S. Navy lieutenant Larry Wood, and various corpsmen ashore first. After reaching the scene about 20 minutes after the attack, Wood and the corpsmen immediately "set up a triage station. . . . and started treating the most seriously injured."[45] The helicopter squadron then started evacuating wounded personnel to a wide number of hospitals within flying distance of the Beirut International Airport, including a British facility on the island of Cyprus.

As rescue crews dug through the debris, Geraghty found the unconscious and ash-covered body of BLT 1/8 commanding officer, Lieutenant Colonel Howard Gerlach. The blast had apparently thrown Gerlach through the concrete wall. At first, Geraghty thought Gerlach had died. However, although knocked out, Gerlach had survived, but suffered severe wounds to his head and body. The rescuers placed Gerlach in a Lebanese ambulance that evacuated him to an Italian multinational force medical facility. Eventually, Gerlach, as well as 20 other seriously wounded Marines, were transferred to local hospitals in Beirut. According to Geraghty, Robin B. Wright, a correspondent of the *Sunday Times* of London, helped positively locate the severely injured BLT commander and assisted in arranging to have him transferred to a U.S. military hospital in Germany, where he eventually recovered. Captain Berry Ford, Geraghty reported, tracked down and recovered the other wounded personnel taken to the local hospitals.[46]

Just as one suicide bomber attacked the Marine Barracks, a second bomber struck the compound of the French multinational forces. The French had 59 paratroopers killed and scores of personnel wounded in the second blast. Until these truck bomb attacks, the French had suffered the highest number of soldiers killed among the multinational

---

[45] Geraghty, *Peacekeepers at War*, 101.
[46] Geraghty, *Peacekeepers at War*, 100–1.

forces, losing 16 servicemembers in action.[47] While the United States also bore substantial casualties before the bombing, the French losses illustrated the significant sacrifice of that nation as well. Incredibly, within 24 hours of the terrorist attacks, French president Francois Mitterrand, minister of defence Charles Hernu, and chief of the defence staff General Jeannou Lacaze arrived to visit their own country's shattered compound as well as extend their condolences to the Americans. After arriving at the Beirut International Airport, Mitterrand surprised Geraghty when he expressed interest in paying a visit to the temporary morgue. By then, surviving Marines had gathered nearly 200 bodies of their fallen comrades and placed them in stacked aluminum transfer cases in preparation for relocating them to Germany for further identification. Geraghty noticed the shocked look on Mitterrand's face when he saw the hundreds of shining coffin cases that extended across the entire floor of the temporary morgue. Mitterrand requested permission to say a private prayer in front of each pallet of caskets. Escorted by Geraghty, Mitterrand's heartfelt expression of sorrow was appreciated by all who were there.[48]

The U.S. military leaders recognized the extent of the major catastrophe that had taken place at the Marine barracks. At least 241 U.S. servicemembers had been killed. The Marines of BLT 1/8 made up the vast majority of deaths, with 225—including Sergeant Major Frederick B. Douglass, a much-beloved senior staff noncommissioned officer with 29 consecutive years of service—killed. In the immediate aftermath of the assault, the U.S. military struggled to identify the killed and wounded personnel because the explosion destroyed all of the BLT's service record books. It was some time before the Marines could positively identify who had been killed, been wounded, or survived the suicide bomb. For example, it took several days before the U.S. military knew the whereabout of Captain Michael P. Marletto. Marletto had been an instructor at the U.S. Army Field Artillery School at Fort Sill, Oklahoma, when he was assigned to the Army's target acquisition battery, which accompanied the 24th MAU to Beirut to support its mission. Marletto had been staying in

---

[47] Hallenbeck, *Military Force as an Instrument of U.S. Foreign Policy*, 107.
[48] Geraghty, *Peacekeepers at War*, 107–8.

the BLT headquarters building most nights, but the day before the suicide attack, he was outside the airport perimeter assisting the LAF with learning how to survey in gun positions near the Chouf mountains. By the early evening of 22 October, Marletto remained with the LAF units. When he radioed the BLT headquarters by radio, he received permission to stay with the LAF overnight and return the following morning due to an increased danger of car bomb attacks on U.S. convoys. Just three days before, a remotely detonated car bomb attack wounded four Marines in one convoy. By the time of the suicide attack, Gerlach was one of the few people who knew Marletto was away from the airport. Being left unconscious and barely alive after the explosion, Gerlach could not provide any information and rescuers initially believed Marletto could be entombed in the rubble before the MAU staff positively confirmed his survival. He was fortunate to be alive.[49]

The bombings of the Marine barracks and the U.S. embassy in Beirut did not end the attacks against the United States in the region that year. On 12 December 1983, a suicide bomber detonated a device similar to the one used against the Marine barracks in October outside the U.S. embassy in Kuwait. Subsequent attacks throughout Kuwait City targeted French forces, Americans working for the Raytheon Company, and some natural gas facilities. Fortunately for the Americans, the bomber drove past the ambassador's office and consular section, where a throng of people were lined up to get visas. Instead, it exploded directly in front of the main administration building. Moreover, the bomb did not function properly, preventing it from causing maximum damage but still leaving behind significant destruction. Similar to the U.S. embassy in Beirut, the explosion tore off the face of the building, but it remained standing for another hour and a half before collapsing. As a result, casualties were

---

[49] Capt Michael P. Marletto to Capt Charles P. Neimeyer, 24 October 1983, in possession of the author. In his letter, Marletto quoted the legendary football coach John E. Madden and stated that he had indeed "dodged a bullet" and was fortunate to be breathing still. Back in the United States, due to the devastation at the bombing site and because service record books had been destroyed, finding and confirming survivors became an issue. Due to difficulties in carefully digging through the rubble for survivors, it took some time to construct a finalized list of those who had died in the bombing, making it a horrible time for the families as they waited for information about their loved ones.

minimal and no Americans were killed. According to Pintak, "Luck had saved the United States; security expertise had nothing to do with it."[50]

In a later investigation conducted by the Federal Bureau of Investigation and the Kuwaiti government, they positively identified the remains of the U.S. embassy bomber, which consisted of a charred finger and part of his skull. The bomber, Raad Aqueel al-Badran, was an Iraqi who had joined the Iranian-backed al-Da'wa underground Islamic group. Being in the midst of a violent war with Iraq, Iran supported the group, which was created to overthrow Iraq's secular president, Saddam Hussein. In addition to fighting to depose Hussein, al-Da'wa had connections to Islamic fighters in Lebanon and the new Iranian-backed organization, Hezbollah.[51]

Incredibly, just two days after the Marine barracks bombing, while recovery efforts were still ongoing, the United States conducted yet another contingency operation thousands of kilometers away on the Caribbean island of Grenada. The nation had long been a sore spot for the Reagan administration. Its socialist prime minister, Maurice R. Bishop, was the leader of the Marxist-Leninist New Jewel Movement, which had overthrown the Grenadian government of Eric M. Gairy in 1979. Bishop quickly aligned the nation with neighboring Communist Cuba, immediately making his government an enemy of the vehemently anti-Communist Reagan administration. He, along with his deputy prime minister, Bernard Coard, and People's Revolutionary Army of Grenada general Hudson Austin, had invited in hundreds of Cuban military construction workers and soldiers to assist them with extending the airport runway at Point Salines, near the capital city of St. George's. However, Bishop apparently was not radical enough for the Revolutionary Military Council under Austin's command. Coard and Austin deposed Bishop, placing him under house arrest and then later murdered him, allegedly during an escape attempt, on 19 October 1983. Austin also ordered the house arrest of Governor General Sir Paul G. Scoon, who was

---

[50] Pintak, *Seeds of Hate*, 188.
[51] Pintak, *Seeds of Hate*, 188–89.

a mostly ceremonial holdover from Grenada's colonial association with the United Kingdom.[52]

Following Bishop's death, Austin placed Coard under arrest. Consequently, chaos broke out across the island, and soldiers began threatening students, including U.S. students, then in residence at the St. George's University medical school. Austin imposed an island-wide curfew, and any violators who got caught were subject to summary execution. The U.S. State Department tried to conduct an orderly evacuation of all U.S. nationals still on the island, but it failed. Scoon allegedly sent "a secret message to the Organisation of Eastern Caribbean States [OECS] and to the governments of neighboring states, appealing for help to restore order on the island."[53] This was a courageous move on his part, considering what had happened to Bishop, and it caused the United States to begin planning immediately for an invasion and restoration of order there. This time, the Joint Chiefs of Staff envisioned a much larger ground operation than anything they ever contemplated in Beirut. Unlike in Lebanon, Grenada would be a combat operation, at least until U.S. forces fully secured the island.

A day prior to the Joint Chiefs of Staff's execute order for what became called Operation Urgent Fury, Secretary of State George P. Shultz requested that the vice director of the Joint Chiefs of Staff, Marine Corps major general George B. Crist, accompany the U.S. ambassador to Costa Rica, Francis J. McNeill, to Bridgetown, Barbados. Crist went to advise McNeill on "U.S. capabilities for military action in the Caribbean." For five days, Crist assisted OECS, Barbados, and Jamaica establish a "small peacekeeping force" while also coordinating their actions with "U.S.

---

[52] Reynold A. Burrowes, *Revolution and Rescue in Grenada: An Account of the U.S. Caribbean Invasion* (Westport, CT: New Greenwood Press, 1988), 29, 31, 35–38, 79–81.

[53] LtCol Ronald H. Spector, *U.S. Marines in Grenada, 1983* (Washington, DC: Marine Corps History and Museums Division, 1987), 1. After he retired from office, Scoon denied that he had written a letter asking for help, but Dominica prime minister Mary Eugenia Charles stated that she had received requests for support from Scoon before the invasion of Grenada. Scoon later admitted that he had made these verbal appeals to Charles. In her role as chairperson of the Organization of Eastern Caribbean States, however, Charles appealed directly to the United States, Barbados, and Jamaica for assistance in resolving the growing chaos there. Although there were several later attempts on her life, Charles remained a stalwart anti-Communist leader and reliable friend of the United States throughout her long time as the Dominica leader.

forces, the Central Intelligence Agency, and the State Department."[54] The operation commander, U.S. Navy admiral Wesley L. McDonald excluded the mostly untrained OECS constabulary force from the assault phase of the operation, but they nonetheless played a vital role in demonstrating to the international community that the invasion went beyond being an U.S.-only operation.

Fortunately for the Marine Corps, the 22d MAU under Colonel James P. Faulkner, which included BLT 2d Battalion, 8th Marines (BLT 2/8), under former Vietnam War Co-Van Lieutenant Colonel Ray L. Smith, had already boarded amphibious ships at Morehead City, North Carolina, on 17 October. Initially, the 22d MAU had been headed toward the Mediterranean to eventually relieve the 24th MAU. With the new development in Grenada, however, the 22d MAU, embarked with Amphibious Squadron 4 (PhibRon 4), commanded by U.S. Navy captain Carl R. Erie, were vectored to the Caribbean to await further orders. The Marine Corps had recently restructured BLT 2/8 per a new table of organization that provided its fighting forces with greater mobility and firepower. Consequently, the personnel in the BLT infantry battalion had been reduced by 10 percent but received "an increase in firepower," including having a total of 134 grenade launchers and 32 M47 Dragon antitank missile systems as well as gaining 8 M2 .50-caliber machine guns.[55] The number of jeeps for the battalions also doubled. More than 40 percent of the Marines in BLT 2/8 had at least two years of experience with the unit, and many had gone through earlier deployments in Lebanon. Moreover,

---

[54] Ronald H. Cole, *Operation Urgent Fury: The Planning and Execution of Joint Operations in Grenada, 12 October–2 November 1983* (Washington, DC: Joint History Office, Office of the Chairman of the Joint Chiefs of Staff, 1997), 28. MajGen George B. Crist was later appointed as the second commanding general of U.S. Central Command (USCENTCOM), and became the first-ever Marine general to lead a unified combatant command. He was also the first full Marine general who had not served as Assistant Commandant or Commandant. Crist relinquished command of USCENTCOM and retired from active duty in 1988, setting the stage for his successor, Gen H. Norman Schwarzkopf Jr., USA, to later conduct Operation Desert Storm in 1990–91.

[55] Spector, *U.S. Marines in Grenada*, 1. LtCol Ray L. Smith, known throughout the Service as "E-Tool" (entrenching tool), was a Marine's Marine. A persistent rumor was that he got his nickname because he allegedly dispatched an enemy soldier with an entrenching tool during the Vietnam War, although Smith has consistently denied this story. During the Vietnam War, however, he was the recipient of the Navy Cross, two Silver Stars, the Bronze Star, and three Purple Hearts. He was a superb choice to lead BLT 2/8 in the assault on Grenada.

"all of the squad leaders and more than a third of the fire team leaders had completed the 2d Marine Division's squad leaders' course."[56] Considering the quality of the leadership team, including Lieutenant Colonel Granville R. Amos commanding HMM-261 in direct support, few MAUs were better prepared to go to war in the era of the modern Marine Corps.

Most of the MAU staff, including Faulkner, did not believe that they would have to conduct a landing at first, but they began to make contingency plans for one. The MAU operations officer, Major Earnest A. Van Huss, envisioned a combined seaborne and vertical assault. The helicopter assault force had the objective of taking the Point Salines airfield to establish a suitable evacuation site for U.S. civilians on the island. The Marines faced a serious problem due to the lack of meaningful information on the situation once they got ashore, especially the actual number of Americans who needed rescuing. Additionally, a dearth of maps of the island caused a significant problem for the Joint forces involved in the operation. Fortunately, PhibRon 4's chief of staff, Commander Richard W. Butler, was an avid sailor who had sailed through the area six years before. This experience made him quite familiar with the beaches, tides, and waters in and around Grenada, knowledge that proved invaluable to the MAU planning staff.[57]

Late on 22 October, changes occurred that threw most of the planning into a state of confusion. At that time, Vice Admiral Joseph Metcalf III, commander of the U.S. Atlantic Command's (USLANTCOM) Second Fleet, received overall command of Operation Urgent Fury, quickly influencing its overall purpose. Instead of simply establishing an evacuation point, Metcalf assigned the amphibious task force that included the 22d MAU—designated Task Force 124—with capturing the smaller Pearls Airport and the port of Grenville while also "neutralizing any opposing forces in the area."[58] All this activity portended that the operation would go beyond landing the MAU to evacuate civilians. Urgent Fury was the largest Joint military operation the United States had con-

---

[56] Spector, *U.S. Marines in Grenada*, 2.
[57] Spector, *U.S. Marines in Grenada*, 2.
[58] Spector, *U.S. Marines in Grenada*, 3–5.

ducted since Vietnam. However, the U.S. military would have significant problems operating jointly.

Metcalf marked 25 October as the mission's D-day, instructing the commanders that the landings should occur after 0400. Task Force 121, which included U.S. Army Rangers and elements of the 82d Airborne Division, were charged with securing the Point Salines airfield and the adjacent capital city of St. George's. A Navy Sea, Air, and Land (SEAL) team had the mission of securing Scoon and the island's radio station. A U.S. Navy carrier battle group built around the aircraft carrier USS *Independence* (CV 62), as well as U.S. Air Force assets based in the United States, would provide general support of the entire operation. Early that morning, most of the initial Marine Corps vertical assault force, made up of 21 helicopters, departed from the amphibious assault ship USS *Guam* (LPH 9). Escorted by a of Bell AH-1 Cobra attack helicopters, this force landed on an unused racetrack near the Pearls airfield—giving the landing zone (LZ) the codename "Buzzard"—rather than setting down directly on the airfield due to concerns over antiaircraft positions around the airport.[59]

Nevertheless, the helicopters still ran into a plethora of obstructions. Major Melvin W. DeMars Jr. noted that the numerous trees around LZ Buzzard made it seem like "you were landing in front of the Palms Springs Inn." Except for a Marine who broke his arm disembarking from a helicopter and some damage to a tube-launched, optically tracked, wire-guided (TOW) missile mounted on a jeep that rolled over into a ditch, the helicopter-borne landing mostly went off without a hitch. Smith ordered Company E to take the critical high ground that overlooked the airport. The Marines ran off token Grenadian resistance without any problem. Company F arrived an hour later when HMM-261 dropped the unit into Grenville, where they established LZ Oriole. At both Pearls and Grenville, the Marines benefitted from a local population that primarily reviled the thuggish behavior of the Bishop-Coard government in St. George's and happily assisted them, even identifying government henchmen and hiding spots for weapons.[60]

---

[59] Spector, *U.S. Marines in Grenada*, 5–8.
[60] Spector, *U.S. Marines in Grenada*, 8–9.

Figure 50. D-day map of Operation Urgent Fury

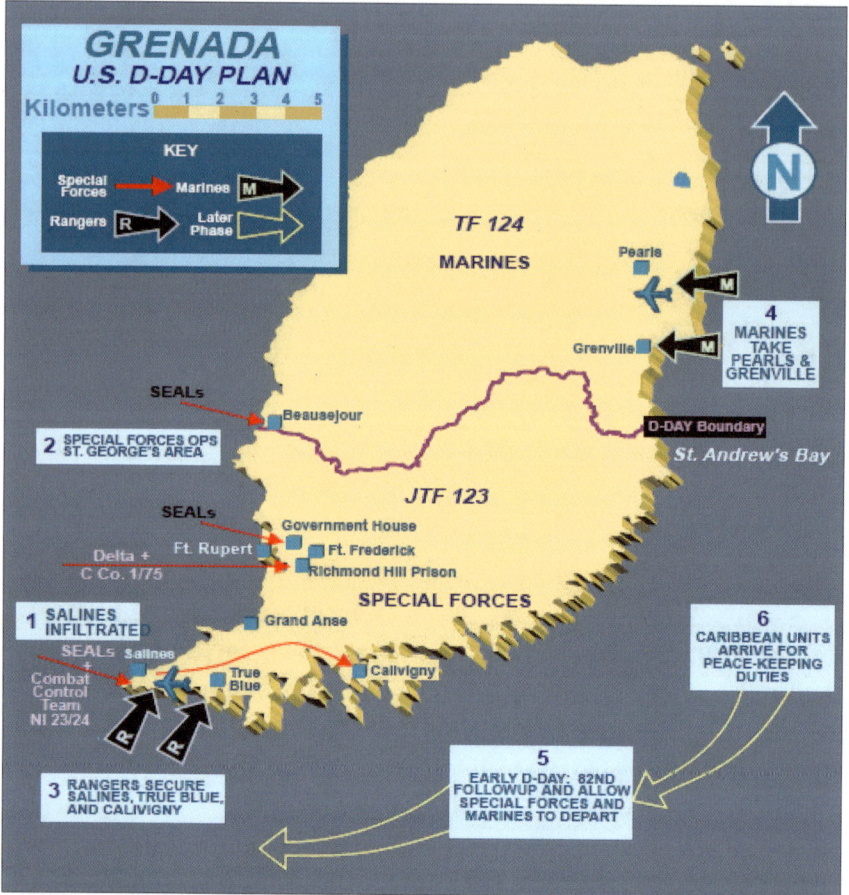

Source: Joint Chiefs of Staff.

Despite the initial success of the Marines and Army Rangers, an amazing lack of intelligence hampered the entire operation. All the ground commanders possessed only a sketchy idea as to the location of U.S. civilians and whether or not they were under guard. The intelligence community made a significant error made when it initially determined that all the U.S. medical students were physically near the "True Blue" campus. It missed more than 230 students who were at a second medical campus at a location called Grand Anse, with the True Blue campus housing "only about a third of the students." Yet, neither Metcalf nor any of the deployed U.S. soldiers knew about Grand Anse "until after their arrival at True Blue itself." Moreover, numerous students lived off-campus, scattered among "houses or apartments on the Lance aux Epines peninsula and elsewhere." Further, the U.S. forces only received vague details on the size, weaponry, and location of potential Grenadian and Cuban combatants.[61]

Beyond the missing intelligence of the students, a distinct lack of Joint warfighting familiarity and planning between the Services added to the operation's problems. British Army major Mark Adkin, who later went ashore with the Barbados constabulary forces, vividly illustrated the Joint planning problems at USLANTCOM during the main planning meeting on 23 October. "As a high-level planning conference," Adkin wrote, "it was not a great success." The issues started when "many officers arrived late" with some of them, such as U.S. Air Force brigadier general Robert B. Patterson, who represented the military airlift command, not arriving until after the meeting concluded. Without communication between Patterson and U.S. Army and special operations forces leadership, they could not coordinate the airborne planning. Additionally, the most significant Marine Corps leadership were already at sea with the amphibious force, meaning that the Navy "dominated the proceedings."[62]

While the Navy and Marine Corps, being in direct contact the whole time, got on extremely well throughout the operation, Metcalf had substantial difficultly coordinating the activities of the Army and Air Force units. The Army had an existing plan that would have brought in the commanding general of the XVIII Airborne Corps as the overall on-scene commander, which would have made sense for Urgent Fury due

---

[61] Maj Mark Adkin, UKA, *Urgent Fury: The Battle for Grenada* (Lexington, MA: Lexington Books, 1989), 128, 130. Adkin served for five years as a staff officer with the Barbados Defense Force. He participated in Operation Urgent Fury as part of the combined forces of the Barbados and Jamaica allies who supplied a small number of troops toward the overall combined international effort. Amazingly, while U.S. planners were unaware of this campus, the parents of the medical students at Grand Anse had telephone communications with their loved ones before, during, and after the operation.

[62] Adkin, *Urgent Fury*, 132.

Figure 51. A combat artist's rendition of Marine Corps helicopters launching from USS *Guam* (LPH 9)

Source: Marine Corps History Division.

to most of the operational ground forces coming from the Army. Instead, the Joint Chiefs of Staff placed the mission under Admiral McDonald's USLANTCOM, based out of Norfolk, Virginia, which included Metcalf's Second Fleet. Most notably during Urgent Fury, Metcalf's planning staff lacked a senior Army officer until the Army detailed Major General H. Norman Schwarzkopf Jr. from his command of the 24th Mechanized Infantry Division at Fort Stewart, Georgia, at the last minute. Schwarzkopf ultimately became Metcalf's deputy task force commander.[63]

Moreover, the commanders had little to no discussion about the logistics necessary for Urgent Fury. Leadership assumed that the entire operation would not take more than a few days, meaning that the operating forces theoretically should have had the ability to personally carry

---

[63] Adkin, *Urgent Fury*, 135.

ashore what they initially needed. If necessary, due to the airfields being prime objectives, the Air Force could later land critical supplies for the ground units. The MAU was in a much better situation because it could operate ashore in a combat environment for a full 30 days thanks to its established naval support. Amazingly, the USLANTCOM's logistics chief, Rear Admiral Neil P. Ferraro, was notified about the operation "only 22 hours before H hour and consequently had no time to achieve anything."[64]

To ensure inter-Service deconfliction, USLANTCOM decided that the Marines would focus on the north end of the island while the broader objectives of the Point Salines Airfield, St. George's, and the southern end of the island fell to the U.S. Army. The Army Rangers, however, would face a significant hurdle to their objective. Moreover, during insertion early in the morning on 25 October, four Navy SEALs who were supposed to inspect the beach area around the Point Salines Airfield "vanished in unexpectedly rough seas."[65] The loss of this SEAL crew deprived the Rangers of critical intelligence about the situation at Point Salines, causing McDonald to delay H-hour until 0530, postponing the insertion of the Rangers until closer to dawn. This decision, in turn, required that the fully loaded Ranger element make their airborne drop at the dangerously low altitude of 500 feet. As soon as the Grenadians and their Cuban allies observed the approach of the specially configured Lockheed MC-130H Combat Talon aircraft, they opened fire with antiaircraft and heavy automatic weapons. A lower jump altitude meant less prolonged exposure in the harness for the Rangers, but also that they could expect an exceptionally hard landing, which indeed happened. Miraculously, only a few Rangers were seriously injured during the high-risk parachute jump. Nevertheless, Cuban troops, supported by numerous armored personnel carriers, put up a surprisingly vigorous resistance at Point Salines.

---

[64] Adkin, *Urgent Fury*, 132.
[65] Cole, *Operation Urgent Fury*, 35.

Still, by 0900, the Rangers had secured "138 American medical students at the True Blue campus adjacent to the airfield."[66]

The siege at the governor general's residence, known as the Government House, in St. George's was especially confusing. Consequently, that evening Metcalf ordered the 22d MAU to land Company G, BLT 2/8, commanded by Captain Robert J. Dobson, along with some of the MAU's M60 main battle tanks and TOW weapons carriers on a beach at Grand Mal on the western side of the island near Fort Frederick.[67] The landing of Marine armor at Grand Mal was especially telling. In his after-action report on the operation, the PhibRon 4 commander, Captain Erie, noted that "the use of armor . . . proved to be very effective," causing a tremendous shock to the Grenadian forces. "Once the Grenadian forces realized the Marines were ashore with armor," Erie reported, "they quickly broadcast this to these units and basically told them it was all over." Erie believed that this action proved that amphibious night assaults "should be the desired option whenever feasible." Finally, the tremendous flexibility that the Navy–Marine Corps amphibious assault force demonstrated its place as an ideal force choice in the furtherance of a Joint operation. Due to the rapidly changing nature of the fighting ashore, Metcalf and Schwarzkopf adjusted the Marine's mission time and again. The Service adapted to the situation and completed any required new tasking. Nevertheless, naval historian Frank Uhlig Jr. recorded that these actions demonstrated "how difficult it is to run a multiservice operation without either a plan, a staff appropriate to the task, or a command center adequate to the need."[68]

The Grand Mal landing force made it to Scoon's quarters by 0712 the day after the landing. Scoon and his wife, Esmai McNeilly, 9 civilians,

---

[66] Cole, *Operation Urgent Fury*, 42. The resistance of the Cubans was indeed a tactical surprise to the Joint Chiefs of Staff. Moreover, the Cuban "construction workers" were well-trained as combat troops. They were led by an experienced officer, Col Pedro Tortola Comas, who had been handpicked for this job by Cuban dictator Fidel Castro. The Rangers later uncovered additional weapons caches that could have easily equipped several Cuban rifle battalions.

[67] Spector, *U.S. Marines in Grenada*, 13–16.

[68] Capt Carl R. Erie, USN, quoted in Frank Uhlig Jr., "Amphibious Aspects of the Grenada Episode," in *American Intervention in Grenada: The Implications of Operation Urgent Fury*, ed. Peter M. Dunn and Bruce W. Watson (Boulder, CO: Westview, 1985), 95–96.

and 22 special operations personnel were quickly lifted by helicopter to the *Guam*.[69] Crist soon arranged for the eventual return of Scoon to St. George's so that he could start to form the nucleus of an interim government. The Marine Corps landing force later "moved a few miles east to the Grenadian stronghold at Fort Frederick in St. George's suburbs," capturing the citadel 10 hours later.[70]

Even with all this success, U.S. forces had to make their way to the Grand Anse campus to rescue any U.S. nationals who remained there. By the second day, the issue of the students at Grand Anse had become critical for Metcalf. At that time, nearly all the available Marine Corps ground forces were engaged in combat on the island. Task Force 121, commanded by Army major general Edward L. Trobaugh and consisting predominately of the Army's 82d Airborne Division, moved out from the True Blue campus and advanced toward Grand Anse. The task force ran into stiff Cuban resistance at a place called Frequente, and Trobaugh requested aid from Metcalf after seeing that Grand Anse was more heavily defended than planners had previously supposed.[71] Schwarzkopf recommended that Army Rangers make a vertical assault directly onto the Grand Anse campus, but no Army or special forces helicopters were available for such an attack. However, there were some helicopters from HMM-261 noticeably present on the deck of *Guam*. To employ the HMM-261 equipment, Schwarzkopf extended the Marine's tactical area for the second time in two days, adding the Grand Anse to their section. Despite later stories that credited the Marines with the liberation of Grand Anse, the Army

---

[69] Cole, *Operation Urgent Fury*, 47; and Spector, *U.S. Marines in Grenada*, 16.

[70] Cole, *Operation Urgent Fury*, 47.

[71] Cole, *Operation Urgent Fury*, 47–48. It was around this time that Gen John W. Vessey, USA, then chairman of the Joint Chiefs of Staff, was supposedly growing frustrated with the slow progress of the 82d Airborne Division in its movement toward Grand Anse. Reportedly, Vessey sent a message to Trobaugh stating, "We have two companies of Marines running all over the island and thousands of Army troops doing nothing. What the hell is going on?" See Rick Atkinson, *The Long Gray Line: The American Journey of West Point's Class of 1966* (New York: Henry Holt, 1989), 485. If true, this was a bit unfair to Trobaugh because he did not have his full division ashore yet. By the conclusion of the second day of fighting, two more 82d Airborne battalions and a brigade headquarters "were landed at Point Salines airfield, increasing the number of airborne troops in Grenada from about two thousand to nearly five thousand." See, Cole, *Operation Urgent Fury*, 49.

Rangers actually liberated the campus and secured the Americans who lived there after a 30-minute firefight. Soon after, four HMM-261 helicopters arrived to evacuate 224 medical students.[72] HMM-261 commander Lieutenant Colonel Amos, being a Virginia Military Institute classmate of this Army Ranger battalion's commander, Lieutenant Colonel Ralph L. Hagler Jr., quickly carried off the Joint force planning to ensure the Rangers' extraction on Marine equipment.[73]

Fighting on the island continued into 27 October, but resistance was rapidly falling off. Earlier fierce fighting around Point Salines and Grand Anse caused the 82d Airborne to adopt a systematic process for identifying and rooting out any more Grenadian or Cuban diehards. Yet, this mission took on a secondary importance to locating and securing threatened U.S. nationals still at large on the island. This primary mission, in turn, required all U.S. forces to adopt a cautious approach to the use of force.[74]

The U.S. forces learned numerous lessons from the action in Grenada. The difficulty of interoperability between various elements of the Joint task force, especially when dealing with communications equipment, emerged as an early lesson for the entire force. At times, this factor led to several sensational calls for fire from troops engaged with the enemy to Fort Bragg, North Carolina, through commercial telephone lines. Further, due to operational security matters, the U.S. military decided early on to neither provide details to the media nor allow reporters to accompany the invasion force. This choice made sense on 25 October but less so after the U.S. forces secured the American medical students. The initial exclusion of reporters and details caused an increasingly skeptical media to question the motives of the Reagan administration for invading Grenada in the first place. This suspicion against the administration dissipated when Reagan's office eventually granted the press access to the returning medical students who, to a person, effusively thanked the U.S. government for sending the military to their rescue. At least one stu-

---

[72] Cole, *Operation Urgent Fury*, 47–48.
[73] Spector, *U.S. Marines in Grenada*, 17.
[74] Spector, *U.S. Marines in Grenada*, 23–25.

dent, in front of an entire bank of media representatives, literally kissed the ground after setting foot on U.S. soil.[75]

Even after the fighting on the island of Grenada had ended, the U.S. force's operation against that nation continued. On 1 November 1983, the Marines of Task Force 124 conducted a dawn raid on the tiny near-by Grenadian dependent of Carriacou. They landed on the island un-opposed, but captured 17 Grenadian Army personnel and "a quantity of military equipment including rifles, radios, explosives, ammunition, jeeps, a truck, and a generator." Task Force 124 did not find any Cuban soldiers. The following day, paratroopers of the 82d Airborne Division relieved the Marines, at which point they reboarded their ships. The Marines, PhibRon 4, and the *Independence* battle group then proceeded on their original mission to the Middle East.[76]

At the same time, the Marine Corps made a change to the 22d MAU command relationship, assigning yet another former Vietnam War-era Co-Van, the 2d Marine Division's assistant division commander, Brig-adier General James R. Joy, as the new MAU commander. Faulkner be-came Joy's chief of staff. This change made sense in the context of the situation in Beirut, as general officers commanded both the French and Italian contingents as well. On 3 November, McDonald turned over op-erational control of all forces on Grenada to Trobaugh, and by 12 De-cember, all U.S. Army airborne battalions on Grenada had returned to the United States.[77]

Meanwhile, the battered but resilient 24th MAU prepared to turn over its peacekeeping duties at the Beirut International Airport to the arriving 22d MAU. President Reagan appointed the recently promot-ed Commandant Kelley as his representative on the scene. Kelley, along with congressional representatives, arrived at the airport on the after-noon of 25 October. For Kelley, who had earlier visited the Marine Corps wounded in a military hospital in Frankfurt, Germany, it was an emo-tional experience. As Commandant, he was determined to ensure that

---

[75] Uhlig, "Amphibious Aspects of the Grenada Episode," 109–12, 119, 124.
[76] Cole, *Operation Urgent Fury*, 60.
[77] Cole, *Operation Urgent Fury*, 60–61.

24th MAU received all the help they needed. On 31 October, Kelley testified before several congressional committees that held hearings on the Marine barracks bombing. The committee members grilled Kelley about whether or not the 24th MAU had provided itself with adequate security. This question was difficult for the Commandant to answer. Although car bombings had taken place, most notably against the U.S. embassy in April, and had occurred with increasing frequency in the weeks leading up to the suicide attack, he implied that no security would have been "adequate to protect the occupants against a 5-ton Mercedes truck carrying 5000 pounds of explosives at high speed." He also saw little evidence that "the commander should have known, given the explosion at the Embassy in April. . . . Both instances involved a terrorist bombing from a motor vehicle," he argued, "but there the similarity ends," as all the aspects of the two assaults, including the delivery system, were "totally different."[78]

The DOD's Long Commission found that numerous mistakes had been made up and down the chain of command, including the damaging effect of mission creep that emanated from the Reagan administration. The commissioners labeled the Marine barracks attack "a terrorist act . . . sponsored by sovereign states," specifically Iran and Syria, "and organized political entities for the purpose of defeating U.S. objectives in Lebanon." Further, Secretary of Defense Caspar W. Weinberger dissected the report and sent it to various Service secretaries. He also requested that DOD subagencies address how their respective organizations were implementing the Long Commission's recommendations, wanting

---

[78] Paul X. Kelley testimony, quoted in Frank, *U.S. Marines in Lebanon*, 103–4. In 1983, vehicle-borne improvised explosive devices were a relatively new phenomenon in the world of terrorism. The use of suicide bombers to deliver such devices was also recent. Before this event, the multinational forces had experienced mainly unmanned command-detonated car bomb attacks or offset explosive devices that did not require the immolation of its perpetrator. Historically, beginning with Japanese kamikaze attacks in World War II, it has been difficult for most Americans to fathom the willingness of a suicide bomber to kill themselves in an attack. In hindsight, it is easy to say that the MAU should have had better security. However, the location of the MAU in its vulnerable airport location, mission creep, and especially the forced transition of the MAU from that of neutral peacekeeper to belligerent in the growing and politically complex Lebanese Civil War must also be taken in context. Geraghty certainly sensed that trouble would ultimately emanate from this transition.

their answers by 9 January 1984. Most notably, the chairman of the Joint Chiefs of Staff, General Vessey, was asked to outline what he had done to tackle the commission's recommendations on "military responses to terrorism; casualty reporting; the chain of command and effective command supervision of the USMNF security positions; tailored intelligence; rules of engagement; and post-attack security."[79] All these issues had hampered Geraghty's 24th MAU throughout its difficult sojourn in Beirut.

After the 22d MAU relieved the 24th MAU toward the latter part of November 1983, Joy immediately went to work improving force protection measures around the airport, which included spreading out any concentrations of Marines who might become potential targets for terrorist bombings. All this activity did not mean that the Muslim militias stopped targeting the airport. At 0700 on 4 December, for example, Navy jets from *Independence* and USS *John F. Kennedy* (CV 67) launched an airstrike against Syrian positions in the Beqaa Valley. Anticipating a possible retaliatory strike against the Marines, Joy put the 22d MAU on its highest state of alert. The day turned out especially bad for the Navy-Marine Corps team. Syrian antiaircraft batteries shot down two of the attacking aircraft, killing one of the pilots while the other bailed out, leading to his capture by the Syrians. Later that evening, the combat outpost Checkpoint Seven, which sat on the road from Ash Shuwayfat to the airport that ran past the Pepsi Cola bottling plant, received small arms fire.[80] The observation posts sat on top of two different two-and-a-half-story buildings that "provided the best observation of all the small buildings in the area." At 2204 that night, a 122-millimeter Katyusha rocket scored a direct hit on the observation post, wounding two and killing eight Marines. No stranger to combat, Lieutenant Colonel Smith believed that his Marines were targeted due to the earlier airstrike. No longer placed under the highly constrained rules of engagement that his

---

[79] Frank, *U.S. Marines in Lebanon*, 109.

[80] Joseph B. Treaster, "2 U.S. Jets Downed by Syrian Gunfire; Navy Staged Raid," *New York Times*, 4 December 1983; and Frank, *U.S. Marines in Lebanon*, 122. The captured pilot, Lt Robert O. Goodman, USN, later gained his release through the vigorous efforts of Rev Jesse Jackson. This was not the first or last time that Jackson would help secure the release of U.S. prisoners held by foreign powers.

predecessors dealt with before 23 October 1983, he subsequently used significant force against the Amal and Druze gunners whenever he felt it was justified. Basically, if Smith's Marines came under fire, the militias could expect a "vigorous response."[81]

By mid-December, Joy brought in additional engineers from Camp Lejeune and even U.S. Naval Construction Force members known as Seabees from Rota, Spain, to complete work on a tank ditch, complete with concrete obstacles, around Beirut International Airport. Extensive hardening of the area took place every day. Joy arranged for Smith's line companies to employ a mobile defense supplemented by mutually supporting strongpoints. Both officers gave BLT 2/8's company commanders extensive leeway to create open fields of fire for their company frontages. Smith told the commanders of Company F and Company G to "use their imagination," which they fully embraced, including literally moving the earth around them.[82] By Christmas, attacks slackened enough for entertainer Bob Hope to stage one of his famed USO shows aboard *Guam* and USS *New Jersey* (BB 62). Throughout January 1984, the Marines continued to face attacks and suffered occasional casualties, but everything changed the following month. By February, the LAF was experiencing significant desertion among its Muslim enlisted personnel, who were receiving encouragement to desert by Muslim military, political, and religious leaders. Many of the Muslim LAF enlistees left to join one of the militias or simply went home. By 6 February, the LAF forces around the airport were in a state of virtual collapse as "Lebanese soldiers, with their tanks and other rolling stock, sought a safe haven within U.S. positions at the airport, or continued to the north to join up with other government forces." On 11 February 1984, in what must have been reminiscent of Operations Frequent Wind and Eagle Pull, the Marines began evacuating U.S. embassy personnel, American nationals, and their dependents. On 7 February, Joy heard in a radio broadcast from the British Broadcasting Corporation that Reagan had announced the withdrawal of all shore-based U.S. forces to Navy shipping that would re-

---

[81] Frank, *U.S. Marines in Lebanon*, 123–24.
[82] Frank, *U.S. Marines in Lebanon*, 125.

main just offshore for the time being. Shortly after, Joy received official orders to withdraw. That same day, the small British multinational force contingent also left Beirut.[83]

While the Marines in Beirut began their withdrawal, some fallout from the terrorist bombing the previous October continued. On 8 February 1984, Geraghty, in perhaps the unkindest cut of all, received a non-punitive administrative letter of instruction from Secretary of the Navy Lehman, who sent it under the orders of Secretary of Defense Weinberger. In the letter, Lehman laid the responsibility for the security measures of his subordinates directly on Geraghty. Lehman's letter concluded that those measures "did not provide adequate security" despite the Long Commission recognizing that the Marines had no options to prevent the assault at the time. Still, Geraghty received the letter as "a nonpunitive reminder that your actions, as commander, were not sufficient to prevent this tragedy." Incredibly, Lieutenant Colonel Gerlach, who remained in a bed at the Boston Veterans Administration hospital with blurred vision and diminished hearing, while also being "virtually a quadriplegic," received a similar letter from Lehman.[84] Neither Geraghty nor Gerlach needed any reminders of their duty at the Beirut International Airport.

Later that month, the long ordeal of the MAU deployments to Beirut was about to end. On 20 February, the commander of USEUCOM, Army general Bernard W. Rogers, reassigned Joy to command Joint Task Force Lebanon, which placed Faulkner back in command of the 22d MAU. Six days later, the 22d MAU was scheduled to hand control of the Beirut International Airport to the LAF. The LAF's problem with desertions, however, left it without the necessary personnel to hold the airport. Major William J. Sublette, the 22d MAU air officer, and Major Earnest A. Van Huss traveled to the LAF liaison office at the airport to retrieve the American colors. As they folded the U.S. flag, a LAF colonel handed the Lebanese national flag to Faulkner, telling him "you may as well take our flag too." According to Marine Corps historians, the 22d MAU on 26 February "left behind more than one million filled sandbags and a lot of deep

---

[83] Frank, *U.S. Marines in Lebanon*, 132–34.
[84] Geraghty, *Peacekeepers at War*, 164.

holes, which . . . the Amal very quickly occupied." Six minutes after the last Marine transport departed, the Amal forces raised their flag "over the watchtower at Black Beach." Soon, Amal flags appeared "all over the airport."[85] On 5 March 1984, the Lebanese government repealed the 17 May Agreement and, just short of one month later, it made a request to France, Italy, Great Britain, and the United States to disband the multinational force, bringing "the unhappy episode of U.S. military involvement in Lebanon" to an end.[86]

By 10 April 1984, the 24th MAU, now commanded by Colonel Myron C. Harrington Jr., relieved the 22d MAU while at sea. The 24th MAU's new mission was to serve as a floating "reaction force to rescue the American ambassador, if necessary, or in other contingency operations in Lebanon or elsewhere in the Mediterranean." Elements of the MAU also provided external security at the U.S.-UK embassy ashore but were no longer needed by 31 July 1984. The United States established a new embassy site in safer East Beirut, and the ambassador, embassy staff, and upgraded Marine Corps security guard detachment quickly moved in. In the 18 months that the MAUs spent in Lebanon, the Marine Corps suffered 436 total casualties, with 238 killed and 158 wounded in action and another 47 injured in nonbattle or accidental situations.[87] On 8 November 1985, Lebanon was inscribed on the Marine Corps War Memorial in Arlington, Virginia.

After months of recriminations about the Beirut bombing, the Navy-Marine Corps team and the U.S. military returned its focus to an increasingly maritime-oriented Soviet Union, which seemed to be growing in capability each year. Commandant Kelley's recapitalization efforts for the Marine Corps were starting to pay some dividends in the critical areas of equipment modernization and readiness. Anticipating greater emphasis on special operations and low-intensity conflicts in the future, Kelley ordered Major General Alfred M. Gray Jr., the II Marine Expeditionary Force commander, to "become better prepared for maritime-

---

[85] Frank, *U.S. Marines in Lebanon*, 137.
[86] Hallenbeck, *Military Force as an Instrument in U.S. Foreign Policy*, 132.
[87] Frank, *U.S. Marines in Lebanon*, 138–40.

based raids and hostage rescue in the Middle East and Caribbean." Originally called the "MAU Special Operations Capable (MAU/SOC) program," Gray supported it with great enthusiasm. Kelley was adamant that the MAU/SOC units would not become part of the JSOC but could provide any Joint task force commander with the option of using amphibiously oriented Marine Corps forces for special missions.[88] The MAU/SOC program did require that MAUs receive extensively more specialized predeployment training before being certified and deployed to carry out such missions.

Beginning in 1976, elements of the 2d Marine Division, most notably the 4th MAB, had been regularly deploying to Norway either independently or as part of a more substantial combined NATO exercise, which continued throughout the 1980s. For instance, Operation Anorak Express in 1980 marked the first time the Marine Corps had "operated north of the Arctic Circle."[89] Consequently, except for some serious shortcomings in cold-weather operations in the late 1970s and early 1980s, the Service had made steady improvements in equipment and doctrine. From 28 February to 22 March 1984, just days after the 22d MAU redeployed to the offshore Navy shipping near Lebanon, the 4th MAB under Brigadier General Norman H. Smith, participated in "one of the biggest naval defence exercises ever conducted by the western allies." Called Operation Teamwork 84, it encompassed landing 12,000 Marines, with support from 38 naval vessels, "beyond the Arctic Circle in northern Norway in severe weather conditions." They were then "expected to hypothetically fight themselves ashore in temperatures down to 14 degrees." Before deploying, the 4th MAB went through extensive cold-weather training at Camp Ripley, Minnesota, which included hiring 28 merchant ships "at a cost of around $2 million [dollars] to simulate realistic convoy conditions." Amazingly, the Soviet military, per an international agreement, was allowed to send a senior servicemem-

---

[88] Allan R. Millett, *Semper Fidelis: The History of the United States Marine Corps*, rev. ed. (New York: Free Press, 1991), 630.
[89] Maj Joseph A. Crookston, "Marine Corps Roles and Missions: A Case for Specialization," GlobalSecurity, accessed 14 August 2019.

ber to observe the exercise, but it violated the agreement by sending numerous uninvited observers.[90] Additionally, several Soviet navy vessels shadowed the NATO fleet, and at least six four-engine Soviet Tupolev Tu-95 "Bear" strategic bombers attempted to make photographic reconnaissance flights.

The Soviet bombers became such a nuisance that the NATO naval task force commander, Vice Admiral Metcalf of Grenada fame, requested that Smith launch his Hawker Siddeley AV-8A Harrier attack aircraft to run them off, which created a conundrum for Smith. Before Smith left for the exercise, Lieutenant General John H. Miller, the commanding general of FMFLANT, instructed Smith that all organic Marine Corps aviation was reserved for the exercise's Marine air-ground task force (MAGTF) and should not be used for any other purpose. Making this issue even more difficult was that the Navy carrier battle group that had been initially assigned to the operation, and which could have easily driven off the bombers, remained in the Mediterranean, most likely due to the volatile situation related to the late-February redeployment of the 22d MAU. Without a supporting aircraft carrier, Smith made the tough decision to launch the Harriers of Marine Attack Squadron 231 from USS *Inchon* (LPH 12). Armed with air-to-air missiles, the Harriers pulled off this mission without a hitch despite lacking onboard airborne radar. The Harriers proved that they could fill in for carrier aviation in an emergency and play a key role in sea control, especially in the absence of a traditional large-deck aircraft carrier.[91] While Metcalf was the commander of the amphibious task force and in charge of evaluating any overarching threats, the actual final decision to launch the Harriers rested with Smith as the commander of the landing force.

---

[90] John Jones, "NATO Ships Plying North Atlantic in Maneuvers," *UPI Archives*, 7 March 1984.
[91] LtGen Norman H. Smith, USMC (Ret), interview with the author, 24 May 2019, Quantico, VA. Smith noted that when he returned to the United States, he paid a visit to LtGen Miller as soon as he could to explain why he allowed the Navy to use his Harriers. To his credit, Miller did not criticize Smith's on-the-scene decision. According to Smith, the Marine Corps had a total of six to eight Harriers on USS *Inchon*. The BLT helicopters were moved to other amphibious ship decks so that the Harriers had a dedicated flight deck. Smith also noted that the Harriers had an Australian exchange pilot assigned to them, but for political reasons, he was not allowed to fly the intercept missions.

Because of the last-minute nature of the mission, the task force had to improvise its support elements. To compensate for a lack of airborne radar, the task force designated the destroyer USS *Kidd* (DDG 993) as the primary control ship for Harrier operations. In a further complication, USS *Inchon* personnel had to be "trained to control AV-8s in an instrument environment performing air-to-air intercepts. The most important joint training conducted was the integrated preparation of the USS *Inchon* and squadron crews to conduct the high tempo deck operations expected for the evaluation." The idea was that *Kidd*'s radar would track an inbound and potentially hostile aircraft before it reached the amphibious task force, and then the Harriers would deploy from *Inchon* to intercept the hostile aircraft within five minutes of notification, before it could fly over the task force. In fact, the task force was only in its second day of deployment in the Atlantic when a sidewinder-armed Harrier successfully intercepted a Tupolev Tu-95 Bear bomber flying out of Cuba. Thanks to *Kidd*'s radar, the Harriers could be vectored to "neutralize" potentially hostile surface targets as well.[92]

Once off the coast of Norway, the Teamwork 84 force began an air command and control subexercise known as Busy Eagle 84. Busy Eagle allowed Smith to practice integrating the Marine Air Command and Control System (MACCS) into Norway's network of these systems. The planners incorporated a significant challenge in this portion of the exercise by purposely operating in places where terrain masking commonly disrupted the radars. This aspect required the MACCS to find suitable radar sites that could overcome the rugged terrain and mountainous peaks that habitually interfered with the electronic line of sight communications, which was far easier said than done. Nevertheless, Smith noted that land-based Marine aviation launched 95 excursions from northern

---

[92] BrigGen Norman H. Smith, "Arctic Maneuvers 1984," *Marine Corps Gazette* 68, no. 12 (December 1984): 31.

Norwegian air facilities, specifically Bodø Main Air Station and Evenes Air Station, during which they accomplished "all exercise objectives."[93]

From 5 to 22 March 1984, the amphibious task force conducted the Teamwork 84 landing exercise. Smith was convinced that the Marine Corps needed "the capability to conduct a combined landing in north Norway" for three reasons. First, he considered north Norway "a door" that "if forced open by the Soviets" would both threaten "the security of transatlantic reinforcement" and help "secure the flank of a central European front." He believed that the Bardufoss, Andoya, Evenes, and Banak air stations and "the ports of Tromso, Narvik, and Sorreisa" were the "keys to the door." Second, possessing north Norway, he argued, "favors the defender as it allows NATO air forces to intercept Soviet aircraft over the Norwegian Sea before the enemy aircraft reach either convoy shipping or bases in Iceland, the United Kingdom, or Denmark." Finally, by controlling the Norwegian Sea, he thought that NATO would have "the ability to restrict the movement of the Soviet Northern Fleet through the Greenland–Iceland and Norway gap en route to their interdiction missions."[94]

For the exercise, the Marine Corps landing force had the objective capturing both Bardufoss Air Station and the port of Sorreisa. Smith's primary assault force consisted of Regimental Landing Team 2, made up of three large battalions—2d Battalion, 2d Marines; 3d Battalion, 8th Marines; and 1st Battalion, 25th Marines, in reserve—and a combined British and Dutch landing team. The assault focused on the Malangen and Bals fjords, with the U.S. contingent centering on Malangen and the British and Dutch force concentrating on Bals. The exercise would also, at least partially, test the land–based prepositioned equipment program, by which the Marine Corps had readied a substantial amount of combat gear,

---

[93] Smith, "Arctic Maneuvers 1984," 31–33. Smith noted that deploying MACCS radars on mountaintops in northern Norway was especially tricky due to winds "in excess of 75 [miles per hour]" and "temperatures below minus 50 degrees Fahrenheit" that rendered them nearly unusable most of the time. Consequently, he stressed the need to consider the use of multiple integrated radars and "maximum use of AEW [early warning] aircraft," Smith, "Arctic Maneuvers 1984," 32–33.

[94] Smith, "Arctic Maneuver 1984," 33.

vehicles, food, water, and maintenance parts for its fly-in forces. In this case, Battery K, 4th Battalion, 10th Marines, flew into Norway and fell in on their prepositioned equipment, including new M198 155-millimeter howitzers. This action resulted in the participants gaining some valuable lessons, especially relating to depreservation and preservation of pre-staged equipment. The battery also got operationally ready long before it would have if it had been required to land via amphibious shipping.[95]

The NATO forces used this exercise to test some equipment for the cold-weather environment as well. All NATO tracked vehicles employed pad cleats to improve their maneuverability. The Marines found that German-made removable track cleats substantially increased the mobility of all their tracked vehicles in the snow, naming them the best option for their equipment. The units also experimented with a new tactical field kitchen that allowed 4th MAB personnel to receive 1,300 hot tray pack meals while in the field. In an improvement over their ancient and bulky Korean War-era cold-weather gear, Marines wore a layered Gore-Tex clothing system that "received overwhelming praise from the Marines who wore it in the field."[96]

Teamwork 84 was a resounding success on several levels. With the exception of the challenge that using towed artillery in the Norwegian permafrost created, exercise observers noted improvements in both cold-weather equipment and doctrine. The ability of the MACCS to link with other NATO and inter-Service radar systems was especially noteworthy. Finally, as Smith wrote, Teamwork 84 "served as a graduation exercise for Marine Corps participation in cold weather exercises in Norway." For years, the Marines had fallen short of their NATO allies in operating in such inhospitable environments, but Teamwork 84 had illustrated their improved capabilities. It also demonstrated the tremen-

---

[95] Smith, "Arctic Maneuver 1984," 33–34, 36. Smith noted there were also minor problems with host-nation support of the prepositioning program.

[96] Smith, "Arctic Maneuver 1984," 36. The U.S. version was a "bolt-on" cleat. The Marines overwhelmingly approved of the German "snap-on" variety. Even so, all the tracked vehicles, cleated or not, were hard on the Norwegian road surfaces. Smith also noted that in such rugged and frozen conditions, his ground forces were mainly "road-bound" for most of the operation.

dous utility and flexibility of the MAGTF in such a climate, making the Marines the force of choice for NATO's Northern Flank.[97]

Throughout the 1980s, the Marine Corps continued participating in several European-oriented exercises, such as the usually biennial Northern Wedding/Bold Guard operations that typically took place on Scandinavian and northern German soil. Doing so provided opportunities for the Marine Corps to improve its ability to contribute to NATO operations. In the fall of 1986, for example, BLT 3d Battalion, 8th Marines (BLT 3/8), under Lieutenant Colonel David F. Bice, took part in this exercise. Prior to their deployment, Bice and BLT 3/8 went to the Marine Corps Mountain Warfare Training Center for about a week. Soon after, they deployed for the Northern Wedding portion of the operations near Oslo, Norway. During the exercise, a Boeing Vertol CH-46 Sea Knight helicopter of HMM-162 crashed at sea, killing eight Marines who were aboard. While this accident was highly regrettable, Bice noted that it resulted in a Service-wide policy change related to improving training and rescue procedures for helicopters. BLT 3/8 then redeployed to Denmark and northern Germany for Bold Guard. Bice noted that his battalion was fully mechanized, meaning that he used his organic Landing Vehicle, Tracked, Personnel 7 assault amphibious vehicle as armored personnel carriers and commanded a Germany Army panzer company for the duration of the operation. His Marines, with the assistance of a German bridging company, even conducted a practice crossing of the famous Kiel Canal. During this crossing, Bice noted that they got 60 vehicles across in rapid succession. Despite receiving assistance from the Germans, Bice stated, some Germans did not appreciate the NATO troops tearing up their roads and fields, with some even being outwardly antagonistic toward the NATO mission in and around the Kiel Canal. Bice, who had previous experience with foreign Services, having completed a tour with the

---

[97] Smith, "Arctic Maneuver 1984," 36–37. One of the odd issues that Gen Smith ran into with the Norwegian locals was the new Kevlar helmet adopted by all U.S. forces in 1983. Better suited to protect one's head from shell bursts, it was also unfortunately shaped nearly identical to the Nazi Germany combat helmet from World War II. The Nazis had occupied Norway throughout the war, and its citizens even in the 1980s did not appreciate seeing military personnel wearing this new helmet, which brought back many painful memories.

British Royal Marines, recognized the exceptional difficulty that multi-Service/multinational force arrangements caused during Joint and combined NATO operations, but which also made conducting such exercises necessary. He also stated that it was important that regional or host nation commanders, such as those he encountered in Norway, would not always cede control of their sovereign territory even during peacetime, which NATO plans officers should always consider.[98]

By 1986, the U.S. inter-Service requirement for more effective Joint warfighting was clearly growing. Consequently, the passage of the Goldwater-Nichols Department of Defense Reorganization Act of 1986 would become the most important piece of defense legislation since the National Security Act of 1947. The legislation marked the final push for improving Joint warfighting. Both the Navy and Marine Corps strongly opposed the reorganizational fallout over the Goldwater-Nichols Act. Further, the powerful Secretary of the Navy Lehman believed that the legislation diminished his authority as a service secretary, which indeed it did.[99]

Eventually, even newly appointed Army and Air Force chiefs of staff joined the counter-reform ranks. By early 1986, it appeared that only the new chairman of the Joint Chiefs of Staff, Navy admiral William J. Crowe Jr., remained in favor of reform, but several events in 1985 shifted the odds toward the reform camp. First, Senator Barry M. Goldwater (R-AZ) formed an alliance with Senator Samuel A. Nunn Jr. (D-GA). Second, the exceptionally reform-minded Congressman Leslie Aspin Jr. (D-WI) was named chairman of the House Committee on Armed Services. Finally, in response to several procurement scandals that were then making the pages of *The Washington Post*, Robert C. McFarlane, now Reagan's national security advisor, convinced the president to appoint a blue-

---

[98] MajGen David F. Bice, USMC (Ret), interview with the author, 11 October 2021, Woodbridge, VA. During his Royal Marines exchange tour, then–Maj Bice taught Joint warfare planning to NATO officers. In fact, Bice had the good fortune to have a number of NATO-related assignments throughout the 1980s and became a leading Marine Corps expert on Joint and combined operations in Europe. Following his retirement in 2002, Bice assisted the Lithuanian military with their transition into NATO.

[99] Locher, "Has It Worked?," 95–115.

ribbon commission headed by former Deputy Secretary of Defense David Packard to study defense acquisition reform.[100] The Packard Commission ultimately made numerous sweeping recommendations that did not remain confined solely to procurements. Most notably, the commission believed that the unified combatant commanders should report directly to the secretary of defense through the chairman of the Joint Chiefs of Staff and that the chairman's office needed considerable strengthening. As a result, all the pieces seemed to fall fortuitously in place for genuine Joint Service reform to occur in 1986.

James R. Locher III, a brilliant Pentagon staffer long associated with Admiral Crowe, crafted the proper language that ultimately resulted in the Goldwater-Nichols Act. Later, he worked hard for its congressional approval and the Pentagon's acceptance after Reagan signed the bill into law. Locher stated that the previous chairman of the Joint Chiefs of Staff, U.S. Air Force general David C. Jones, originated the reforms in 1982. Just five months from retirement, a despondent Jones testified before Congress that "the system is broken. I have tried to reform it from inside, but I cannot. Congress is going to have to mandate necessary reforms."[101] Locher believed that "the number-one problem plaguing the Department of Defense was an imbalance between service and joint interests." Because the Services "absolutely dominated," he argued that "they had weakened the unified commanders."[102] The next closest problem was that the position of the chairman of the Joint Chiefs of Staff was too weak. Before 1986, the Service chiefs had the ability to ensure they included their dissent in any military advice being passed to the secretary of defense and even the president. Consequently, as former Secretary of Defense Schlesinger sarcastically noted, the combined advice of the Joint Chiefs of Staff was "generally irrelevant, normally unread, and almost always disregarded."[103]

---

[100] James R. Locher III, "Has It Worked?: The Goldwater-Nichols Reorganization Act," *Naval War College Review* 54, no. 4 (Autumn 2001): 95–115.
[101] Gen David C. Jones, USAF, quoted in Locher, "Has It Worked?," 101.
[102] Locher, "Has It Worked?," 95–115.
[103] James Schlesinger, quoted in Locher, "Has It Worked?," 104.

Due to this terrible situation for the Joint Chiefs of Staff, subordinate officers wanted to avoid any appointments to it. In fact, Locher noted, any naval officer in the early to mid-1980s considered service on the Joint staff "a kiss of death—meaning that one's career was over." General Crist later testified that while commander of U.S. Central Command, there was "not a single volunteer for any of the thousand billets on his headquarters staff—all of them joint billets." Any officer who received an unfortunate assignment to "joint duty got orders out of it as soon as they could," making their time on them "dysfunctionally short."[104] Overall, the Goldwater-Nichols Act seemed to finally address the chronic problem of the ineffective Joint Staff, diminished the power of the Service chiefs to interfere with Joint operations and planning, invigorated the chairman of the Joint Chiefs of Staff's office; and created the office of vice chairman of the Joint Chiefs of Staff. To assist with Service interoperability, the Services would now share defense procurement programs.

Although Commandant Kelley still opposed the Goldwater-Nichols Act, in a nod to Jointness, he agreed to sign on to White Letter 4-86, titled "Omnibus Agreement for Command and Control of USMC Tactical Aviation." Essentially, the agreement established that the Joint force air component commander (JFACC), presumably an Air Force general, would have "final authority under the theater or joint task force commander for the design of the air campaign."[105] This arrangement meant that a commander from another Service could "dictate the targets for Marine and Navy aviation not immediately involved in a naval campaign, including amphibious operations." This idea had been long resisted within the halls of Headquarters Marine Corps, by past Commandants, and by nearly all Marine Corps aviators. Most Marines believed that retain-

---

[104] Locher, "Has It Worked?," 95–115.

[105] Millett, *Semper Fidelis*, 630–31. On the surface, the Omnibus Agreement and the Commandant's accompanying White Letter 4-86 seemed more threatening to Marine Corps tactical aviation than it actually was. The issue revolved around "excess sorties." What this meant was that if and when Marine Corps tactical aviation was not in the direct support of the MAGTF commander, its aviation assets were liable to be given mission taskings from the JFACC. The Marines were not really losing total control over their aviation component. However, the actual meaning of the word "excess" brought serious concern to some. When was a Marine Corps aircraft deemed in excess to the direct support of the MAGTF?

ing full-time command and control over MAGTF aviation assets ensured that its aviation was not available to participate in "a futile interdiction campaign to the detriment of engaged ground forces." At that time, the Goldwater–Nichols Act, the DOD and Joint Staff reorganizations, and the JFACC omnibus agreement all seemed "an ill omen that Marine Corps positions no longer brought much respect in Congress, the traditional protector of Marine Corps interests."[106] While Kelley's signing of the omnibus agreement was a major concession toward Jointness, the Marine Corps and the Navy still remained skeptical of its utility.

The year 1986 proved a watershed moment for U.S. maritime strategy as well. By the late 1980s, the Navy added USS *Theodore Roosevelt* (CVN 71) to its fleet, giving it 15 carrier battle groups at its disposal. In 1984 alone, the Navy and Marine Corps participated in various Joint exercises that included 55 other countries while visiting 108 countries as well. The CNO, Admiral Watkins, referred to this time as the "era of violent peace."[107] Clearly, the new strategy developed for the Navy–Marine Corps team was oriented to greater offensive capabilities and to enable maritime forces to operate on both northern and southern flanks of NATO to further complicate a possible Soviet armored attack across the central German plain.

The Marine Corps faced another difficult year in 1987, which started off poorly for the Service. Lehman, perhaps unhappy with the outcome of the fight over the Goldwater–Nichols Act, resigned as secretary of the Navy in April and returned to the private sector. That same month, Soviet intelligence agents had penetrated the Marine security guard detachment at the U.S. embassy in Moscow. At that same time, multiple Reagan administration officials, including Marine Corps lieutenant colonel Oliver L. North, were required to testify before Congress and, ultimately, a special prosecutor about their roles in the Iran–Contra Affair, the covert actions related to the facilitation of senior Reagan administration officials of selling weapons to Iran in exchange for money to support the Con-

---

[106] Millett, *Semper Fidelis*, 630–31.
[107] Adm James D. Watkins, USN, "The Maritime Strategy," U.S. Naval Institute *Proceedings* 112 no. 1, Supplement (January 1986).

tra group in Nicaragua. This issue became the lead domestic news story for months, which Democratic Party opponents to Reagan exploited to its fullest. On the bright side, Lehman's replacement as secretary of the Navy was the enigmatic award-winning author and decorated infantry Marine James H. Webb. That summer was also the time for the appointment of a new Commandant of the Marine Corps. As usual during such occasions, the rumor mill was rife about who would replace Kelley. Simultaneously, Congress, for the first time in years, slashed the proposed Weinberger defense budget for fiscal year 1987. The glory days of significantly increased defense spending appeared to reach an abrupt end.[108]

As usual, the scramble for Commandant was not without its theatrics. While Kelley strongly desired his steady and capable Assistant Commandant, General Thomas R. Morgan, to succeed him, Webb seemed determined to create another path for the Marine Corps. Morgan, an outstanding general officer and combat veteran of Vietnam, would have been a strong choice for the post. His previous service primarily as a naval aviator most likely made him an unappealing selection to Webb, however. As a former Marine infantry officer, Webb may have fallen into the traditional Marine Corps bias against naval aviators becoming Commandant. Nevertheless, many observers believed that the Marine Corps had benefited from the personnel and equipment renaissance that had taken place under Wilson, Barrow, and Kelley. Webb now wished for the Service to look toward improving its operational doctrine.[109]

Based on this aspiration, Webb looked for a new Commandant that possessed substantial leadership experience in the field. He desired appointing a general that had significantly more time with the Fleet Marine Forces than with positions at Headquarters Marine Corps. One of Webb's early favorites for the position was Lieutenant General Ernest C. Cheatham Jr., who had recently served as commander of the 1st Marine Division at Camp Pendleton, California, and was then serving as Kelley's deputy chief of staff for manpower.[110] Cheatham had played football as a

---

[108] Millett, *Semper Fidelis*, 631.
[109] Author's personal recollections; and Millett, *Semper Fidelis*, 631–32.
[110] Millett, *Semper Fidelis*, 631.

defensive tackle at Loyola Marymount University and was drafted out of college by the Pittsburgh Steelers of the National Football League (NFL). He put his playing career on hold to join the Marine Corps, briefly serving during the Korean War. He returned to the NFL in 1954 and played a total of six games for the Pittsburgh Steelers and Baltimore Colts. After retiring, Cheatham resumed his Marine Corps career, famously commanding the 2d Battalion, 5th Marines, during its struggle in Hue City against the People's Army of Vietnam during the Tet Offensive in 1968. For his extraordinary performance at Hue City, Cheatham was awarded the Navy Cross. Although being the frontrunner for Webb, Cheatham had major health concerns after years of hard service in the field. Likely, this issue led him to not enthusiastically pursue the job.[111]

Consequently, Secretary Webb became interested in the ubiquitous and famously field-oriented Lieutenant General Alfred M. Gray. Gray made a career out of avoiding service at Headquarters Marine Corps, an aversion that was legendary. Moreover, Gray started his career as an enlisted Marine before being commissioned as a second lieutenant in 1952. During the Vietnam War, he served numerous tours of duty as a decorated combat officer. Following that conflict, Gray served in various leadership roles with the infantry in the 3d Marine Division on Okinawa and especially with the 2d Marine Division at Camp Lejeune, North Carolina, commanding the unit during the Beirut crisis. After promotion to general officer, he became even more famous for his leadership of the 4th MAB. Gray was a strong advocate of military education and possessed one of the world's best collection of volumes on the ancient Chinese military philosopher Sun Tzu. In 1984, he was promoted to lieutenant general and appointed commanding general of FMFLANT. During his time as a general officer, "Gray had developed strong views about military careerism" as well as "bureaucratic sloth," both attributes that Webb greatly appreciated.[112] On 1 July 1987, Gray became the 29th Commandant of

[111] John H. Cushman Jr., "Activist General in Line for Top Marine Post," *New York Times*, 5 June 1987; David Vergun, "Sports Heroes Who Served: Pro Football Player to Formidable Marine," U.S. Department of Defense, 15 September 2020; and Millett, *Semper Fidelis*, 631–32.
[112] Millett, *Semper Fidelis*, 632.

the Marine Corps. Although 58 years old at the time, he had the energy level of officers half his age. He was determined to shake the Marine Corps out of its operational lethargy, and he wanted to get started as soon as possible.

CHAPTER SIX

General Alfred Gray and Maneuver Warfare

Award-winning Marine Corps historian Allan R. Millett wrote that "from the start of his tenure" as Commandant of the Marine Corps, Lieutenant General Alfred M. Gray Jr. "acted like a man possessed, a man who heard a ticking clock (or bomb) behind him and who could not do enough fast enough to suit himself."[1] The 29th Commandant would be an action-oriented leader and was on an urgent mission to reinvigorate the operational doctrine of the Marine Corps. Until the 1980s, few, if any, senior Marine Corps leaders had as much operational experience as Gray. After taking command of the newly established 4th Marine Amphibious Brigade in 1978, Gray had been primarily in charge of East Coast-based Marines. He later successfully led both the 2d Marine Division and Fleet Marine Force, Atlantic (FMFLANT), throughout the early to mid-1980s, commanding the latter until becoming Commandant in 1987.

Based on his experience, Gray wanted to drive operational decision making down to its lowest levels so that line commanders could take advantage of battlefield opportunities as they emerged rather than overly rely on massive firepower and material strength to compensate for poor operational doctrine. He realized since his days in Vietnam at Firebase Gio Linh that unimaginative planning was rarely a recipe for success. Consequently, Gray would become the Corps' greatest advocate for

---

[1] Allan R. Millett, *Semper Fidelis: The History of the United States Marine Corps*, rev. ed. (New York: Free Press, 1991), 632–33.

Figure 52. Gen Alfred M. Gray Jr.

Source: official U.S. Marine Corps photo.

the maneuver warfare concept—in which an intellectually reinvigorated Marine Corps could move quickly, take advantage of battlefield opportunities, and win battles despite the enemy potentially outnumbering and outgunning them—and he was going to do all this "in the face of a declining budget."[2]

Early into Gray's term, the Rand Corporation published a pamphlet that questioned whether the President Ronald W. Reagan administration's defense buildup had come to an end. Reagan's eight years in office saw the U.S. defense budget rise to unprecedented levels. As the 1980s neared their end, even Reagan sensed that the Soviet Union was verging on collapse. In May 1989, newly elected President George H. W. Bush addressed the North Atlantic Treaty Organization (NATO) and proposed a substantial reduction of military forces in Europe. That October, Hungary renounced Communism, declaring itself an independent republic. On 4 November, a combined 1 million citizens in both East and West Germany demonstrated against the Soviet Union and the corrupt Communist regime in East Germany. Just five days later, East and West Berliners spontaneously tore down the infamous Berlin Wall, and the Soviet Union did nothing to stop it. By December, the Communist Party of Czechoslovakia relinquished its one-party power after the so-called "Velvet Revolution."[3] By 1990, most of the former Eastern Bloc states had driven their Communist parties out of office. No one in the U.S. intelligence or defense community could have predicted such a rapid series of events unfolding.[4] The Warsaw Pact meltdown created the early beginnings of what was then commonly called the Peace Dividend. It also created a tremendous conundrum for the U.S. defense establishment, as it now

---

[2] Millett, *Semper Fidelis*, 632–33. Around this same time, the U.S. Army also partially embraced maneuver warfare. However, they called their concept AirLand Battle. The general assumption was that on most potential NATO battlefields, the Soviet Union would always outnumber and outgun U.S. forces. Consequently, the AirLand Battle concept would allow NATO commanders to utilize the benefits of maneuver and combined Joint forces to retain the strategic initiative against the Warsaw Pact powers and thereby prevail against them in the end.

[3] Bart Brasher, *Implosion: Downsizing the Military, 1987–2015* (Westport, CT: Greenwood Press, 2000), 16.

[4] Brasher, *Implosion*, 14–16; and Kevin N. Lewis, *What If the Reagan Defense Buildup Is Over?* (Santa Monica, CA: Rand, 1987).

Figure 53. The fall of the Berlin Wall

Source: Senate of Berlin.

had to figure out its role after its most existential enemy of the past 60 years had been seemingly swept into the dustbin of history. Now, with the Cold War coming to an end, the United States and its Western allies could stop spending so heavily on defense and redirect this surprise dividend toward other things, such as education and long-deferred internal infrastructure improvements.[5]

In the fall of 1989, U.S. Army general Colin L. Powell became the new chairman of the Joint Chiefs of Staff. Powell went on to become one of the most influential and important chairmen in the position's history. Many on the Joint Staff, including Powell, had already begun to see that any future threats would likely come from "indigenously caused

---

[5] Peter Passel, "The Peace Dividend's Collateral Damage," *New York Times*, 13 September 1992, A3.

conventional regional conflicts with little likelihood of direct Soviet intervention."[6] With Soviet military power in a serious ebbtide in Eastern Europe and around the globe, the Joint Chiefs of Staff paid more attention to regional issues such as the insurgency of Communist Sandinistas in Nicaragua and the dictatorship of General Manuel Antonio Noriega in Panama. Powell believed that major force realignments, including yet another review of Service roles and missions, were in order. He argued that the United States should respond to these transformations not only by making significant cuts to "its conventional forces and chang[ing] the pattern of their deployment" but also by reducing "its strategic nuclear arsenal." Having the United States decrease "its forward deployments in Europe and Korea" would reduce the U.S. Army from its "18-division active strength of 760,000 to 10–12 divisions totaling 525,000." Similarly, the Navy could "plan for 400 ships, including 12 carriers" for its deployments rather than the "551 ships, including 15 carriers" it had currently distributed. The changes would allow for "its active strength" being "reduced from . . . 587,000 to 400,000." Although Powell had not set a "projected size" of the Air Force, he hoped to decrease the Marine Corps' "congressionally-mandated three division/wing teams" from an "active strength of 197,000 to between 125,000–150,000."[7]

Consequently, the Office of the Joint Chiefs of Staff advocated for sharply reduced defense manning levels to coincide with declining defense dollars. Powell envisioned a general overall force reduction, which he referred to as the Base Force, to achieve this goal. To make these shocking cuts palatable to the Service chiefs, Powell ensured that they would be phased in over a number of years. He ordered the Joint Staff to conduct a study on what they wanted for the new post–Cold War force, calling it *Quiet Study II*. Completed in October 1989, *Quiet Study II* contradicted an earlier study that assumed that the end of the Cold War would not bring significant changes in strategic direction. Instead, *Quiet Study II* "postulated a shift in focus from the East-West confrontation in Europe

---

[6] Lorna S. Jaffe, *The Development of the Base Force, 1989–1992* (Washington, DC: Joint History Office, Office of the Chairman of the Joint Chiefs of Staff 1993), 2.
[7] Jaffe, *The Development of the Base Force*, 14–15.

to regional contingencies."[8] On 15 November 1989, at the invitation of a skeptical Secretary of Defense Richard B. "Dick" Cheney, Powell briefed President Bush on the details of a potential new national military strategy based on data that the study generated. With the Goldwater–Nichols Act of 1986 now in place, Powell had no obligation to include the Service chiefs in the process and went straight to the president with his ideas. He sought little to no input from the other chiefs because he knew of their unanimous opposition to a changed strategy for the 1990s and the Base Force in general. After Powell returned to the Pentagon, he was adamant that the Services would have to incur some significant cuts in force manning levels.[9]

All the Service chiefs, including Gray, strongly pushed back against Powell's proposal. Powell countered that the Base Force was akin to a "floor," a point "below which we dare not go." The chiefs shot back that they had witnessed similar changes during the immediate post–Vietnam War years and feared that such deep cuts would affect morale, training, and readiness once again. Although the Navy Department's program objective memorandum proposed reducing the Marine Corps' active strength to 159,000, Gray insisted that the Service could not reduce its strength below 180,000. He also argued that none of the suggested personnel numbers for the Base Force were sufficient for the Marine Corps, especially for the Service's traditional role and mission of responding to regional contingencies. In private meetings with Powell, Gray argued that the Joint staff had "no justification for cutting his service since geography, not the Soviet threat, had determined its mission and hence its strength." In a concession to Gray and the Marine Corps, Powell raised their Base Force level to the 159,000 that the Navy suggested in its program objective memorandum. Although "still well below Gray's objective of 180,000," the Marine Corps was the only Service "to which Powell made such a concession."[10] In 1988, Gray emphasized the necessity of a larger repurposed Marine Corps when he relabeled Marine amphibi-

---

ous forces and Marine amphibious units as Marine expeditionary forces (MEFs) and Marine expeditionary units (MEU)s.[11]

Like previous Commandants, Gray sought greater tactical mobility so that forces landing onshore could maneuver rapidly. He championed the development of an advanced amphibious assault vehicle, a tracked amphibious vehicle that could rapidly operate over great distances from U.S. Navy amphibious support vessels to troops ashore as a mechanized troop carrier. This role was asking a lot of the technology, but Gray believed it could be done. He also became Commandant shortly after the Navy had completed its development of a new landing craft, air cushioned (LCAC) vehicle. This water transport would act as a platform that carries heavy loads, including tanks and towed artillery, at nearly 40 knots.[12]

At the same time and despite considerable developmental delay, the new Bell Boeing V-22 Osprey tiltrotor aircraft appeared to near its testing completion in the late 1980s. This unique and increasingly controversial hybrid aircraft used two powerful tiltrotor engines that combined the vertical lift of a helicopter with the in-flight prop-driven capacity of an airplane. It flew twice as fast and five times as far as the Vietnam War-era Boeing Vertol CH-46 Sea Knight helicopter it was meant to replace while also having the ability to refuel in flight to extend its operational range further. Like the past Marine Corps experience with the Hawker Siddeley AV-8A Harrier attack aircraft, the V-22 had several downsides. It was difficult to fly, had a horrible crash record early on, and was becoming increasingly expensive as compared to standard medium- and even heavy-lift helicopters. Consequently, Secretary of Defense Cheney moved to cancel the V-22 program, much to Gray's dismay.[13] Fortunately for the Marine Corps, key congressional representatives kept the platform alive. For instance, Representative W. Curtis Weldon (R-PA) organized broad support for the V-22 program within the House of Rep-

---

[11] Brasher, *Implosion*, 28.

[12] Gen Alfred M. Gray, *Annual Report of the Marine Corps to Congress, 1990* (Washington, DC: Department of the Navy, 1990), 15.

[13] Gray, *Annual Report of the Marine Corps to the Congress, 1990*, 15. In his report, Gray stated that "no Marine requirement is more pressing than the need to identify a solution for the replacement of our medium-lift assault capability."

resentatives by demonstrating that parts of the aircraft were made in 48 of the 50 states.[14]

Gray was also a strong proponent of the eight-wheeled Light Armored Vehicle-25 (LAV-25) made by General Dynamics. The transport could carry six combat-equipped Marines while mounting a 25-millimeter M242 Bushmaster chain gun. Certain variants had the ability to fire antiarmor missiles or could even assist with air defense and command and control. The LAV-25 proved remarkably versatile, having the ability to play a variety of roles on any modern battlefield. It was exceptionally fast as compared to the tracked armored amphibious vehicles that the Marine Corps used as a mechanized troop carrier at the time, while also providing greater mobility and firepower to maneuver elements. The LAV-25 entered active operational service in 1983, and the Marines could not have been happier with it.[15]

Coupled with the new McDonnell Douglas AV-8B Harrier II, all this new technology would allow amphibious task forces to avoid assaulting defended and predictable beachheads. Instead, beaches that the Marine Corps once considered inaccessible to traditional seaborne assault forces now could act as potentially undefended landing sites that the Navy-Marine Corps team could exploit. The new technology coupled with a philosophy of rapid maneuver foreshadowed a winning combination that finally put to rest the idea that amphibious forces were too slow and light to prevail against a numerically superior and mechanized enemy. This same combination seemed to restore the strategic initiative to amphibious task force commanders. In the future, an enemy would have to worry about powerful and offensively minded amphibious forces landing anywhere along a littoral or even well inland.[16]

During Gray's term as Commandant, the Marine Corps also benefited from the Maritime Prepositioning Force (MPF) finally coming into its own. Based on lessons learned from past exercises that included the em-

---

[14] Author's personal discussion with Representative W. Curtis Weldon; and Melissa Healy, "Warplane Survives Attack," *Los Angeles (CA) Times*, 29 November 1990.

[15] Author's personal discussion with Weldon; and Healy, "Warplane Survives Attack."

[16] Author's personal discussion with Weldon; and Healy, "Warplane Survives Attack."

ployment of the MPF sets, the Navy-Marine Corps team embraced a force module concept that vastly improved MPF effectiveness and demonstrated its flexibility. From 1990 to 1993, the Navy-Marine Corps team conducted four major exercises or operations for the MPF, resulting in numerous other lessons learned. First and foremost, the Navy and Marine Corps created a roll-on, roll-off discharge facility (RRDF) that was essentially "a platform constructed by lighterage that allows for . . . the offload of rolling stock in-stream."[17] The RRDF enables the MPF to shorten its offload and backload times by several days. Yet, it became quickly apparent that establishing an RRDF was considerably dependent on the forces operating on a calm sea. In smooth seas, the RRDF greatly assisted the MPF, but the Navy and Marine Corps struggled to establish a RRDP in rougher seas. Nevertheless, early in Operation Restore Hope (1992–93), a United Nations (UN)-led humanitarian operation in Somalia, four MPF ships managed to unload their cargo despite the fact that the piers and harbor facilities in the port of Mogadishu were a disaster with no electricity, broken vehicles, abandoned cranes, and "trash and excrement covered the piers and warehouse floors."[18] In fact, the MPF vessels "offloaded and staged" 759 containers and 2,033 vehicles in the harbor during 17 days of operations. While the U.S. forces faced a small problem keeping looters out of the portside staging areas, each of the MPF "offloaded its vehicles in just 36 hours."[19] In another instance, at the start of Operations Desert Shield and Desert Storm (1990–91), Maritime Prepositioning Squadron 2 (MPS-2), was stationed at the Navy Support Facility Diego Garcia, located in the British Indian Ocean Territory. Being deployed to the Persian Gulf, it arrived "within seven days" delivering the "heavy tanks, self-propelled artillery and the sustainment pack-

---

[17] Mark B. Geis and William A. D. Wallace, *Improving MPF Operational Effectiveness: Lessons Learned from Past Exercises and Operations* (Quantico, VA: U.S. Marine Corps, 1993), 28.
[18] Geis and Wallace, *Improving MPF Operational Effectiveness*, 16.
[19] Geis and Wallace, *Improving MPF Operational Effectiveness*, 11–12.

age for a 16,000-man Marine Expeditionary Brigade."[20] Despite these success, many observers of the MPF program, including former Commandant Robert H. Barrow, noted that the most significant concern for the ships is that "they are not warships; they cannot adequately defend themselves or take substantial punishment."[21]

As Commandant, Gray was integral to the operational art renaissance taking place at Marine Corps Schools in Quantico, Virginia, at the same time that a group of like-minded defense analysts worked diligently to change the status quo inside the Pentagon. In fact, defense reform in the 1980s had become a sort of cottage industry. Numerous analysts, including William S. Lind and retired U.S. Air Force colonel John R. Boyd, among others, all sensed that the defense establishment required changes as soon as possible.[22]

Only a few of these analysts contributed to the concepts of maneuver warfare, however. Lind, as a national security advisor to Senator Gary W. Hart (D-CO), provided two studies on the subject. In 1986, Lind and Hart coauthored the book entitled *America Can Win: The Case for Military Reform*. The volume centered entirely on the need for all the U.S. Services to embrace the maneuver warfare concept. According to Hart and Lind, the Marine Corps had a positive start to embracing the concept initial-

---

[20] LtCol Douglas O. Hendricks, *Maritime Prepositioning Force in Theater Level Campaigning* (Fort Leavenworth, KS: U.S. Army School of Advanced Military Studies, 1991), 5–6. Hendricks noted that while MPF ships provided a task force commander with tremendous flexibility, they usually required a friendly port for offloading. Moreover, since most MPF ships were of the breakbulk roll-on/roll-off variety and designed for rapid unloading, they "were easy to kill" by submarines since water-tight compartmentalization found on most U.S. Navy purpose-built amphibious ships was noticeably lacking.

[21] Hendricks, *Maritime Prepositioning Force in Theater Level Campaigning*, 22.

[22] The analysts involved in developing ideas for defense reform came from both the private and government sectors. In addition to Lind and Boyd, they included Dr. Edward N. Luttwak, a recent émigré and national security expert from Romania and the United Kingdom; James Fallows, a former speechwriter for President James E. "Jimmy" Carter Jr. and national security writer for *The Atlantic Monthly*; Pierre M. Sprey, a NATO and aviation expert; Charles C. Moskos, a military sociologist and U.S. Army veteran; David Packard, cofounder of the Hewlett-Packard Corporation, former U.S. deputy secretary of defense, and chairman of President Ronald W. Reagan's Blue Ribbon Commission on Defense Management; and Thomas P. Christie, an Air Force mathematician and career defense analyst. For more on these and other contributors, see Maj Michael J Leahey, "A History of Defense Reform since 1970" (master's thesis, Naval Postgraduate School, 1989), 7–24.

ly. Yet, following General Barrow's retirement, Marine Corps leadership "moved to suppress" maneuver warfare "where it had been flourishing." While the authors may have been overstating the case, they noted that Gray's former command, the 2d Marine Division, marked the one exception to this policy. Gray, as the commander of FMFLANT at the time, ensured that the doctrinal situation for Marines stationed along the East Coast did not change. Hart and Lind further criticized the Service for its field training, which they called "rigid and mechanical," and its tactics, which they thought were "taught as recipes and formulas." Additionally, they argued that its "education and doctrinal development are at a standstill, except at the Amphibious Warfare School [AWS], where some progress is still being made."[23]

Throughout the 1980s, Lind enjoyed unprecedented access to Gray from the time he was at Camp Lejeune, North Carolina, through his command of FMFLANT and even after he became Commandant, gaining knowledge that he would use in *America Can Win*. Lind later remarked that two active-duty Marine Corps officers led the way toward the adoption of the maneuver warfare concept. Initially, he kept their identities anonymous, trusting that doing so would prevent opponents of the concept—of which there were many at the time—from derailing their military careers. After Gray became Commandant, Lind believed that he could safely reveal the two officers, Gray and Colonel Michael D. Wyly, in an interview from 1989. Wyly incorporated the concept into the curriculum of the AWS as the head of the school's tactics department. Gray, as com-

---

[23] Gary Hart and William S. Lind, *America Can Win: The Case for Military Reform* (Bethesda, MD: Adler & Adler, 1986), 39.

manding general of the 2d Marine Division, made maneuver warfare the division's official operational doctrine starting in 1981.[24]

Before his work alongside Hart, Lind had authored the *Maneuver Warfare Handbook* in 1985. Not intended as an "academic monograph," he wanted to create a "ready reference for field Marines" serving in the operational forces.[25] He argued that "history suggests God is on the side of the bigger battalions"—such as the Warsaw Pact forces in Europe—"unless the smaller battalions have a better idea." When dealing with a larger, stronger force, Lind contended that it was "never very promising" if the two forces got into "a slugging match." Even if the smaller units somehow won the fight, he argued, "the cost is usually high." Consequently, Lind urged the Marine Corps to think of maneuver warfare as a form of judo, where a smaller force could use what a larger force believes are its inherent advantages against them. "It is a way of fighting smart," he wrote. Maneuver warfare allows a smaller battalion to outthink an enemy that it "may not be able to overpower with brute strength."[26] Essentially, Lind was advocating that the Marines eschew mass in favor of movement. The concept was not solely about moving a smaller force into an advantageous position to defeat a larger one. It

---

[24] Leahey, "A History of Defense Reform since 1970," 43n6. Wyly was later promoted to the rank of colonel and was one of the most forward-thinking military strategists in the Marine Corps at the time. Yet, even he was not without his detractors. As Wyly noted, the more he published, the more trouble he seemed to get in. Other reformers outside traditional Department of Defense circles included St. Anselm's College professors Richard A. Gabriel (a former career Army intelligence officer) and Paul L. Savage and Brookings Institute scholars Jeffrey Record and Martin Binkin. Binkin and Record's sharp critique of the Marine Corps amphibious mission likely spurred Gray, Wyly, and others to think about alternative ways in which an expeditionary-minded Marine Corps can be best employed in the future. See Richard A. Gabriel and Paul L. Savage, *Crisis in Command: Mismanagement in the Army* (New York: Hill and Wang, 1978); and Martin Binkin and Jeffrey Record, *Where Does the Marine Corps Go from Here?* (Washington, DC: Brookings Institution, 1976).

[25] William S. Lind, *The Maneuver Warfare Handbook* (Boulder, CO: Westview Press, 1985), 2. A few senior Marines were turned off by Lind's views on maneuver warfare concept. Some argued that as a civilian with no experience in the Armed Services, he could not possibly know what they knew. Instead, Lind used examples from World War II and other conflicts since then to support why maneuver was a better option than attrition, especially for forces that expected to be outnumbered on the battlefield. Fortunately, thanks in large part to the insistence of Gen Gray, maneuver warfare ultimately became Marine Corps-wide operational doctrine just in time for Operation Desert Storm.

[26] Lind, *The Maneuver Warfare Handbook*, 2.

was about maneuvering to strike at the enemy's center of gravity. Lind's work in both these books provided the basis for the Marine's development of maneuver warfare.

One of Gray's early acolytes, Colonel Anthony C. Zinni, came to embrace the maneuver warfare concept as fervently as his commanding general. In 1986, Zinni was selected for the Chief of Naval Operations' prestigious Strategic Studies Group at the U.S. Naval War College in Newport, Rhode Island. From 1987 to 1989, Zinni performed in a dual role as commanding officer of the 9th Marine Regiment and the 35th MEU with the 3d Marine Division on Okinawa. At that time, Zinni decided that the 9th Marine Regiment should experiment with the maneuver warfare concept, similar to what the 2d Marine Division had instituted a few years earlier. As he later observed, this push was difficult to implement. Zinni noted that maneuver warfare and mission-type orders made a lot of his staff officers especially uncomfortable. Many of them preferred the highly complex and micromanaged operational plans of the past. For those who did eventually embrace maneuver warfare, they came to see its vast superiority compared to the previous operational system in the 3d Marine Division. When Zinni returned to the United States, he had a brief assignment at Quantico as the chief of staff of the Marine Air-Ground Training and Education Center. After this post, Zinni, under Gray's orders, went on a lecture tour to select installations to discuss the maneuver warfare experience of the 9th Marine Regiment.[27] He was selected for flag rank shortly thereafter. It was clear that Zinni's sterling combat record and willingness to embrace the concept of maneuver warfare played a strong role in his promotion to this coveted rank in the Marine Corps.

During the debates about maneuver warfare, Gray, amazingly, even considered the opinions of junior officers. A highly influential writer for the *Marine Corps Gazette* was Captain John F. Schmitt, who was later instrumental in cooperating with Gray in the writing and production of

---

[27] Col Anthony C. Zinni, Maneuver Warfare Lecture, U.S. Naval Academy, late fall 1989. The author attended this fascinating lecture, but the exact date and month cannot be accurately remembered.

Warfighting, Fleet Marine Force Manual 1 (FMFM 1) in 1989. This small book became one of the most profound doctrinal publications that the Marine Corps ever produced, being on an equivalent level as the *Tentative Manual for Landing Operations* (1934) and the *Small Wars Manual* (1940) and completing the holy trinity of significant Marine Corps doctrinal publications.[28] In the foreword to *Warfighting*, Commandant Gray wrote that "he expected every officer to read—and reread—this book, understand it, and take its message to heart." He argued that the concept in the manual represented "not just guidance for actions in combat, but a way of thinking in general."[29] Divided into four succinct chapters, *Warfighting* starts with a discussion on the nature and theory of war and ends with how to prepare for and conduct warfare in general. Gray' was heavily influenced by the work of classical strategists, such as Chinese military theorist Sun Tzu's *The Art of War*, which was "the only theoretical book on war that [Air Force colonel John R.] Boyd did not find fundamentally flawed."[30] Moreover, *Warfighting* encapsulated a decade of debate between "maneuverists" and "attritionists," making a strong argument for why maneuver warfare heralded a winning warfighting operational strategy for the Marine Corps.[31]

The ideas of another little-known (at that time) advocate of maneuver warfare undoubtedly influenced *Warfighting*. Colonel John R. Boyd was an unusual individual. He had enlisted in the U.S. Army Air Forces toward the end of World War II and deployed to Japan as part of the U.S. occupation forces after the war ended. After graduating from the Uni-

---

[28] *Warfighting*, Fleet Marine Force Manual (FMFM) 1 (Washington, DC: Headquarters Marine Corps, 1989); *Tentative Manual for Landing Operations, 1934* (Washington, DC: Headquarters Marine Corps, 1934); and *Small Wars Manual*, Fleet Marine Force Reference Publication 12-15 (Washington, DC: Headquarters Marine Corps, 1940).

[29] Gen Alfred M. Gray, "Foreword," in *Warfighting*, FMFM 1. This manual was later updated in 1997 and renamed Marine Corps Doctrinal Publication 1 to incorporate post-Cold War lessons learned about the nature of warfighting. Gray later remarked that he was happy to see the document updated.

[30] Robert Coram, *Boyd: The Fighter Pilot Who Changed the Art of War* (Boston, MA: Little, Brown, 2002), 320–21, 323–24, 326.

[31] *Warfighting*, FMFM 1, 61–65. It should be noted that the terms *maneuverist* and *attritionist* were not generally liked by most people involved in the debate. They seemed to unnecessarily create a narrowly defined dichotomy of "us" versus "them," while their actual positions were much more nuanced.

versity of Iowa in 1951, he was commissioned as a pilot in the recently formed U.S. Air Force but only flew 22 total combat missions during the Korean War.[32] However, his theories on combat decision making later became part and parcel of the maneuver warfare movement.

Boyd's entire career—both military and civilian—was more about accomplishing something rather than attaining higher rank or public recognition. He occasionally received less than positive evaluation reports for being somewhat defiant. Yet, other senior evaluators found him to be one of the most promising officers in the Air Force. Following Korea, Boyd was selected for the Air Force's prestigious Fighter Weapons School (FWS) at Nellis Air Force Base in Nevada. Boyd was not overly impressed with the instruction at the time, later deriding it as a "gunnery school."[33] He did well enough as a student to get an appointment as a fighter tactics instructor. He became fixated with revising the FWS curriculum and later published a pathbreaking manual called the "Aerial Attack Study," which the Air Force eventually adopted. In his role as FWS instructor, he became known as "40 second Boyd."[34] He bet his students that through a series of rapid-fire maneuver decisions, he could, within a 40-second window, get on his opponent's tail for a kill shot or "pay the victor 40 dollars." Allegedly, "no pilot ever collected on the bet."[35]

After retiring as a colonel in 1975, most of Boyd's few remaining friends at the Pentagon and in the Air Force, called "acolytes" by his biographer Robert Coram, became part of the growing post-Vietnam War defense reform movement. At the same time, Boyd became infatuated with learning theory. In 1976, he produced an 11-page paper titled

---

[32] Coram, *Boyd*, 30–33.

[33] Coram, *Boyd*, 70–72; and Maj Ian T. Brown, *A New Conception of War: John Boyd, the U.S. Marines, and Maneuver Warfare* (Quantico, VA: Marine Corps University Press, 2018), 6–7, 9, 12, https://doi.org/10.56686/9780997317497. Brown's book has already found its way onto the Commandant's Professional Reading List.

[34] Brown, *A New Conception of War*, 11–12, 14–16; and Coram, *Boyd*, 114, 116.

[35] Brown, *A New Conception of War*, 11–12.

Figure 54. U.S. Air Force Col John R. Boyd

Source: official U.S. Air Force photo.

"Destruction and Creation."[36] Although several of his acolytes urged Boyd to publish this extraordinary document, he never did. While Coram described reading the study as "tough sledding," it first suggested a significant element of the maneuver warfare concept with "Boyd's elaboration on the idea that a relationship exists between an observer and what is being observed."[37] Out of this article, Boyd established his now-famous observe, orient, decide, act (OODA) loop theory, which Coram has called "the intellectual heart of the new war doctrine so craved by elements within the U.S. military."[38]

Boyd's OODA loop theory is often mischaracterized and misunderstood even today. Taking a page from Sun Tzu, Boyd had designed the theory to "shatter cohesion, produce paralysis, and bring about collapse of the adversary by generating confusion, disorder, panic, and chaos." Basically, Boyd believed that "if someone truly understands how to create menace and uncertainty and mistrust, then how to exploit and magnify the presence of these disconcerting elements, the Loop can become vicious, a terribly destructive force, virtually unstoppable in causing panic and confusion." Boyd and his acolyte at the Pentagon, Franklin C. Spinney, later developed a 185-slide briefing on this subject titled "Patterns

---

[36] Coram, *Boyd*, 323. Boyd was fortunate in that a few of his acolytes or professional friends became senior-level decision makers in the Pentagon after his retirement from active duty in 1975. One such individual was Thomas P. Christie, who became deputy assistant secretary of defense. Others, such as Pierre Sprey and Franklin C. Spinney, played key roles in promoting Boyd's often wide-ranging ideas on the ways the U.S. military needed to change inside the Pentagon. Consequently, Boyd's philosophy spread beyond just the Air Force and had already been embraced by his most fervent Marine Corps acolyte, Col Michael D. Wyly.

[37] Coram, *Boyd*, 323–24.

[38] Coram, *Boyd*, 326.

Figure 55. Lockheed P-80 Shooting Star jet fighter

Source: official U.S. Air Force photo.

of Conflict."[39] Boyd famously refused to present this briefing in less than five hours, much to the ire of a number of Service chief executive assistants who strongly desired their bosses to get the briefing. At that length, however, the chiefs simply would not attend. Over time, Boyd and Spinney presented it to larger audiences, including ones on Capitol Hill. This amazing series of slides began with an analysis of maneuver warfare in Greek and Roman conflicts and then progressed rapidly through eras of history that emphasized other examples, both successful and unsuccessful. Boyd even took on the issue of using maneuver warfare to counter

---

[39] Coram, *Boyd*, 332–34.

guerrilla operations, which the U.S. military had been struggling with since the Korean War concluded.[40]

Lind provided a deeper understanding of what Boyd meant with the concept of maneuver in maneuver warfare. In *The Maneuver Warfare Handbook*, he noted that "the Boyd Theory defines what is meant by the word 'maneuver' in the term 'maneuver warfare.' Maneuver means Boyd cycling the enemy, being consistently faster through however many OODA loops it takes until the enemy loses his cohesion—until he can no longer fight as an effective, organized force."[41] Time and movement, "or more precisely speed and maneuver," were two of the most important components of Boyd's theory. He believed that "if one moves and constantly presents an opponent with a changing situation and does so quickly, one has a tremendous advantage. If one cannot do this . . . the chances for success in almost any kind of combat are seriously degraded."[42]

Another Boyd acolyte, Air Force lieutenant colonel James G. Burton, was an early convert to Boyd's reformist ideas. From 1979 to 1982, Burton served three different secretaries of the Air Force. He described the distrustful and dysfunctional inter-Service atmosphere that existed within the executive ring of the Pentagon during that time as one that made Machiavelli "a rank amateur in some of the contests waged there."[43] Burton further noted:

Coalitions form and dissolve overnight between the strangest of bedfellows. Dire enemies momentarily join forces to battle someone else, then resume their old fight as if nothing had happened. The only way to get a decision to stand is to "shoot the losers"—line up everyone who opposes the decision and [metaphorically] shoot them down. Otherwise, they begin to

---

[40] Col John R. Boyd, USAF, "Patterns of Conflict," ed. Col Chet Richards, USAF (Ret) and Chuck Spinney (PowerPoint presentation, Defense and the National Interest, 2005), accessed 30 August 2019.

[41] Lind, *The Maneuver Warfare Handbook*, 6.

[42] Grant T. Hammond, *The Mind of War: John Boyd and American Security* (Washington, DC: Smithsonian Institution Press, 2001), 151.

[43] Hammond, *The Mind of War*, 113. LtCol James G. Burton paid a price for his association with Boyd, being passed over the first time he was eligible for promotion to the rank of colonel. He was eventually selected on a later board.

undermine the decision before the ink is dry on the paper. Quite often, the real debate begins only after a major decision is made. Time and again, I have listened to senior officials express total frustration when issues they thought were settled suddenly reappeared.[44]

While the more managerially focused Navy and Air Force partially resisted adopting many of Boyd's ideas, the two Services that felt the strongest effects from the Vietnam War, the Army and Marine Corps, moved to embrace them. Boyd's concept became so popular with the Marines that they essentially made him an honorary Marine. In 1980, Wyly brought Boyd to Quantico to present a lecture to his AWS students. Wyly did so before knowing about Boyd's self-imposed requirement that he present the entire lecture or nothing at all. This required Wyly give Boyd a huge chunk of valuable educational time—something that rarely, if ever, happened at Marine Corps Schools for any presenter. Fortunately, Wyly got a supportive Brigadier General Bernard E. Trainor, then the deputy for education and director of the education center at Marine Corps Development and Education Command, to allow Boyd the time he required. Consequently, Boyd delivered one of the most impactful lectures ever to AWS, which ultimately lasted nearly seven hours. The young officers, according to Boyd's biographer, found it riveting and "could not get enough of what he had to say." Wyly later remarked that he "would remember this day for the rest of his life." Boyd had provided his stu-

---

[44] LtCol James G. Burton, USAF, quoted in Hammond, *The Mind of War*, 113–14. Essentially, during the cash-flush days of the 1980s, lousy procurement ideas and faulty doctrine never truly went away and were given continuously new leases on life, making the work of defense reform extremely frustrating. Things got so bad that reformers such as Burton, Boyd, Spinney, and Sprey "had to be careful about being seen with each other." They even resorted to using a system of aliases when they did make contact. See, Hammond, *The Mind of War*, 114.

dents "ideas that could be translated into tactics that worked on the modern battlefield."[45]

Wyly's role in introducing Boyd and the maneuver warfare concept to Quantico and eventually to Gray should not be underestimated. Bringing a controversial retired Air Force officer with no ground combat experience to the Marine Corps School was a risky move but one that eventually paid huge dividends. By the 1980s, Wyly had become known as somewhat of a maverick who had gone through two tours in the Vietnam War. In 1969, he commanded Company D, 1st Battalion, 5th Marines. Before Wyly took command, the company had received the nickname "Dying Delta" because it had lost so many Marines on fruitless search and destroy missions. Wyly, however, had different ideas about how to fight a war where only the enemy seemed to control when and where the forces would initiate contact. Albeit on a smaller scale, Wyly's initial steps toward maneuver warfare seemed to help him change the tempo of fighting for his company and enabled his Marines to be the ones to initiate combat against the enemy vice the other way around. Interestingly, President Reagan's final secretary of the Navy, James H. Webb Jr., was one of Wyly's platoon commanders. Webb later received the Navy Cross for gallantry in combat due to Wyly's recommendation.[46]

While in command of the 2d Marine Division in the early 1980s, Gray formed a group of officers and labeled them the maneuver warfare board. The officers who comprised the board ranged in rank from captain to lieutenant colonel, and several of them had been in Wyly's AWS maneuver warfare seminar before reporting to the 2d Division. Gray gave these officers wide latitude and did not interfere with their deliberations. The board's notes eventually evolved into a manual with subsections that fo-

---

[45] Coram, *Boyd*, 379. Boyd soon became a regular lecturer at the Amphibious Warfare School at Marine Corps University in Quantico, VA. Wyly's uncle was none other than Capt Donald F. Duncan, commanding officer of the famed 96th company, 2d Battalion, 6th Marines, which included in its ranks a future Commandant of the Marine Corps, Lt Clifton B. Cates. Duncan was killed on 6 June 1918 at Belleau Wood. Wyly recalled how family stories about his late uncle influenced his later decision to become a Marine. See, Coram, *Boyd*, 373.
[46] Coram, *Boyd*, 375.

cused on specific military occupational specialties, making it "the key to understanding how to implement maneuver warfare."[47]

According to historian Fideleon Damian, interrelated factors influenced the progress of this doctrine. Damian found a "direct connection" between the activities of the maneuver warfare board, Wyly's AWS, and doctrinal debates taking place in the pages of the *Marine Corps Gazette*. After the class of 1981 graduated from AWS, two graduating Marine Corps captains who had gone through Wyly's maneuver warfare seminar received assignments to the 2d Marine Division. When they arrived, they approached Gray at the Camp Lejeune officer's club and invited him to a meeting of their informal maneuver warfare group. Gray, however, was already predisposed to the concept, having earlier heard one of Boyd's "Patterns of Conflict" lectures. According to William Lind, "Gray was also an avid reader of military history and was known for an open mind" and that "he was also rare among the senior leadership of the Marine Corps in that he was receptive to maneuver warfare."[48]

Lind answered one of the biggest questions regarding maneuver warfare. If the vital element to warfighting success was to think, act, and move faster than the adversary, then the new doctrine had to address the question of "how can one consistently be faster?"[49] Lind believed that the answer to this question resided in the effective use of operational art. This idea meant "using tactical events—battles and, equally, refusals to give battle—to strike directly at an enemy strategic center of gravity." If a force is able to destroy this " 'hinge' in the enemy's system," then it would "bring it down." With operational art "permitting its practi-

---

[47] Anthony J. Piscitelli, *The Marine Corps Way of War: The Evolution of the U.S. Marine Corps from Attrition to Maneuver Warfare in the Post-Vietnam Era* (El Dorado Hills, CA: Savas Beatie, 2017), 48–49. The author of this superb book had unprecedented access to numerous post-Vietnam "maneuverists" of the Marine Corps, such as Wyly and Gray.

[48] Fideleon Damian, "The Road to FMFM 1: The United States Marine Corps and Maneuver Warfare Doctrine, 1979–1989" (master's thesis, Kansas State University, 2008), 84–85.

[49] William S. Lind, "The Theory and Practice of Maneuver Warfare," in *Maneuver Warfare: An Anthology*, ed. Richard D. Hooker Jr. (Novato, CA: Presidio Press, 1993), 9.

tioner to avoid unnecessary fighting," the units could become faster by "fighting only where and when necessary."[50]

While this explanation seems somewhat simplistic, during the 1980s, due to his contact with Gray and the 2d Marine Division and Boyd in Washington, Lind became the major advocate for the maneuver warfare concept with the Marine Corps. Yet, when briefing senior Marine Corps leadership on the subject, he leaned toward the caustic side in his delivery. Some of these leaders, many of whom had experienced extensive combat in Korea and Vietnam, struggled to listen to a mere civilian academic with virtually no military experience telling them how they had essentially been doing their jobs all wrong.[51] To make matters worse for the messenger, Lind sometimes showed up to observe a Marine Corps field exercise "wearing an inverness and a deerstalker" hat, making him the "most incongruous figure" the Marines had ever seen.[52] In reality, Lind was quite informed about the art of war, but that did not stop Marine Corps colonel John C. Studt from humorously stating in the foreword to Lind's *The Maneuver Warfare Handbook* that Lind would "have difficulty passing" the physical fitness test.[53] For a Service that had turned physical fitness into one of its holiest sacraments, Lind's ultra-civilian personal appearance made his message an even tougher sell.

In 1985, Lind added fuel to the fire when teamed with longtime Marine Corps critic Jeffrey Record for a scathing critique of Marine Corps leadership in the 28 July edition of *The Washington Post*. Lind and Record accused senior Marines of "ineptitude" in Beirut, Lebanon, and during

---

[50] Lind, "The Theory and Practice of Maneuver Warfare," 9. An excellent counterpoint to maneuver warfare enthusiasts was Daniel P. Bolger's "Maneuver Warfare Reconsidered." Bolger is particularly critical of Lind's claim that the Germans early in World War II were "the greatest practitioners of maneuver warfare" as well as his frequent use of German words to describe the concept., especially because they had been defeated in two world wars. Even so, Bolger argued that the idea was still something that needed consideration but was mostly overhyped. See Bolger, "Maneuver Warfare Reconsidered," in *Maneuver Warfare*, 26–29. Lind later admitted that German military doctrine and practices "were not always the best." See Brown, *A New Conception of War*, 135.

[51] Piscitelli, *The Marine Corps Way of War*, 49. The author engaged contemporary sources on this issue to include Lind, Wyly, and LtGen Bernard E. Trainor.

[52] Coram, *Boyd*, 383; and Brown, *A New Conception of War*, 146–48.

[53] Brown, *A New Conception of War*, 146.

"at least a dozen Marine Corps peacetime field exercises." They claimed that the Service's leadership had knowingly long covered up operational shortcomings. The authors firmly believed that the Marine Corps' way of war had "remained basically unchanged since the Pacific campaigns of World War II," which affected the Service's progress. For instance, Lind and Record noted that one Marine Corps student reported having "failed an exercise for outmaneuvering a dug-in enemy beach defense instead of assaulting it head on." Supposedly, the student conveyed, the faculty had rejected his actions "because he had a 3-1 numerical superiority over the enemy," meaning he should have attacked head-on instead of flanking the position. Apparently, they had ignored the student's "concern about casualties rendering his force ineffective for further actions." Lind and Record pushed the Marine Corps to immediately and enthusiastically embrace maneuver warfare to remedy this malaise. Comparing the Marine Corps reform efforts to that of the Army, Lind and Record feared that the old "argument for letting the Army absorb the Corps may gain overpowering strength."[54] A rebuttal to Lind and Record's critique of the Corps was provided by none other than Lieutenant General Victor H. Krulak in *The Washington Post*'s op-ed pages several months later. Lind and Record both wished for the Marine Corps to remain a separate Service but feared it would not survive as an independent entity unless it changed its warfighting doctrine.[55] Needless to say, Lind found himself increasingly unwelcome aboard many Marine Corps installations.

Despite this issue, Lind continued to have occasional access to Gray. Although Gray did not agree with Lind's indictment of senior Marine Corps leadership, he supported continuing the intellectual debate about maneuver warfare. After Gray became Commandant in 1987, changes related to maneuver warfare doctrine at Quantico and throughout the Service picked up speed. Trainor, now a lieutenant general, believed that opposition to Lind emerged because he publicly blamed senior Marine Corps leadership for allegedly having a "calcified commitment to attri-

[54] William Lind and Jeffrey Record, "The Marines' Brass Is Winning the Battle but Losing the Corps," *Washington Post*, 28 July 1985, B1; and Brown, *A New Conception of Warfare*, 147.
[55] Victor H. Krulak, "The Corps' Critics Are Wrong," *Washington Post*, 27 October 1985.

tion warfare where it did not exist to the degree he claimed."[56] In fact, Lind could not have been "a more polarizing patron for maneuver warfare," but his work allowed Boyd's ideas to remain at the forefront of the doctrinal debate in the Marine Corps during the 1980s.[57]

While Commandant Paul X. Kelley certainly did not appreciate the complaints that Record and Lind lodged, he did not necessarily oppose maneuver warfare as a doctrinal concept. Yet, he believed that Lind's hyperbolic criticism of Marine Corps doctrine generated more enemies than supporters, causing him to keep Lind at a distance. According to Colonel William S. Woods, who held a spot on Gray's maneuver warfare board as a major, he believed that the board members had erroneously faulted Kelley's lack of overt support for the concept to mean that he was against it. Woods later noted that Kelley did not have the ability to simply "condemn the current fighting style."[58] After all, Woods continued, "a retired, foul mouthed fighter jock (Colonel John Boyd), an overweight pompous civilian (William Lind), and a couple of young Captains with no combat experience are not much reason for tossing out all your current operational theories, particularly when that motley group is espousing war fighting methods gleaned from an army that lost its last couple of wars. Kelley was a warrior; but he was also a pragmatist."[59]

By 1989, the Marine Corps experienced major movements related to education and maneuver warfare doctrine. First and foremost, to ensure that the Marine Corps officer corps, including noncommissioned officers, was indoctrinated with maneuver warfare, Gray established the Marine Corps University (MCU), unifying all the Marine Corps schools at Quan-

---

[56] Brown, *A New Conception of Warfare*, 146–48.
[57] Coram, *Boyd*, 383. Col Gary I. Wilson, another maneuver warfare board plank holder from the 2d Marine Division, believed that since many officers only had a cursory knowledge of the concept and a limited grasp of military history, this made them "uncomfortable" with Lind. Instead of intellectually challenging his message, they "attacked him personally, not the issues, and in the meantime, Lind cleaned house by sticking to the issues." See Piscitelli, *The Marine Corps Way of War*, 91.
[58] Col William S. Woods, quoted in Piscitelli, *The Marine Corps Way of War*, 91.
[59] Piscitelli, *The Marine Corps Way of War*, 91.

tico under a single commanding general, who also became the university's president.[60]

Next, Gray wanted to establish a Commandant's Professional Reading List for use within the entire Service. Wyly had long advocated for the development of a professional reading list, having even started a precursor to the Commandant's list at the AWS in the early 1980s.[61] Yet, it took Gray's own indirect intervention for the Marine Corps Schools to follow through on this step.

To get the maneuver warfare concept into all the curriculums at Quantico, Gray sent Colonel Patrick G. Collins, one of his most trusted advisors, to The Basic School (TBS) to initiate the process. According to retired Marine Corps general John F. Kelly, when Collins arrived at Quantico, he told the TBS commanding officer, Colonel Terry J. Ebbert, a distinguished Vietnam War Navy Cross recipient, that the curriculum emphasis on maneuver warfare was "what the commandant wants, and you guys are going to do it." Additionally, Collins informed Ebbert, "he [the Commandant] wants a reading list . . . a reading program."[62] After Collins left the room, a somewhat chagrined Ebbert asked his staff, "Does anyone know what the hell he was talking about? What is maneuver warfare?" Kelly, then a relatively new infantry major on the TBS instructor staff, made the classic mistake of saying that he knew something about this topic. Consequently, Ebbert tasked Kelly with producing a draft reading list. The highly talented Kelly immediately started working, putting dozens of titles on 3-by-5-inch index cards so that he and Collins could see quickly what books were recommended for the

---

[60] "The History of Marine Corps University," Marine Corps University, accessed 27 April 2023.

[61] Damian, "The Road to FMFM 1," 78.

[62] Gen John F. Kelly, interview excerpt, in Piscitelli, *The Marine Corps Way of War*, 102. In the interview, Kelly noted that both BGen Paul K. Van Riper and Col Terry Ebbert got a chance to see and discuss the list before it went forward. During Operation Iraqi Freedom, Kelly served as the assistant division commander for MajGen James F. Mattis. Ultimately promoted to four-star rank, Kelly finished his lengthy active-duty career as the commanding general of USSOUTHCOM. Following retirement, Kelly was appointed by President Donald J. Trump as the secretary of homeland security. He later served as Trump's White House chief of staff before returning to private life in 2018. Col Collins was a legendary "Old Breed" Marine who fought with valor in two wars and served three tours of duty in Vietnam. He passed away in 1997. He was extremely close to Gray and a firm believer in the maneuver warfare concept.

various subdivisions to the list. Gray agreed to much of what emerged from the index cards, which became the first Commandant's Professional Reading List.[63]

Established in 1989, the professional reading list provides Marines of all ranks a select group of professional books related to some form of the military profession of arms for them to read. The books are rank-specific, educationally appropriate, and include a mix of classic volumes and new, cutting-edge scholarship. Since its establishment, every Commandant has adjusted, added to, deleted from, or reorganized the professional reading list. It is a living compendium that was and is seen as lifting the intellectual depth of all Marines. Its effect has spread to where all the Armed Services now have professional reading lists.[64]

That same year, shortly after Gray had established MCU, a regional contingency, something that the Joint Chiefs of Staff had predicted, emerged in turmoil-riven Panama. This new contingency would test the Goldwater-Nichols Act under combat conditions for the first time. At the time, Panama was ruled by Manuel Noriega's dictatorship. Noriega had long been associated with Panamanian military strongman General Omar Torrijos, who had famously held successful negotiations with President James E. "Jimmy" Carter Jr. in 1977 to ensure that the United States would turn over control of the Panama Canal and Canal Zone to Panama in 1999. Two years after Torrijos was killed in a plane crash in 1981, Noriega became the de facto leader of Panama, maintaining his rule by appealing to nationalism.[65]

Long considered a valuable anti-Communist intelligence asset, Noriega's rule became increasingly oppressive as the 1980s progressed. As early as 1986, the Reagan administration had "cut aid to Panama by 85 percent" due to Noriega's unscrupulous activities. Around that time, Reagan also sent his national security advisor, retired U.S. Navy rear admiral John M. Poindexter, to Panama to warn Noriega to stay out of

---

[63] Piscitelli, *The Marine Corps Way of War*, 101–2.

[64] "Military Reading Lists," U.S. Naval Institute, accessed 27 April 2023.

[65] LtGen Edward M. Flanagan Jr., USA (Ret), *Battle for Panama: Inside Operation Just Cause* (Washington, DC: Brassey's, 1993), 6–10.

the drug business. Noriega ignored both messages and charged the United States with once again interfering in Panama's internal affairs. By 1987, even Noriega's former chief of staff, Colonel Roberto Diaz Herrera, accused Noriega of having "manipulated the 1984 election"; having ordered "the grisly torture and murder" of one of his well-known critics, Dr. Hugo Spadafora; and even possibly having "had a part in the plane crash that killed Torrijos." Additionally, Herrera admitted that "he and other Panamanian military officers had become wealthy extracting exorbitant fees from Cubans desiring Panamanian visas." That sum-

Figure 56. U.S. Marshals Service mugshot of Panamanian dictator Manuel A. Noriega

Source: U.S. Marshals Service.

mer, Noriega routinely violated the treaties created in 1977 when he ordered the Panamanian Defense Forces (PDF) to regularly "stop and check American school buses and military dependent's automobiles."[66] This policy resulted in a rising number of ugly incidents between the PDF, U.S. nationals, and U.S. military personnel and their dependents.

By this time, Noriega's illicit activities conflicted with U.S. drug policies, creating a contradictory situation in his nation. He had become deeply involved in money laundering, and additional intelligence sources revealed that he was still heavily involved in the drug trade in Colombia and beginning to turn Panama into a major narcotics transshipment point. Despite his active involvement in the drug trade, Noriega willingly assisted the Reagan administration in its crackdown on these damaging activities when it served his purposes. In addition to Noriega working both sides, he was covertly engaging in trafficking intelligence and

---

[66] Flanagan, *Battle for Panama*, 9–10.

even Panamanian passports to Cuba for its intelligence services to use. However, he still retained the support of the PDF. As further insurance, he created a paramilitary units called "dignity battalions" to assist him with remaining in power. These Noriega loyalists had the barest of military training and discipline, but they served a purpose for the increasingly isolated Noriega.[67]

Starting in 1903, the U.S. Army and Marine Corps had long maintained a substantial presence in Panama. The following year, Major John A. Lejeune established the first Marine barracks in Panama City. At the height of World War II, the Marine Corps had nearly a regiment of Marines assigned to security duties there. More than 40 years later, the Marine force in Panama became officially known as Marine Corps Security Force (MCSF) Panama. MCSF Panama was predominately located at the Rodman Naval Station, the largest of its kind in Panama and located close to the Pacific Ocean entrance to the Panama Canal. The Marines had the primary mission of protecting "local naval installations (especially Rodman)," while the Army units had the role of guarding "the canal itself."[68]

By 1988, heightened political tension inside Panama caused planners in U.S. Southern Command (USSOUTHCOM) to initiate the earliest phases of a larger operations plan called Elaborate Maze. The fully fleshed out Elaborate Maze ultimately arranged for the 6th Marine Expeditionary Brigade (6th MEB) from Camp Lejeune, alongside substantial Army and Air Force reinforcements, to phase in forces to the Canal Zone until getting the entire MEB ashore. The Joint Chiefs of Staff, however, were not ready for such a force buildup in Panama, consequently causing USSOUTHCOM to scale it back. Since the situation started primarily as a security mission, the initial Marine Corps contingent that reinforced the Marines at Rodman "was a platoon from Fleet Anti-Terrorist Security Team (FAST) Company, Marine Corps Security Force Battalion, Atlan-

---

[67] Flanagan, *Battle for Panama*, 8–12; and William R. Doerner, "Lead-Pipe Politics: As Noriega Bloodies His Opposition, the U.S. Sends in More Troops," *Time*, 22 May 1989.
[68] LtCol Nicholas E. Reynolds, USMCR, *Just Cause: Marine Operations in Panama, 1988–1990* (Washington, DC: Marine Corps History and Museums Division, 1996), 1.

tic, based in Norfolk, Virginia."[69] Commanded by Major Eddie A. Keith, the FAST contingent was placed under the control of MCSF Panama and first assigned to guard the Arraijan fuel tank farm located on the far edge of Rodman Naval Station. The situation for the FAST platoon was quite complicated. The tank farm was unfenced, surrounded by thick jungle, and was close to the well-traveled Inter-American Highway. Despite the hesitation from the Joint Chiefs of Staff, by 1 April 1988, the Pentagon "formally announced that 1,300 additional U.S. troops would be sent to Panama, including 300 U.S. Marines."[70]

Even with the increased deployment, the U.S. forces' mission remained to ensure the security of Canal Zone assets. The 6th MEB planned for the reinforcements to deploy as a Marine air-ground task force, but the commanding general of USSOUTHCOM, Army general Frederick F. Woerner Jr., overruled this idea because he deemed the Marine Corps aviation component as unnecessary. A reinforced rifle company—Captain Joseph P. Valore's Company I, 3d Battalion, 4th Marine Regiment—from the 2d Marine Division soon arrived in Panama to help with the operation, including securing the fuel tank farm. These Marines had already been through substantial predeployment training, making them the most prepared for the operation. MCSF Panama ultimately fell under the control of Joint Task Force (JTF) Panama, which was under the command of Army major general Bernard Loeffke. Unlike what took place during Operation Urgent Fury in Grenada in 1983, largely thanks to new command relationships emerging from the Goldwater-Nichols Act, USSOUTHCOM's early establishment of JTF Panama enabled a single commander to coordinate the activities of a wide variety of Joint forces in the region.

Shortly after arriving, Company I faced some trouble within JTF Panama. After the Marines took up their duties of patrolling the fuel tank farm, they believed they had discovered some indications that some force, most likely the PDF, was attempting to infiltrate their position, which led to a tragic incident for the unit. When a flare that did not de-

---

[69] Reynolds, *Just Cause*, 2. Reynolds stated that the FAST platoon was a kind of military SWAT team.

[70] Reynolds, *Just Cause*, 2, 5.

ploy properly was mistaken for a gunshot on 11 April, one element of a Marine patrol fired into another, killing the patrol leader Corporal Ricardo M. Villahermosa. The incident resulted in USSOUTHCOM conducting a field investigation that concluded that the Marines had overreacted to an illusory PDF threat.[71]

However, the unit quickly proved its worth in a scenario the following night. Shortly before 1930 on 12 April, Company I, while employing ground sensors on loan from the Army, picked up the movement of approximately 40 individuals moving toward the tank farm. After receiving additional reports of infiltrators, Valore moved forward to investigate and reported seeing about 12 unknown individuals moving "between the highway and the tank farm." A short firefight ensued, ending in a matter of minutes. Amazingly, Major General Loeffke arrived on scene in civilian clothes around 2200 that evening and assured Valore that no PDF troops were in the vicinity, despite later evidence disputing his statement. Loeffke quickly ordered Valore to stand down and stop all military activity. The next morning, Marines inspected the area. Although they found a bit of debris from the firefight, including "fresh foreign-made battle dressings," they found no bodies or even enemy shell casings. Consequently, the Naval Criminal Investigative Service questioned Valore and his Marines, and those "who fought in the jungle that night" received an order from the Army leadership "to submit to a urinalysis [drug test], which they did, with negative results." The reaction of the Army high command, including the public affairs officer for JTF Panama, frustrated the Marines because they seemed to believe that the Marines had been firing at shadows. The leaderships' opinion changed two days later when elements of the Army's 3d Battalion, 7th Special Forces Group, "operating west of Howard Air Force Base allegedly engaged another group of well-disciplined intruders."[72]

Although the Marines had clearly been in a firefight at the tank farm on 12 April 1988, some of the details remained unclear in its aftermath. Military historian and Army veteran Lieutenant General Edward M. Fla-

---

[71] Reynolds, *Just Cause*, 7–8.
[72] Reynolds, *Just Cause*, 8–9.

nagan Jr. wrote that some intelligence sources indicated that "members of the Cuban Spetsnaz Special Forces—Cuban commandos—attacked the Marine outposts around the tank farm." Flanagan noted that the assault likely originated from the adjacent estate of "Cuban leftist" and Panamanian National Assembly member Rigoberto Paredes.[73]

Meanwhile, things in the United States grew increasingly worse for Noriega. An American drug smuggler named Stephen M. Kalish testified before a Senate committee that "in exchange for help [with] his own drug business in Panama, he had bribed Noriega with millions of dollars." Moreover, federal grand juries returned indictments against Noriega, charging him with a number of drug and money laundering offenses.[74]

Noriega also planned to run his candidate, Carlos Duque, for Panamanian president in the national elections the following spring—an event that would surely draw a substantial amount of international scrutiny. By April 1989, former President Carter, now acting as a U.S. election observer, announced even before he had departed Panama that Noriega "had stolen the elections by fraud." The Catholic Church, a strongly influential organization in Panama, declared that the anti-Noriega candidate, Guillermo Endara, won the election with 74 percent of the vote. Noriega annulled the results and had his dignity battalion thugs violently attack a motorcade of the opposition candidates. These actions increased local Panamanians' anger toward Noriega. Many Panamanians also became increasingly enraged at the United States because they viewed Noriega as an U.S. creation. By mid-May 1989, due to the election fraud, President George H. W. Bush recalled U.S. ambassador Arthur H. Davis Jr. and ordered an additional 2,000 troops into Panama "within the following two to three weeks to protect American citizens and property." Bush strove to isolate Noriega diplomatically as well, successfully getting most of the nearby Central American states, except for Cuba and Nicaragua, to "strongly condemn" the Panamanian leader and his fraudulent government.[75]

---

[73] Flanagan, *Battle for Panama*, 18.
[74] Flanagan, *Battle for Panama*, 11.
[75] Flanagan, *Battle for Panama*, 15–16.

At the same time, operational contingency planning for Panama shifted from USSOUTHCOM, headquartered within the Canal Zone, to the XVIII Airborne Corps at Fort Bragg, North Carolina. The looming contingency operation now seemed to call for the establishment of a robust corps operational headquarters under the overall control of USSOUTHCOM to ensure adequate command and control of functional elements assigned to the mission. The expeditionary minded XVIII Airborne Corps, under the command of none other than Beirut veteran Lieutenant General Carl W. Stiner, nicely fit the operation's need. Yet, USSOUTHCOM commander, General Woerner, soon found himself in trouble after making some critical comments about U.S. policy in Panama. By July 1989, Secretary of Defense Cheney announced that Woerner was going to retire and that Army general Maxwell R. Thurman would replace him.[76]

Soon after Thurman's appointment, a group of Panamanian officers staged yet another military coup against Noriega on 3 October 1989. This effort marked the second coup attempt against him in 18 months. This time, a former Noriega loyalist, Major Moises Giroldi, led the challenge. Giroldi's coup was initially successful, and the rebels even placed Noriega under arrest briefly. Yet, things began to fall apart only a few hours later. Somehow, Noriega's friends had gotten the word about his arrest—allegedly because someone had foolishly allowed Noriega to make a courtesy phone call to his mistress. This leak led to the arrival of PDF Battalion 2000, an elite unit commanded by Noriega loyalist Major Francisco Olechea, which quickly crushed the revolt. Olechea had earlier pledged to not interfere with the coup but apparently changed his mind. Soon government loyalists surrounded the PDF headquarters building, called La Comandancia, where Major Giroldi was holding Noriega at gunpoint. Having been apprised that his friends were on the verge of liberating him, Noriega supposedly screamed at Giroldi, "To be a commander you have to have balls. You don't have balls."[77] Giroldi lost his

---

[76] Lawrence A. Yates, *The U.S. Military Intervention in Panama: Operation Just Cause, December 1989–January 1990* (Washington, DC: U.S. Army Center of Military History, 2013), 27–28, 31–32.

[77] Flanagan, *Battle for Panama*, 25.

nerve and allegedly surrendered to Noriega. He was later taken away, tortured, and executed, along with nine other rebel leaders.

By that December, the situation between Panama and the United States had deteriorated rapidly. On 15 December, the Panamanian General Assembly, at Noriega's direct behest, declared a state of war between Panama and the United States. The next night, the PDF stopped a civilian-owned vehicle carrying four unarmed U.S. military officers at one of its checkpoints. A violent confrontation quickly ensued. With armed soldiers closing in on their vehicle, the driver attempted to speed away. PDF troops then opened fire, which killed Marine Corps first lieutenant Robert Paz. In the aftermath, the PDF detained a junior U.S Navy officer and his wife, both of whom had witnessed the checkpoint incident. Instead of questioning the couple, however, the PDF beat the naval officer so severely that he was hospitalized for two weeks. They proceeded to slam his wife against a wall while out of control PDF soldiers luridly threatened her with sexual assault. All these incidents convinced President Bush that Noriega could no longer remain in power. Bush ordered Powell to initiate Operation Just Cause, which began during the early morning hours of 20 December 1989.[78]

Concerns for the thousands of American citizens still in Panama—far too many to individually identify—led the U.S. plan to speed up from the initial idea of a time-phased buildup of forces. Instead, U.S. leadership rewrote the original operations plan to allow a more rapid buildup of Joint forces, now known as Operation Blue Spoon, to achieve the objective of quickly taking down Noriega and the PDF. Immediately, Powell and Thurman informed Lieutenant General Stiner that he would have the task of completing the objectives of Operation Blue Spoon in a matter of days, rather than the originally envisioned three weeks. Once the U.S. forces landed, the ground units would make a significant effort to capture Noriega as soon as possible. Most importantly, the JTF's primary objective was the "disarming and dismantling of the Panama Defense

---

[78] Ronald H. Cole, *Operation Just Cause: The Planning and Execution of Joint Operations in Panama, February 1988–1990* (Washington, DC: Joint History Office, Office of the Chairman of the Joint Chiefs of Staff, 1995), 14, 27–28.

Force." In Powell's words, "if you're going to get tarred with a brush, you might as well take down the whole PDF . . . pull it up by the roots."[79]

The Marine Corps played a limited role in Operation Just Cause. Soon after the May 1989 Panamanian elections, General Woerner had consented to the addition of Company A, 2d Light Amphibious Infantry Battalion (2d LAIB), which was equipped with LAV-25 amphibious armored infantry fighting vehicles. Prior to Operation Blue Spoon and Operation Just Cause, the company received orders to conduct a series of security operations dubbed Sand Fleas. Meant as a tool to confuse the PDF, the ubiquitous LAVs "seemed to be everywhere at once." According to Marine historian Nicholas E. Reynolds, "Participation in this series of operations signaled a change for Marine Forces Panama." Having previously been "largely confined to the boundaries of Rodham and neighboring military installations," the arrival of the LAVs converted the Marines "from a reinforced security force into a maneuver force."[80]

Following the coup attempt in October 1989, Stiner, rightly believing that the potential deployment of U.S. forces to Panama might happen sooner than expected, had his operations staff rapidly rework the original Operation Blue Spoon plan into the plan for Operation Just Cause. All the Marines forces, including those already on the ground with MCSF Panama, were grouped into Task Force Semper Fi, placed under the command of Colonel Charles E. Richardson. Richardson had arrived in Panama in September 1989 and quickly became concerned with what he perceived as the original plans for Marine Corps participation in potential combat operations. He believed that the original concepts called for the inclusion of the entire 6th MEB, which was not available to him. Instead, he only had Company K, 3d Battalion, 6th Marines; Company D, 2d LAIB, 2d Marine Division; a FAST platoon; and the MCSF Panama detachment at Rodman Naval Station on the ground. Moreover, the number of missions for his forces had increased, with the most notable addition being providing security for Howard Air Force Base, the projected hub for all

---

[79] Cole, *Operation Just Cause*, 14.
[80] Reynolds, *Just Cause*, 14–15; and Col Robert P. Mauskapf and Maj Earl W. Powers, "LAVs in Action," *Marine Corps Gazette* 74, no. 9 (September 1990): 51.

inbound logistics and transportation for Operation Just Cause. Consequently, Richardson requested bringing in more Marines. Stiner steadfastly refused the appeal but promised him a battalion from the U.S. Army's 7th (Light) Infantry Division "within two days after the beginning of hostilities." He also arranged for Richardson to take command of an Army military police company and the 536th Engineer Battalion "at or before H-Hour."[81]

The LAI's Sand Fleas operations were fortuitous for Richardson. After having campaigned along the canal's west bank, his forces now had an excellent working knowledge of the area, including "its road networks, its towns and cities, as well as the people who lived in them." This knowledge included "the location of ranking political and administrative officials in the Noriega regime who might have to be apprehended in the event of hostilities."[82] This invaluable intelligence provided Task Force Semper Fi with the ability to prepare for combat in a highly effective fashion.

Other pre–Just Cause operations assisted in preparing the Marines for the larger campaign. Two months before the operation's initiation, the Marines sent 12 LAV crews on "an excursion into several populated neighborhoods west of Howard Air Force Base," known as Operation Rough Rider, which had "turned confrontational" before its conclusion.[83] The idea behind the expedition was for the Marines of Company D, 2d LAIB to familiarize themselves with the densely populated areas near both Howard Air Force Base and Rodman Naval Station. As the Marines neared the town of Vista Alegre, the LAVs ran into a PDF roadblock. Company D's commander, Captain Gerald H. Gaskins, received permission to crash through the barrier as well as a second one that had been set up on the route back to Rodman. After convincing a local farmer to remove his pickup truck from what happened to be yet another roadblock, it was clear that these episodes and others like them provided the Marines with special operating procedures for "crowd control and the art

---

[81] Yates, *The U.S. Military Intervention in Panama*, 142.
[82] Yates, *The U.S. Military Intervention in Panama*, 143.
[83] Yates, *The U.S. Military Intervention in Panama*, 142–43.

Figure 57. Marine Corps LAV-25 on patrol near the Panama Canal

Source: National Archives and Records Administration.

of removing roadblocks." Due to his acute combat arms troop shortage and an ever-expanding mission list, Richardson, in keeping with the famous leatherneck mantra that "every Marine is a rifleman," organized his support and service troops for possible use in a direct combat role.[84]

In the days before Operation Just Cause, Richardson learned from Stiner and the other Joint leaders the plans they had for his Marines. Yet, Stiner requested such tight secrecy that Richardson could not inform his subordinate commanders about the pending combat operations until nearly H-hour, which Stiner designated as 0100 on 20 December 1989. Task Force Semper Fi was tasked with both defensive and offensive mis-

---

[84] Yates, *The U.S. Military Intervention in Panama*, 142–43.

sions. Defensively, besides providing security for Howard Air Force Base and Fort Kobbe, Richardson's unit had the duty of maintaining perimeter security for the Rodman Naval Station, which included the Arraijan fuel farm and the base ammunition dump, as well as establishing defensive positions at the Bridge of the Americas that spanned the Pacific entrance to the canal at Balboa and "other transit points on the west bank." Offensively, the Marines were ordered to seize multiple Panamanian government locations, including the directorate of traffic and transportation station 2 (DNTT 2), multiple PDF stations and substations, "a Dignity Battalion training facility," a port facility and radio tower, and the Noriega "regime's political headquarters in Arraijan." Essentially, the Marines were supposed to "block the western approach to Panama City to prevent Panama Defense Forces . . . reinforcement."[85] Despite these multiple roles, Stiner, in a move that stunned Marine Corps liaison officers, reassigned Richardson's promised military police company to the Joint Special Operations Task Force (JSOTF) under Army major general Wayne A. Downing. Nevertheless, Richardson seemed to take this last-minute change in stride.

Approximately 10 minutes before H-hour, the Marines were forced to begin their part of the operation. Shortly after midnight on 20 December, U.S. forces spotted multiple PDF V300 armored vehicles near Rodman, which they believed were "en route to Panama City."[86] Without hesitation, Richardson ordered two LAV platoons with and heavily armed counterterrorism Marines embarked to proceed to their blocking position at Arraijan, which sent them directly past the key objective of DNTT 2. As the LAVs approached the building, they came under small arms fire. Under orders that stressed that all Just Cause units should minimize the use of force when and where possible, a squad of eight Marines moved up and breached the building. Fighting at close quarters with the PDF inside, the squad cleared the building room by room. Corporal Garreth C. Isaak breached the final room and was hit by several enemy rounds. He later succumbed to his wounds. After the LAV company secured the build-

---

[85] Yates, *The U.S. Military Intervention in Panama*, 144.
[86] Yates, *The U.S. Military Intervention in Panama*, 145.

Figure 58. U.S. Army map of Operation Just Cause

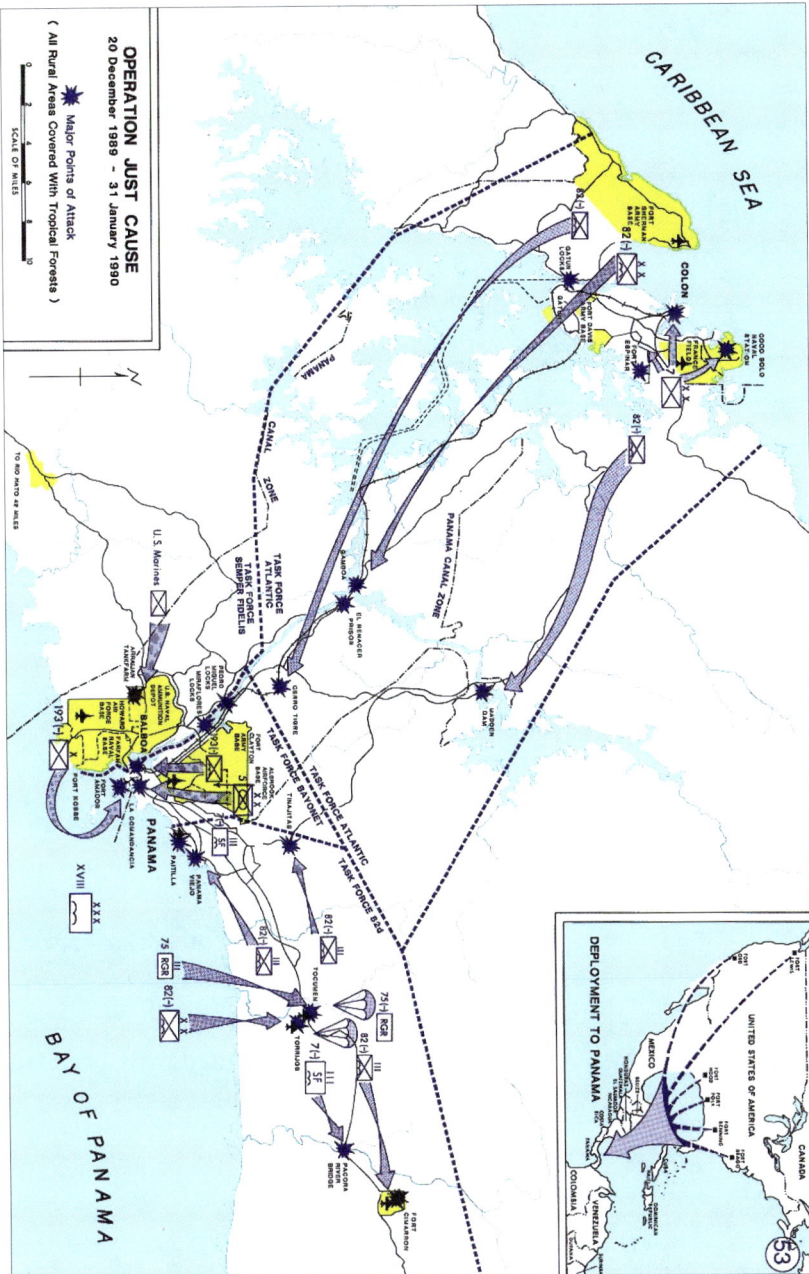

ing and grounds, Richardson moved up his makeshift support troops to control DNTT 2 and the adjacent vicinity while the LAV company continued to Arraijan. Despite the multiple actions of the Marines, Corporal Isaak would be the only Marine killed during Operation Just Cause.[87]

The other U.S. Services faced far worse experiences. Navy Sea, Air, and Land (SEAL) Team 4 was assigned to capture the Paitilla airfield, which housed Noriega's private Learjet. During the early hours of Operation Just Cause, some U.S. leaders had concerns that Noriega might try to escape to Cuba or Nicaragua in the jet. U.S. Navy lieutenant commander Patrick Toohey personally led the ground assault force to seize the Paitilla airfield and Noriega's Learjet. In the initial hours of the operation, SEAL Team 4 used rubber boats with muffled engines to proceed from Rodman to Paitilla. The team did not see the mission as especially difficult, believing they could simply go disable Noriega's jet and then block the runway with vehicles and damaged airplanes to ensure that the PDF could not use the field. Consequently, they did not carry anything more substantial than their small arms, some shoulder-fired antitank weapons, and a mortar.[88]

The SEALs got ashore undetected and on schedule at 1230, 30 minutes before H-hour. Around the same time, Toohey received a radio message that Noriega might be on his way to Paitilla in a helicopter. With that information, the SEALs needed to get to the hangar quickly. A second message soon came from Stiner's headquarters that "reinforced the rules of engagement" meaning that all the forces were to "minimize both collateral damage and casualties." This headquarters broadcast created significant confusion, as it had not been for the SEALs, who "interpreted the message as a change in mission." The order was actually directed toward a unit of Army Rangers who were "confronted with a civilian airliner on the ground at Torrijos International Airport and a possible hostage situation."[89]

---

[87] Yates, *The U.S. Military Intervention in Panama*, 148.
[88] Thomas Donnelly, Margaret Roth, and Caleb Baker, *Operation Just Cause: The Storming of Panama* (New York: Lexington Books, 1991), 114–15.
[89] Donnelly, Roth, and Baker, *Operation Just Cause*, 115–16.

In addition to the confusion, Toohey was informed that three PDF armored cars with 90-millimeter guns were moving toward the north end of the airfield. With only light weapons at their disposal, the SEALs needed to conduct a successful road ambush to stop the vehicles. Toohey ordered the SEAL element on the north side of the airfield to move to the road and set up the position quickly. Their movement took them directly past the airfield hangars, which caught the attention of a PDF soldier inside the hangar housing Noriega's jet. The PDF soldier unleashed a burst of automatic rifle fire that instantly killed two SEALs and wounded six more. A supporting platoon lost another two SEALs killed and two wounded. In less than five minutes, the SEALs had lost Lieutenant Junior Grade John P. Connors, Chief Engineman L. Donald McFaul, Torpedoman's Mate Second Class Isaac G. Rodriguez III, and Boatswain's Mate First Class Christopher T. Tilghman killed in action and eight total wounded.[90] The Panamanian vehicles continued unimpeded into Panama City. To make matters worse, Toohey could not get an immediate medevac helicopter response, most likely due to the enormous air traffic jam at Howard Air Force Base. Although it was a bloody day for the SEALs, they accomplished their goal of disabling Noriega's jet and blocking the runway. Still, of the numerous actions during Operation Just Cause, "the Paitilla mission may be the most controversial among the military."[91] After all, if the mission's purpose was to simply destroy the Learjet and disable the runway, a precision airstrike could have accomplished this mission while freeing these valuable special forces for

---

[90] Donnelly, Roth, and Baker, *Operation Just Cause*, 117; and Yates, *U.S. Military Intervention in Panama*, 91.

[91] Donnelly, Roth, and Baker, *Operation Just Cause*, 117–19. One of the complicating factors noted by the authors was that the SEALs could not communicate with their Lockheed AC-130 gunship support, which might have taken out the hangars and the armored car threat but would have been less precise if used. Others argued that the SEALs were too close to the hangars for the gunships to be used in the first place and, in a worst-case scenario, may have resulted in friendly fire casualties. While the SEALs have long alleged that the second radio message from Stiner's headquarters about minimizing the use of force was a change in mission for them, all the Joint Task Force commanders stated that this was not the case. In hindsight, it remains difficult to see why MajGen Downing, commanding all special forces during Operation Just Cause, selected the SEALs for this mission. Donnelly, Roth, and Baker, *Operation Just Cause*, 119–20.

another mission that made more sense. Clearly, the blending of special operations into the overarching Joint operations plan was an issue that still needed improvement.

In other areas, the Marines received new assignments that expanded their role in Operation Just Cause. Throughout the campaign, the Marines of Task Force Semper Fi roved the roads between Rodman and Arraijan. After accomplishing its H-hour missions on the morning of 20 December, Task Force Semper Fi was unexpectedly ordered to secure the PDF's 10th Military District Headquarters in La Chorrera—a moderate-size city of 80,000 residents. As the unit began its approach that afternoon, a Marine Corps aerial observer, having seen the strength of the PDF position in La Chorrera, requested an airstrike against the PDF compound. Amazingly, the Joint force air component commander (JFACC) initially denied the request due to a legitimate fear of collateral damage including civilian casualties. The JFACC eventually relented and authorized two Virginia Air National Guard LTV A-7 Corsair II attack aircraft to provide close air support, strafing the compound with their 20-millimeter cannons. Their fire was of such "pinpoint accuracy" that the U.S. forces later discovered that "none of the civilian homes near the headquarters had been damaged."[92]

The airstrike had achieved its objective, weakening the defenses around the PDF headquarters. Marine LAVs immediately followed up, crashing into the compound. Except for scattered sniper fire, most of the PDF defenders had fled the area. Due to persistent combat manpower shortage, Richardson and Task Force Semper Fi struggled to keep the PDF from infiltrating back into parts of La Chorrera. Yet, with their roadblock at the Bridge of the Americas, Task Force Semper Fi had indeed achieved its mission of keeping PDF reinforcements from crossing

---

[92] Yates, *The U.S. Military Intervention in Panama*, 150–53. The author went on to describe some behind-the-scenes drama concerning the A-7 airstrike. Apparently, despite having the airstrike request approved by Stiner, the acting JFACC, BGen Robin G. Tornow, USAF, continued to block the mission until he was forcefully confronted by the XVIII Airborne headquarters' Marine Corps liaison officer, who informed him that "the mission was essential and that Marine lives were at stake and that he had better approve it because LtGen Stiner had." Only then did Tornow relent. Even so, it still took more than an hour for the airstrike to be conducted. See Yates, *The U.S. Military Intervention in Panama*, 151.

over to assist their hard-pressed units on the east bank in Panama City. The Marines of Task Force Semper Fi, to include MCSF Panama, had all performed extremely well.[93] Still, Richardson and his subordinates probably wondered what the task force could have accomplished on the west bank had the entire 6th MEB, including its accompanying organic air component, been in place at H-hour.

True to his word, Stiner added the 2d Battalion, 27th Regiment, 7th (Light) Infantry Division, to Task Force Semper Fi by 22 December. Richardson immediately vectored the Army unit to occupy La Chorrera, which his Marines were physically unable to do after having captured the 10th PDF district headquarters. The 7th Infantry Division soldiers swept through the area. They drove any remaining belligerents toward Task Force Semper Fi Marines in blocking positions on the eastern side of the city, resulting in the capture of a considerable number of demoralized PDF soldiers. The captured troops "absorbed a great deal of time and effort," however. By the end of Operation Just Cause, the Marines, despite accounting for approximately 4 percent of the U.S. Joint task force, "had captured some 1,320 Defense Force and Dignity Battalion members," approximately 25 percent of total prisoners that the U.S. Joint task force detained.[94]

By 23 December, the Marines around Howard Air Force Base had secured the Arraijan fuel farm from the PDF. With little possibility of a PDF threat, Major Robert B. Neller, the commander of the Marine guard there and future 37th Commandant of the Marine Corps, asked Richardson to reassign his now largely unoccupied force to an upcoming mission to take out a nearby seaport. Called Vaca Monte, the port was supposedly a base for an antiterrorist PDF unit. Richardson concurred with Neller's request and, to Neller's amazement, put him in charge of the entire operation, codenamed Task Force Bull Dog. Neller was given a company of U.S. Army soldiers from the 27th Regiment, 7th Infantry Division; a company from the 2d LAIB with its FAST Marines; and a detachment of security forces. The Navy screened the task force from the sea and sent in

---

[93] Yates, *The U.S. Military Intervention in Panama*, 150–53.
[94] Reynolds, *Just Cause*, 29.

a SEAL unit to support the operation. That afternoon, the Marine units, approaching from the west, rendezvoused with the Army personnel, approaching from the east, before moving toward Vaca Monte. When they arrived, however, Neller and his task force faced no resistance. Instead, they ran into approximately 400 heavily laden looters attempting to leave, which included numerous "foreign seamen" from Cuba, Germany, Peru, and Nicaragua, among others, who had been docked at Vaca Monte. By midnight, the Task Force Bull Dog had released "all the foreigners" reducing the "number of detainees . . . to 100 Panamanians, 13 of them regular Panamanian soldiers assigned to Vaca Monte, who were held for further processing." Although Neller and his task force searched for the PDF command, Noreiga's troops had already left the area. Later that same day, MCSF Panama was "placed under the operational control of the 7th Infantry Division," remaining in that arrangement until the beginning of February 1990.[95]

The Marines had achieved their mission of keeping PDF forces from crossing over the canal into Panama City. This task, however, did not require them to confront substantial bodies of PDF soldiers. With the Army dominating the planning for Operation Just Cause, most of the U.S. forces engaged in the campaign came from that Service. The soldiers, especially those in the 7th Infantry Division, had performed exceedingly well on short notice. Throughout the operation, the Army planners clearly were not keen on any increased use of Marines. Their reluctance kept most of the 6th MEB out of the campaign, something that hindered the Marines while they had responsibility for nearly all operations on the west bank of the canal. Richardson certainly could have used the additional manpower at the time. This issue ultimately became a moot point when the operation transitioned into a more benign nation-building operation called Promote Liberty in January 1990.

This transformation from military operation to nation-building undertaking emerged due to events that occurred only two weeks into the campaign. During Operation Just Cause, Noriega avoided detection un-

---

[95] Reynolds, *Just Cause*, 30.

til he arrived at the Apostolic Nunciature, the diplomatic quarter of the Holy See, on 24 December 1989, where he was briefly given sanctuary. For the next nine days, the Army conducted some well-publicized psychological operations experiments to force the dictator to surrender. For instance, the Army set up large speakers and blasted loud popular music into the Nunciature 24 hours a day. President Bush personally ordered the Army to stop the practice after receiving a Vatican request to end it, but they continued to bother Noriega as much as possible. When Noriega emerged from the Nunciature, dressed in a wrinkled military uniform and with his ears likely ringing, he was arrested and flown to Miami, Florida, for interrogation, trial, and eventual incarceration. By this time, most Panamanians celebrated his downfall. During Noriega's refuge, a set of Marines conducted house searches on Christmas Day for high-value PDF fugitives. That unit had the honor of capturing Rigoberto Paredes in Nuevo Arraijan. When the Marines arrived in town, "local citizens were literally chasing Paredes down the street."[96]

Operation Just Cause illustrated the success of the congressional efforts to reform Joint warfighting with the implementation of the Goldwater-Nichols Act. As noted by Joint historian Ronald H. Cole, the operation circumvented the mistakes from Lebanon and Grenada in many ways. A combination of President Bush's determination and the augmented authority of both General Powell, as chairman of the Joint Chiefs of Staff, and General Thurman, as commander of USSOUTHCOM, allowed for "specific, readily attainable objectives and responsive and effective command and control while giving the tactical commander considerable operational freedom."[97] Nevertheless, Powell interjected in a few nontactical situations, such as when U.S. Army troops detained the Cuban ambassador and illegally searched the Nicaraguan embassy. Powell's "ready access to both" Secretary of Defense Cheney and President Bush, as well as his ability to directly communicate with "other federal agencies such as the U.S. State Department," enabled him to provide

---

[96] Reynolds, *Just Cause*, 32.
[97] Cole, *Operation Just Cause*, 3.

timely "politico-military guidance to the operational commanders."[98] Powell began to play a more direct role after issues of political sensitivity began to take precedence over strictly military affairs.

In an interview with the *Army Times* on 26 February 1990, Stiner noted how the training related to new Joint efforts had improved the forces' ability to operate. He stated, "Our training program paid off in spades in Panama and that's the reason you saw the discipline, the efficiency, the effectiveness and the proficiency that was demonstrated by our troops." Additionally, he declared that Operation Just Cause was "a joint operation in every sense of the word," with each Service contributing a "unique and important" capability that was "needed to perform this mission." He called the cooperation between the Services "absolutely outstanding," and pointed to numerous factors that assisted in their success. "First of all," he reported, "we received clear guidance from the national command authority level of what was expected. Secondly, we were allowed to prepare a plan in detail to accomplish that. Third, we briefed that plan all the way up through the decision making authority, and that plan was approved. Fourth, we were allowed sufficient time to conduct detailed rehearsals for its execution. And fifth, when conditions dictated that it should be executed, we were allowed to execute it without changes to the plan." All these elements, he stated, were "very germane in the outcome of Operation Just Cause."[99]

The political and military leadership of the United States did not have much time to absorb the lessons from Operation Just Cause. On 2 August 1990, Iraqi dictator Saddam Hussein surprised many Western observers when the armed forces of Iraq brutally attacked the neighboring state of Kuwait. In just two days, Iraqi forces conquered and occupied the small, oil-rich emirate. Hussein did not, however, anticipate the strong adverse reaction of the United States and the international community at large when he claimed Kuwait as Iraq's 19th province. Shortly after his invasion, the UN Security Council convened an emergency session and con-

---

[98] Cole, *Operation Just Cause*, 74.
[99] Cole, *Operation Just Cause*, 71.

demned Hussein's actions. It was soon apparent that Hussein had few friends there or anywhere else in the world.[100]

For years, the Iraqis had been dissatisfied with Kuwait's high oil production levels and the slant drilling practices of its petroleum industry, especially related to the sizable Rumaila oil field that ran beneath both countries. Nearly one-third of all 615 Iraqi national oil drills worked Rumaila alone. While oil fields crossed political boundaries in many situations, the affected parties usually had a preset agreement in place for revenue sharing and production levels. In 1990, no such agreement existed between Iraq and Kuwait. That May, the League of Arab States held an emergency meeting, during which Hussein announced that due to issues related to the Rumaila field, Kuwait now owed Iraq $2.4 billion (USD). At the time of his invasion, Hussein was also deeply in debt to Kuwait, owing the emirate $14 billion (USD) that he had borrowed to fund the costly Iran-Iraq War (1980–88). Yet, he also claimed during the meeting that excessive Kuwaiti oil production was costing his nation $14 billion (USD) a year, "far more than Kuwait had lent Iraq during Iraq's confrontation with Iran," and warned that the league could not "tolerate this type of economic warfare." Others argued that Hussein's claims acted as a "smokescreen" for his real desire to plunder Kuwait's wealth and bring the productive Rumaila oil field under his sole control.[101]

Although this study does not include a detailed analysis of Marine forces during Operation Desert Shield and Operation Desert Storm, some of the more salient aspects of the Service's participation in the largest deployment of U.S. Joint forces since the Vietnam War demonstrates the further incorporation of maneuver warfare.[102] Indeed, many observers of these campaigns saw Desert Storm as the successful culmination of the 1980s maneuver warfare debate, which it was. Despite this fact, it be-

---

[100] "The 1991 Iraq War—The Battle at the UN," Association for Diplomatic Studies and Training, 20 October 2015.

[101] Thomas C. Hayes, "Confrontation in the Gulf; The Oilfield Lying below the Iraq-Kuwait Dispute," *New York Times*, 3 September 1990.

[102] The most definitive operational history to date is Paul W. Westermeyer, *U.S. Marines in the Gulf War, 1990–1991: Liberating Kuwait* (Quantico, VA: Marine Corps History Division, 2014), hereafter *Liberating Kuwait*. It is an excellent reference on U.S. Marine Corps activity during the conflict.

came clear within the first moments of ground combat that most of the Iraqi Army was not motivated to fight, although U.S. leadership initially expected mass casualties due to Hussein maintaining the fourth-largest army in the world, and maneuvering against them did not prove particularly difficult. While the U.S. forces faced pockets of intense combat, the most nettlesome problems that emerged were over the handling of hordes of dispirited Iraqi prisoners of war and friendly fire issues.

The first Marines that the Iraqi invasion physically affected were the embassy guard forces in Baghdad, Iraq, and Kuwait City, Kuwait. Things did not go smoothly for the embassy guards in Kuwait City. Because the Iraqis were able to so rapidly overrun Kuwait, chaos ruled just outside the embassy gates, and by 4 August Iraqi troops sat just outside the embassy grounds. The Marines prepared to defend the embassy, but a garbled order on 7 August had them change into civilian clothes and directed the destruction of their small arms. Yet, three days later, the Marines managed to get back in uniform and even scrounged up a few weapons. Eventually, Hussein allowed the Marines and embassy personnel, as well as a number of U.S. civilians caught up in the Kuwait City maelstrom, to proceed in a convoy to the U.S. embassy in Baghdad. However, to score further political points, Hussein would not allow the Americans to leave Baghdad until December 1990.[103]

The UN Security Council later approved the United States using force, if necessary, to eject the Iraqis from Kuwait.[104] In January 1991, after much rancorous debate, Bush received congressional approval, garnering a surprisingly narrow margin in the Senate, to use force against the Iraqi forces. While numerous Democrats and a few Republicans in both the House of Representatives and the Senate opposed the measure primarily due to imagining yet another open-ended Vietnam scenario, Congress demonstrated little inclination to deny Bush's request. Imme-

---

[103] Westermeyer, *Liberating Kuwait*, 25–29.
[104] United Nations Security Council, Resolution 678, Authorizing Member States to Use All Necessary Means to Implement Security Council Resolution 600 (1990) and All Relevant Resolutions (29 November 1990).

diately, Bush ordered Powell to begin USCENTCOM preparations for offensive action in Kuwait.[105]

The commanding general of U.S. Marine Corps Forces Central Command (USMARCENT) was Lieutenant General Walter E. Boomer, who had previous experience as a Co-Van during the Vietnam War. As part of USCENTCOM, I MEF was also under Boomer's command. Boomer first heard about the Iraqi invasion of Kuwait on his way to his new duty station at Camp Pendleton, California. He recognized that the conflict likely meant that I MEF would find itself in the thick of things in the weeks to come.

One of the first problems for Marine Corps forces arriving in theater had to do with an issue of command and control. Thanks to the timely activity of the Diego Garcia-based MPS-2, the Marine Corps established "the first combined-arms task force in Saudi Arabia," complete with tanks, artillery, and even armored amphibious troop carriers. The 7th MEB, under the command of Major General John I. Hopkins, flew into Saudi Arabia, linking with its equipment by mid-August 1990. The immediate mission of the 7th MEB was to find a feasible position within "100 miles of the Kuwaiti border" from where it could defend northern Saudi Arabia from the Iraqis if their elements crossed the border. The problem was that if the Iraqis had decided to continue their offensive, the 7th MEB would have been forced into temporarily assuming total control over the entire Marine Corps effort, something that its smaller command and control element was not set up to handle. Consequently, Hopkins and other military leadership feared that the 7th MEB would have to maintain an entirely defensive stance before the I MEF forward command and control echelon arrived. The 7th MEB's missing air component especially concerned Hopkins, as his Marines would rely on air power to assist against any potential Iraqi invasion. Without the overall MPS being "tied together at the Joint Chiefs of Staff level," however, the 7th MEB's critical aviation assets had been delayed at Marine Corps air stations at Beaufort, South Carolina, and Cherry Point, North Carolina,

---

[105] Adam Clymer, "Confrontation in the Gulf; Congress Acts to Authorize War in Gulf; Margins Are 5 Votes in Senate, 67 in House," *New York Times*, 13 January 1991.

due to the Air Force not allocating the necessary tankers to ensure that the Marines' McDonnel Douglass F/A-18 Hornet fighter aircraft could "get across the Atlantic."[106] All this worry was ultimately for naught, as I MEF headquarters took control of all Marine Corps forces in Saudi Arabia by early September 1990.

Additional Marine Corps forces soon began to arrive. Many of them were already aboard Navy ships when they received orders to deploy to Saudi Arabia as soon as possible. The ubiquitous 4th MEB, now commanded by Major General Harry W. Jenkins Jr., had been preparing to take part in yet another Bold Guard training exercise with NATO forces in Europe when they were ordered to the Middle East on 10 August 1990. Jenkins quickly expressed concerns that the Navy seemed to have little interest in conducting amphibious operations in the Persian Gulf, especially the Navy's CENTCOM component commander, Vice Admiral Henry H. Mauz Jr., who apparently did little "in developing a naval campaign that went beyond the level of presence."[107] Still, during the early fall of 1990, I MEF and the Army's XVIII Airborne Corps provided most of the initial deterrent force against any further movements by Iraqi troops. By November 1990, nearly one-quarter of the active-duty Marine Corps—approximately 42,000 Marines, who accounted for "a fifth of the total U.S. force in Desert Shield"—had been deployed. Of this number, I MEF consisted of more than 31,000 Marines onshore. The other 11,000 Marines, with the 4th MEB and the 13th MEU Special Operations Capable

---

[106] Westermeyer, *Liberating Kuwait*, 37–40. Unless the MEBs maintained a standing headquarters, which many of them did not, the deployment of a robust MEF forward command and control element was usually necessary. Since an existing nearby MPS set allowed a MEB to arrive in theater rapidly, and far more quickly than a MEF forward element could deploy, it seemed that as the 31st Commandant, Gen Charles C. Krulak, once noted, MEBs were eminently "deployable" but were, at least initially, "not employable." Hopkins admitted that "one of the failures of the whole damn war was intelligence." As an example, Hopkins noted that intelligence believed that his MEB was likely going to run into the Iraqi Army's *80th Tank Brigade*, but as things turned out after the fighting began, this unit "wasn't in our sector after all. It had left Kuwait months before, and we didn't know it." See MajGen John I. Hopkins, "This Was No Drill," U.S. Naval Institute *Proceedings* 117, no. 11 (November 1991).
[107] Westermeyer, *Liberating Kuwait*, 44.

(SOC), remained afloat as a potential landing force.[108] More Marines from II MEF, based on the East Coast, were on the way as well. Commandant Gray wryly stated that in 1990 "there were four kinds of Marines: those in Saudi Arabia, those going to Saudi Arabia, those who want to go to Saudi Arabia, and those who don't want to go to Saudi Arabia but are going anyway."[109] To drive home this point, Gray even ordered Colonel Peter Pace, the then-commanding officer of Marine Barracks Washington, DC, who would become a future chairman of the Joint Chiefs of Staff, to deploy a company of his mostly ceremonial command to Saudi Arabia to act as a headquarters security force where needed.[110] While the DC-based Marines were not actually needed in Saudi Arabia, Gray still ordered their deployment to underscore that all Marines, regardless of their post and station, were ready for combat deployment on a moment's notice.

By mid-November 1990, the 2d Marine Division entered its first overseas combat deployment since World War II. The unit, then commanded by Vietnam-era Co-Van and Navy Cross recipient Major General William M. Keys, began to arrive and would continue to bring in forces along with the 2d Force Service Support Group (2d FSSG), then commanded by future 31st Commandant of the Marine Corps Brigadier General Charles C. Krulak. Because the 2d Marine Division had already supplied the 4th MEB and several other MEUs to USMARCENT, the division's main contribution to the buildup was the 6th and 8th Marine Infantry Regiments, the 10th Marine Artillery Regiment, the 2d Light

---

[108] BGen Edwin H. Simmons, USMC (Ret), "Getting Marines to the Gulf," in *U.S. Marines in the Gulf War, 1990–1991: Anthology and Annotated Bibliography*, comp. Maj Charles D. Melson, USMC (Ret), Evelyn A. Englander, and Capt David A. Dawson (Washington, DC: Marine Corps History and Museums Division, 1992), 13.

[109] Simmons, "Getting Marines to the Gulf," 14. The author checked in to the U.S. Naval War College at Newport, RI, as a prospective student in the Naval College of Command and Staff on 1 August 1990. The Iraqis invaded Kuwait the next day. Just a few weeks later, Gray came up to the college to address the faculty, students, and staff, as per the annual tradition for all the Service chiefs. Nearly every Marine in both the junior and senior courses urged the Commandant to allow them to immediately deploy to Saudi Arabia with Marine Corps units getting ready for Operation Desert Shield. From the author's memory of that moment, Gray essentially told everyone to "calm down—that the school was not closing—and that he knew where we were if we were needed." Nevertheless, it was not the response any of the Marines hoped to hear.

[110] Simmons, "Getting Marines to the Gulf," 14.

Armor Reconnaissance Battalion, the 2d Tank Battalion, combat engineers, and, of course, its logistical support from the 2d FSSG. The 2d Marine Division's deployment had a choppy quality in that it had battalions ashore, others floating just offshore, and was even "committed to providing two battalions to maintain Marine expeditionary units in the Mediterranean." Not counting the forces afloat, I MEF was "structured like the III Marine Amphibious Force in Vietnam: two divisions, a very large [air] wing, and a substantial service support command."[111]

One Marine Corps equipment concern became evident soon after the MPS began unloading at the port of al-Jubayl, Saudi Arabia. For years, the Chrysler Defense M60A1 Patton had served as the main battle tank for the Marine Corps. By the start of actual ground combat operations in February 1991, only the 2d Tank Battalion possessed the vastly superior General Dynamics Land Systems M1A1 Abrams main battle tanks also used by the U.S. Army. To offset an apparent armor imbalance for the infantry-centric Marines, which had only three active-duty tank battalions in the entire Service, General Schwarzkopf initially paired the British Army's 7th Armored Brigade, the famous "Desert Rats," with I MEF. Yet, USCENTCOM planners and the commanding general of all British forces, General Sir Peter de la Billière, decided at the last moment to detach the Desert Rats from the Marines and connected them with U.S. Army maneuver units on their planned broad western sweep of Hussein's forces dug-in on the Kuwaiti border. Billière also thought that such a move would allow him to better show off the British-built FV4030/4 Challenger 1 main battle tank. Instead, the U.S. Army's 1st "Tiger" Brigade, 2d Armored Division, called "Hell on Wheels," joined the Marine infantry units. This action was a fortuitous trade because, according to Boom-

---

[111] Simmons, "Getting Marines to the Gulf," 15; and Westermeyer, *Liberating Kuwait*, 60–61. Much of the 2d Marine Aircraft Wing was folded into the 3d Marine Aircraft Wing. Both FSSGs for I MEF and II MEF eventually made their way ashore, and they played a significant role in keeping the large Marine Corps presence on the ground well-supplied. BGen James A. Brabham commanded the 1st FSSG, and once the 2d FSSG arrived under BGen Krulak, the two commanders worked out a system in which Krulak would oversee the close-in direct support logistics effort while Brabham, who had extensive experience working with the Saudi government, performed in a general support role.

er, the U.S. soldiers "fell right in and did a terrific job" when the fighting began.[112]

To bring on the major land engagement he was certain his forces would ultimately win, Hussein ordered elements of his elite armored forces, including a large body of mechanized infantry, to attack Coalition forces in the northern Saudi Arabian town of al-Khafji, just across the Kuwaiti border, on 28 January 1991. The Iraqi assault force possessed a "combination of T-54/55 and T-62 main battle tanks while their mechanized infantry battalions were equipped with BMP-1 armored personnel carriers supported by BRDM scout vehicles."[113] Hussein absolutely believed that Americans had become so casualty adverse in the aftermath of the Vietnam War that they would choose to leave the region as soon as the bodies started to pile up. He thought that the political fallout from the United States losing 58,000 dead during the entire Vietnam War illustrated their unwillingness to sustain loses compared to his own nation, especially when Iraq suffered 51,000 dead in one battle against Iran on the al-Faw peninsula in 1986.[114] Consequently, Hussein assumed that his and his army's willingness to sustain and, in turn, inflict heavy casualties on the United States was the key to his success.

During the early evening of 29 January 1991, Iraqi armored and mechanized infantry forces closed on Outpost 4, located west of al-Khafji. First Lieutenant Steven A. Ross, commanding 2d Platoon, Company A, 1st Marine Reconnaissance Battalion, first spotted the Iraqis. Ross and his Marines had been joined earlier in the day by Captain Roger L. Pollard's Company D, 3d Light Armored Infantry Battalion, which possessed 19 LAV-25s, including a detachment of 7 LAV-ATs that had BGM-71 tube-launched, optically tracked, wire-guided (TOW) antitank missile-firing capabilities. Ross and Pollard planned for the 2d Platoon to deploy forward, then to fall back to the stronger LAV-25 position if

---

[112] Simmons, "Getting Marines to the Gulf," 13; Westermeyer, *Liberating Kuwait*, 72–75; and SSgt Richard E. Osbourne, USMCR, *Sand and Steel: Lessons Learned on U.S. Marine Corps Armor and Anti-Armor Operations from the Gulf War* (Quantico, VA: Marine Corps Combat Development Command, 1993), 14–15, 23.

[113] Westermeyer, *Liberating Kuwait*, 99.

[114] Westermeyer, *Liberating Kuwait*, 21.

it spotted the Iraqis. During the early moments of the confusing engagement, one LAV-AT fired on another after mistaking it for an Iraqi tank. The incident destroyed the vehicle, killing its crew of four Marines. At the time, most believed that an Iraqi tank had hit the LAV-AT, only to discover the following day that the vehicle had been the victim of "friendly fire."[115]

While Pollard's other LAVs kept the Iraqi tanks at bay for a time, their M242 Bushmaster 25-millimeter chain guns were not designed to penetrate tank armor. At this point, they called in supporting Coalition aviation assets, which arrived in the form of two Fairchild Republic A-10 Thunderbolt II attack aircraft. These venerable aircraft were uniquely configured to destroy armored vehicles. In addition to their nose-mounted 30-millimeter rotary cannons, they also carried tank-killing AGM-65 Maverick air-to-ground missiles. In yet another friendly fire incident, an issue that would plague Coalition forces throughout the combat phase of Operation Desert Storm, a malfunctioning Maverick missile hit one of Pollard's LAV-25s, killing seven more Marines. Only the LAV-25 driver, who was ejected from the vehicle by the blast, survived.[116]

On the coast in al-Khafji proper, the main Iraqi thrust in the assault, the fighting was a bit more desperate. Here, the U.S. Marines, sailors—including a Navy SEAL detachment—and an Army special forces team were eventually driven from their observation posts by Iraqi shellfire and mechanized infantry. The Marine Corps Surveillance, Reconnaissance, and Intelligence Group ordered other reconnaissance-connected Marines in town to fall back to al-Khafji's southern outskirts. The Iraqi attack overran some of the reconnaissance units before they could withdraw, forcing them to remain hidden for the time being. The Saudis were directly responsible for this sector. Some bitterness arose when they requested Coalition air support only to find that none was forthcoming at that moment. While USCENTCOM saw the Iraqi occupation of al-Khafji as inconsequential, it meant much more to the Saudis, who "saw it as

---

[115] Westermeyer, *Liberating Kuwait*, 109–13.
[116] Westermeyer, *Liberating Kuwait*, 113–16.

an assault on their own sacred soil."[117] Consequently, the Saudis independently planned a counterattack on the town as early as midday on 30 January.

Near al-Khafji, the 3d Marine Regiment, called Task Force Taro, operating with the 1st Marine Division, held positions around the city. Task Force Taro's commander, Colonel John H. Admire, was quite concerned about the teams still hiding in al-Khafji. They had intermittent communications, reported back intelligence, and even called in airstrikes when the opportunity arose. Their presence, however, confounded any plans for an immediate counterattack. Saudi Arabian Army colonel Turki al-Firmi soon showed up and was emphatic that his forces needed to liberate al-Khafji because the town was in his sector and on his home soil. After discussing the situation with 1st Marine Division commander, Major General J. Michael Myatt, Admire deferred to the Saudis over the al-Khafji counterattack, which he considered "one of the most difficult decisions I ever had to make."[118] Yet, he also deemed it "an opportunity" for the United States to "demonstrate our trust and confidence in the Arab coalition forces."[119]

When the Saudi forces counterattacked against the Iraqis in al-Khafji, the Marines provided fire support and assistance. According to Captain Joseph Molofsky, the Marine Corps liaison to the Saudi brigade, the 7th Battalion of the King Abdul Aziz Brigade drew the assault assignment. The Saudis launched the attack on time, but it soon devolved into chaos due to the lack of an attack plan from the Saudi leadership. Molofsky stated that he and the Saudi officers spent the time from "2200 until about 0400, dawn actually, driving around . . . trying not to get killed." He claimed, "Nobody was in control. The Saudi battalion commander went into shock. He was sitting in his APC [armored personnel carrier] staring straight ahead." Fortunately, an U.S. Army special forces advisor, Lieutenant Colonel Michael Taylor, was there to keep the

---

[117] Westermeyer, *Liberating Kuwait*, 118–19.
[118] Otto J. Lehrack, *America's Battalion: Marines in the First Gulf War* (Tuscaloosa: University of Alabama Press, 2005), 102–3. John H. Admire retired from the Marine Corps in 1998 as a major general.
[119] Lehrack, *America's Battalion*, 102–3.

Saudis moving in the right direction, though this was not an easy task. When the Saudi column entered al-Khafji and began to take fire from the Iraqis, according to Molofsky, they initially did not react. Once "one of them finally opened up," Molofsky stated, "they all opened up in every direction—up in the air, down in the ground." Major Joseph Stansbury, the battalion operations officer for the 3d Battalion, 3d Marine Division, stated, "Colonel Taylor fought the war for Khafji. The Saudis planned, but Colonel Taylor would bump them and say they needed to do this or that."[120] After the poor start, the Saudis casually regrouped just south of town before relaunching the assault, in conjunction with Qatari forces, and recapturing al-Khafji the next day. Admire called the effects of this battle "profound." He argued that it led to an increase in "the confidence and morale of the Arab coalition forces" and highlighted that "the Iraqi army had no resolve."[121]

The fight for al-Khafji provided an early turning point in the operation and exposed potential issues. The battle resulted in the destruction of hundreds of Iraqi armored vehicles and hundreds of Iraqi casualties. After initially stumbling, the Arab Coalition forces, specifically the Saudi and Qatari forces, performed credibly enough to defeat Saddam Hussein's vaunted armored forces. With this tremendous boost in confidence, the Arab Coalition, as part of Joint Forces Command East, would receive a more active role in ground combat operations. After al-Khafji, Boomer, Myatt, and Keys recognized that the Iraqi Army, which had a fearsome prewar reputation, was not as motivated to fight the Coalition as had been previously supposed. The issue of friendly fire casualties during the fight caused Boomer to order a full investigation into the matter. With more friendly fire incidents to come, Boomer later lamented that the "technological marvels that helped the coalition forces defeat Iraq, sometimes fail, and with disastrous results."[122]

Coalition airpower was another issue that emerged in Joint warfare efficiency. On 22 January, a Joint Surveillance Target Attack Radar Sys-

---

[120] Lehrack, *America's Battalion*, 136–37.
[121] Lehrack, *America's Battalion*, 151.
[122] Westermeyer, *Liberating Kuwait*, 132, 134.

tem aircraft identified 320 Iraqi armored vehicles that had entered the al-Wafrah oil fields in Kuwait, less than 13 kilometers from the border with Saudi Arabia. Yet, the Coalition air forces did not respond to the Iraqis' action. Instead, U.S. Air Force general Charles A. Horner, the JFACC for USCENTCOM, was required—mostly for political reasons—to deploy his airpower in fruitless hunts for Soviet-made R-11 Scud A tactical ballistic missiles that were then provocatively hitting the state of Israel or U.S. troops formations seemingly at will. When not looking for Scud missiles, Horner's aviation assets were vectored toward striking the vaunted Iraqi Republican Guard force concentrations. Schwarzkopf strongly desired that his forces degrade the Republican Guard "to 50 percent of [its] original combat strength before the invasion of Kuwait." In fact, Schwarzkopf seemed fixated only on the destruction of the Iraqi Republican Guard. He ordered Horner to use more and more attack aircraft against these particular forces. As a result, the Coalition air forces flew "literally thousands of air missions" to try to achieve his objective.[123]

During the Battle of al-Khafji, I MEF had its own aviation in the area. Yet, it did not use it against the Iraqi mechanized units that eventually overran the town. Saudi Arabian general Prince Khalid bin Sultan al Saud thought he knew why the Marines held back their aviation. He bitterly claimed that at that same time, "the Marines were fighting their own battles farther west" that the Service considered "of greater significance" than defending al-Khafji. The Marines, according to Khalid bin al Sultan, stockpiled "all the supplies they needed for the coming attack into Kuwait" at a logistics base at "al-Kibrit about 70 miles west of R'as al Mishab." He believed that protecting this logistics base was "the prime concern of Lieutenant General Boomer," which caused him to commit the Marine airpower "to the battles in their sector." Doing so, Khalid bin al Sultan stated, meant "the Marines starved [the Saudi units] of the air support [they] needed and expected to get."[124] Khalid bin al Sultan likely overstated the issue because Coalition aircraft rap-

---

[123] David J. Morris, *Storm on the Horizon: Khafji—the Battle that Changed the Course of the Gulf War* (New York: Free Press, 2004), 79.

[124] Morris, *Storm on the Horizon*, 179.

idly delivered highly effective airstrikes in support of U.S. forces still in hiding inside al-Khafji despite missing this lucrative target initially. Regardless, it would have been better to have engaged the Iraqis before they broke into al-Khafji.

After weeks of air strikes on other targets, which resulted in some Coalition aircrew casualties, the time approached for the ground campaign to begin on 24 February 1991. USCENTCOM planners originally developed a relatively lackluster concept of operations that simply called for Coalition forces to frontally attack across the Kuwaiti border. Schwarzkopf, unhappy with the first plan, famously brought in more innovative planners from the U.S. Army's School of Advanced Military Studies in Fort Leavenworth, Kansas. Known as the "Jedi Knights," these officers envisioned a bolder plan that launched select Coalition forces on a sweeping left hook around Hussein's Kuwait border defenses and, most notably, vast belts of lethal mines. This plan was exactly the type that Schwarzkopf desired.[125]

The mission for Boomer's I MEF largely remained unchanged due to the Marines not being a part of the grand maneuver element. Instead, both the 1st and 2d Marine Divisions would breach the Iraqi mine barriers in column, then proceed to seize the Kuwait International Airport and other nearby objectives before entering Kuwait City.[126] Keys was not happy with the plan, however, because it called for his 2d Marine Division to pass through the 1st Marine Division after it breached the mines. He believed that passage of lines operations on this scale rarely went well in actual combat. Consequently, he recommended to Boomer that each Marine division make their own breaches. Keys' confidence in creating a second breach emerged from his division possessing the necessary engineering equipment that fortuitously came with the attached Army Tiger Brigade. As for Boomer, he eventually agreed with Keys' suggestion, believing the Marines "could move much faster" allowing them to

---

[125] David Evans, "Schwarzkopf's 'Jedi Knights' Praised for Winning Strategy," Chicago Tribune, 1 May 1991.

[126] "Rolling with the 2d Marine Division: Interview with Lieutenant General William M. Keys, USMC," in U.S. Marines in the Persian Gulf, 1990–1991: Anthology and Annotated Bibliography, 149–52.

cover "much more ground" "to kill that many more Iraqis."[127] Brigadier General Krulak put any concerns about the double breach at ease when he informed Boomer that he could create another large forward logistics base that would support the entire I MEF operation.

On 2 February 1991, Boomer met with Schwarzkopf and Vice Admiral Stanley R. Arthur, the Navy's USCENTCOM component commander, aboard USS *Blue Ridge* (LCC 19). Due to the significant number of sea mines along the Kuwaiti coast and the amount of time it would take to clear them, the Navy strongly recommended against conducting any amphibious operations. Schwarzkopf concurred with Arthur but asked Boomer if he required an amphibious assault to support the ground attack. When Boomer responded negatively, he added a caveat that he wished for the amphibious task force to demonstrate off the Kuwaiti coast as part of a planned deception that would cause the Iraqis to worry about both a land and an amphibious attack. Although not directly involved in attacking the Iraqi forces, the afloat Marines played a critical role in the overall USCENTCOM deception plan, seizing some largely undefended islands and processing many prisoners of war. The 5th MEB, commanded by Brigadier General Peter J. Rowe, eventually came ashore to form the I MEF reserve force.[128]

As a further deception, Major General Jenkins's 4th MEB made plans to possibly raid Failaka Island, located about 32 kilometers off the coast of Kuwait City. Jenkins stated that the raid force was meant to establish a "base for operations against the Kuwaiti coast" on the island before eventually withdrawing "under the cover of darkness."[129] Although Schwarzkopf was concerned that the diversion of Navy carrier aviation to support the raid would detract from the overall air campaign, Arthur assured him it would not affect that element, and planning for the raid proceeded ahead. Still, the sea mines issue remained "the single most

---

[127] Westermeyer, *Liberating Kuwait*, 135–36; and "Rolling with the 2d Marine Division," 149–52.
[128] LtCol Ronald J. Brown, USMCR, *U.S. Marines in the Persian Gulf, 1990–1991: With Marine Forces Afloat in Desert Shield and Desert Storm* (Washington, DC: Marine Corps History and Museums Division, 1998), 152, hereafter *With Marine Forces Afloat in Desert Shield and Desert Storm.*
[129] MajGen Harry W. Jenkins, USMC (Ret), *Challenges: Leadership in Two Wars, Washington DC, and Industry* (London: Fortis Publishing, 2020), 252.

important argument against an amphibious assault."[130] Sea mines caused substantial superstructure damage to two American ships, USS *Princeton* (CG 59) and USS *Tripoli* (LPH 10). Mine sweeping had never been given much priority in the Navy, to the regret of all during Operation Desert Storm. Jenkins said he "would never forget the sight of a lone sailor sitting on a high-back chair on the bow of his ship with a helmet, flak jacket, and binoculars, looking for mines as we sailed by." Consequently, on the morning the of raid, the operation was canceled. Jenkins lamented that "the way events had been progressing, I was not really surprised."[131]

As the 1st and 2d Marine Divisions readied for the coming ground assault, the Marine leadership focused on solidifying their logistics elements. Boomer admitted that the "long pole in the tent remains logistics, as has been the case for every force in this theater." Fortunately for Boomer, he had two dynamic subordinates in charge of this effort, Krulak and Brigadier General James Brabham. Krulak established a forward supply base named al-Khanjar, the Arabic name for a dagger worn by Saudi men. Krulak's combat service support troops rapidly built up a massive 11,000-acre supply base, complete with a working field hospital, an expeditionary airfield, ammo bunkers, and a large bulk fuel storage area. Due to the scorching conditions of the Saudi desert, potable water was always a concern. Much to the amazement of everyone, Krulak's troops discovered a lone water pipe sticking out of the ground. No one seemed to know why it was there, but the Marines immediately incorporated it into their system. This "miracle well," as Krulak referred to it, pumped out 100,000 gallons of water per day.[132]

In preparation for the assault, the 1st and 2d Marine Divisions pushed through the border sand berm and moved to the edge of the vast Iraqi minefield. At 0400 on 24 February, the fully mechanized Task Force Ripper, commanded by Colonel Carlton W. Fulford of the 1st Marine Division, began the division's assault. Two other 1st Marine Division task forces—Grizzly and Taro—soon joined in the breaching operations. Know-

---

Figure 59. Navy officials inspect mine damage to the USS *Tripoli* (LPH 10)

Source: official U.S. Navy photo.

ing that the Iraqis had used chemical weapons against the Iranians in the past, the Marines wore components of their mission-oriented protective posture clothing that provided them with some protection in case the Iraqis launched a chemical weapons attack. Bulky and not conducive to a hot desert environment, the clothing forced Marine leadership to prepare for the potential of heat casualties after the sun came up.[133]

The 2d Marine Division launched their assault an hour and a half after the 1st Marine Division. The 6th Marine Regiment, commanded by another Vietnam War-era Co-Van and Navy Cross recipient Colonel Lawrence H. Livingston, made the breach through the minefields, but several chemical weapons scares hindered its progress. The Army's Tiger Brigade then proceeded through the lanes that the Marines created and broke out into open battlespace.[134] As the two divisions pushed forward, they captured significant numbers of prisoners—more than they initially expected, which became a problem. They soon eased the problem after creating two temporary prisoner of war camps "at the flanks of the breach area."[135] In this initial step of ground action, the two divisions "suffered fewer than 50 casualties," and lost "less than 10 vehicles."[136]

The minefield breakthrough had been greatly assisted by a deception plan devised by the 1st Marine Division's assistant division commander, Brigadier General Thomas V. Draude. Once the decision was finalized to make the double-breach, the 2d Marine Division had to move to a position west of the 1st Marine Division. Draude had to make the Iraqis believe that the 2d Marine Division was "still there as the division pulls out behind us and gets over to the west of us in preparation for the breach." Given the name Task Force Troy, in reference to the famous Trojan Horse in Greek mythology, Draude used 200 Marines to replicate the activity of the entire 2d Marine Division. Navy Seabees as-

---

[133] Westermeyer, *Liberating Kuwait*, 163–66.

[134] LtCol Dennis P. Mroczkowski, USMCR, *U.S. Marines in the Persian Gulf, 1990–1991: With the 2d Marine Division in Desert Shield and Desert Storm* (Washington, DC: Marine Corps History and Museums Division, 1993), 43, 48–49, hereafter *With the 2d Marine Division in Desert Shield and Desert Storm*.

[135] Mroczkowski, *With the 2d Marine Division in Desert Shield and Desert Storm*, 52–53.

[136] James F. Dunnigan and Austin Bay, *From Shield to Storm: High Tech Weapons, Military Strategy, and Coalition Warfare in the Persian Gulf* (New York: William Morrow, 1992), 273.

Figure 60. U.S. Air Force chart shows various levels of mission-oriented protective posture equipment

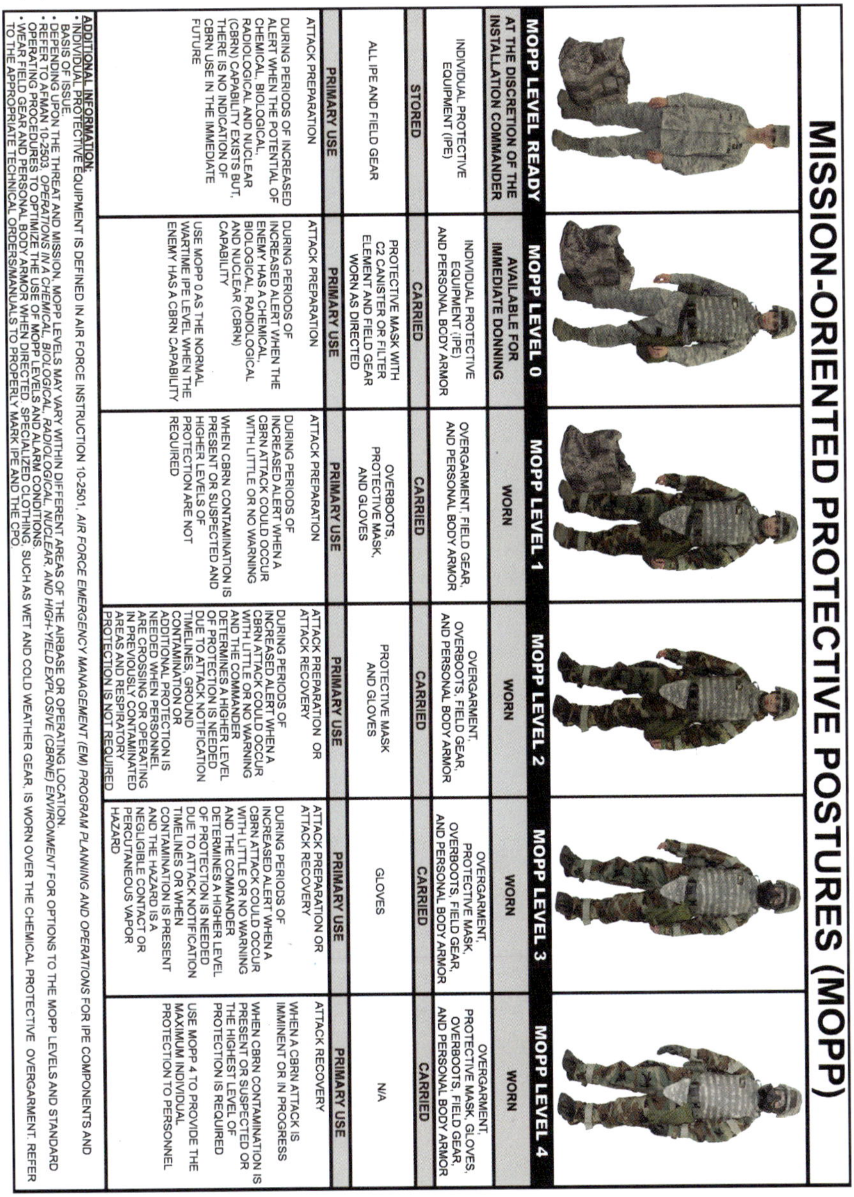

MISSION-ORIENTED PROTECTIVE POSTURES (MOPP)

	MOPP LEVEL READY	MOPP LEVEL 0	MOPP LEVEL 1	MOPP LEVEL 2	MOPP LEVEL 3	MOPP LEVEL 4
	AT THE DISCRETION OF THE INSTALLATION COMMANDER	AVAILABLE FOR IMMEDIATE DONNING	WORN	WORN	WORN	WORN
	INDIVIDUAL PROTECTIVE EQUIPMENT (IPE)	INDIVIDUAL PROTECTIVE EQUIPMENT (IPE) AND PERSONAL BODY ARMOR	OVERGARMENT, FIELD GEAR, AND PERSONAL BODY ARMOR	OVERGARMENT, OVERBOOTS, FIELD GEAR, AND PERSONAL BODY ARMOR	OVERGARMENT, PROTECTIVE MASK, OVERBOOTS, FIELD GEAR, AND PERSONAL BODY ARMOR	OVERGARMENT, PROTECTIVE MASK, GLOVES, OVERBOOTS, FIELD GEAR, AND PERSONAL BODY ARMOR
	STORED	CARRIED	CARRIED	CARRIED	CARRIED	CARRIED
	ALL IPE AND FIELD GEAR	PROTECTIVE MASK WITH C2 CANISTER OR FILTER ELEMENT AND FIELD GEAR WORN AS DIRECTED	OVERBOOTS, PROTECTIVE MASK, AND GLOVES	PROTECTIVE MASK AND GLOVES	GLOVES	N/A
	PRIMARY USE	PRIMARY USE	PRIMARY USE	PRIMARY USE	PRIMARY USE	PRIMARY USE
	ATTACK PREPARATION	ATTACK PREPARATION	ATTACK PREPARATION	ATTACK PREPARATION OR ATTACK RECOVERY	ATTACK PREPARATION OR ATTACK RECOVERY	ATTACK RECOVERY
	DURING PERIODS OF INCREASED ALERT WHEN THE POTENTIAL OF CHEMICAL, BIOLOGICAL, RADIOLOGICAL AND NUCLEAR (CBRN) CAPABILITY EXISTS BUT, THERE IS NO INDICATION OF CBRN USE IN THE IMMEDIATE FUTURE	DURING PERIODS OF INCREASED ALERT WHEN THE ENEMY HAS A CHEMICAL, BIOLOGICAL, RADIOLOGICAL AND NUCLEAR (CBRN) CAPABILITY. USE MOPP 0 AS THE NORMAL WARTIME IPE LEVEL WHEN THE ENEMY HAS A CBRN CAPABILITY	DURING PERIODS OF INCREASED ALERT WHEN A CBRN ATTACK COULD OCCUR WITH LITTLE OR NO WARNING	DURING PERIODS OF INCREASED ALERT WHEN A CBRN ATTACK COULD OCCUR WITH LITTLE OR NO WARNING AND THE COMMANDER DETERMINES A HIGHER LEVEL OF PROTECTION IS NEEDED DUE TO ATTACK NOTIFICATION TIMELINES GROUND CONTAMINATION OR ADDITIONAL PROTECTION IS NEEDED WHEN PERSONNEL ARE CROSSING OR OPERATING IN PREVIOUSLY CONTAMINATED AREAS AND RESPIRATORY PROTECTION IS NOT REQUIRED	DURING PERIODS OF INCREASED ALERT WHEN A CBRN ATTACK COULD OCCUR WITH LITTLE OR NO WARNING AND THE COMMANDER DETERMINES A HIGHER LEVEL OF PROTECTION IS NEEDED DUE TO ATTACK NOTIFICATION TIMELINES OR WHEN CONTAMINATION IS PRESENT AND THE HAZARD IS A NEGLIGIBLE CONTACT OR PERCUTANEOUS VAPOR HAZARD	WHEN A CBRN ATTACK IS IMMINENT OR IN PROGRESS. WHEN CBRN CONTAMINATION IS PRESENT OR SUSPECTED OR THE HIGHEST LEVEL OF PROTECTION IS REQUIRED. USE MOPP 4 TO PROVIDE THE MAXIMUM INDIVIDUAL PROTECTION TO PERSONNEL

ADDITIONAL INFORMATION:
- INDIVIDUAL PROTECTIVE EQUIPMENT IS DEFINED IN AIR FORCE INSTRUCTION 10-2501, AIR FORCE EMERGENCY MANAGEMENT (EM) PROGRAM PLANNING AND OPERATIONS FOR IPE COMPONENTS AND BASIS OF ISSUE.
- DEPENDING UPON THE THREAT AND MISSION, MOPP LEVELS MAY VARY WITHIN DIFFERENT AREAS OF THE AIRBASE OR OPERATING LOCATION.
- REFER TO AFMAN 10-2503, OPERATIONS IN A CHEMICAL, BIOLOGICAL, RADIOLOGICAL, NUCLEAR, AND HIGH-YIELD EXPLOSIVE (CBRNE) ENVIRONMENT FOR OPTIONS TO THE MOPP LEVELS AND STANDARD OPERATING PROCEDURES TO OPTIMIZE THE USE OF IPE.
- WEAR FIELD GEAR AND PERSONAL BODY ARMOR WHEN DIRECTED. SPECIALIZED CLOTHING, SUCH AS WET AND COLD WEATHER GEAR, IS WORN OVER THE CHEMICAL PROTECTIVE OVERGARMENT. REFER TO THE APPROPRIATE TECHNICAL ORDERS/MANUALS TO PROPERLY MARK IPE AND THE CPO.

Source: official U.S. Air Force photo.

sisted Draude's work by making "dummy artillery pieces, tanks, and so forth." He convinced helicopter pilots flying through the area to temporarily touch down, and he had LAVs run around kicking up dust. Draude even had an Army psychological warfare unit blast vehicular noise complete with clanking tank treads through loudspeakers.[137] By all accounts, Task Force Troy worked perfectly.

After clearing the two major minefields, both divisions made significant progress. The situation for nearly all the attacking task forces on "G-Day" initially "developed into a confusion of surrendering Iraqis and intermittent engagements against determined defenders."[138] Large numbers of surrendering Iraqis clogged the cleared lanes through the minefields and began to affect the forward progress of the combat elements. The Marines employed ad hoc solutions to remove the surrendered Iraqis from the lanes and to temporary prisoner of war assembly points.

After the first day of fighting, both divisions had broken through the major obstacle belts and were now poised to attack Kuwait International Airport. The following day, the Tiger Brigade met some resistance but mainly took in more prisoners of war. About the same time, in what has been called the "Reveille Counterattack," a "battalion-sized Iraqi unit of tanks and mechanized infantry collided with the 1st Battalion, 8th Marines."[139] This resulted in the destruction of dozens of Iraqi tanks and armored vehicles. At that moment, however, the biggest issue for the Marines (besides any resisting Iraqis) throughout the 100 hours of ground combat operations—the total amount of combat time for all Coalition ground units during Desert Storm—were the boundaries between the two divisions. Fortunately, "close coordination" between Generals Myatt and Keys prevented any serious incidents from taking place.[140]

---

[137] "Brigadier General Thomas V. Draude," in *Desert Voices: An Oral History Anthology of Marines in the Gulf War, 1990–1991*, ed. Paul W. Westermeyer and Alexander N. Hinman (Quantico, VA: Marine Corps History Division, 2016), 109–10.

[138] LtCol Charles H. Cureton, USMCR, *U.S. Marines in the Persian Gulf, 1990–1991: With the 1st Marine Division in Desert Shield and Desert Storm* (Washington, DC: Marine Corps History and Museums Division, 1993), 69, 77. Cureton accompanied the 1st Marine Division during Operation Desert Shield and Operation Desert Storm.

[139] Mroczkowski, *With the 2d Marine Division in Desert Shield and Desert Storm*, 53.

[140] Mroczkowski, *With the 2d Marine Division in Desert Shield and Desert Storm*, 61.

Figure 61. Kuwaiti oil field burns during Operation Desert Storm

Source: U.S. Army Corps of Engineers.

All the Marines involved in Operation Desert Storm performed well beyond everyone's already high expectations. The Joint planning process had worked even better than in Operation Just Cause. Much of this credit is due to Schwarzkopf, who had rejected the original conservative plan for one that featured maneuver warfare on a grand scale. For the Marine Corps, the entire operation was also a testament to the new doctrine that the Service's leadership had struggled with in the classrooms of Quantico and during combined arms exercises in the training areas of Twentynine Palms throughout the 1980s. Using mission-type operational orders, the 1st and 2d Marine Divisions fervently embraced maneuver warfare, which paid great dividends for them in the end. Boomer came away with a similar conclusion. He further "attributed his success to his junior Marines" and noted that "the young lance corporal would take a look and see something 75 or 100 meters out in front that needed to be done and go out and do it without being told."[141]

---

[141] Westermeyer, *Liberating Kuwait*, 223.

Schwarzkopf was no less effusive in his praise of the Marines. In a news conference held shortly after the cessation of fighting, he stated:

I can't say enough about the two Marine divisions. If I use words like brilliant, it would really be an under description of the absolutely superb job that they did in breaching the so-called impenetrable barrier. It was a classic, absolutely classic military breaching of a very, very tough minefield, barbed wire, fire trenches type barrier. They went through the first barrier like it was water. They went across into the second barrier line, even though they were under artillery fire at the time. They continued to open up that breach. And then they brought both divisions streaming through that breach. Absolutely superb operation, a textbook, and I think it'll be studied for many, many years to come.[142]

As the Marine Corps prepared to enter the new post–Cold War era of the 1990s, it discovered that current political events and military technology were progressing so rapidly that all the U.S. Services were struggling to keep up with the changes. Change and uncertainty became the watchwords for the U.S. military in that decade. New weapons and new paradigms, such as the advent of precision-guided munitions, the internet, the proliferation of antiaccess/area-denial (A2/AD) weapons, the rise of terrorist organizations, and even cyber warfare, all emerged. These aspects portended to make future amphibious operations a challenging proposition. Moreover, after 1986, Joint operations were the new name of the game, and rarely was a single Service contingency event ever going to occur again. Moving forward, the Marines needed to think hard about these pieces and possibly investigate further adjustments to their warfighting doctrine.

---

[142] Gen H. Norman Schwarzkopf Jr., USA, "Commander's Briefing; Excerpts from Schwarzkopf New Conference on Gulf War," *New York Times*, 28 February 1991.

CHAPTER SEVEN

To the Crucible and Beyond

The U.S. Central Command (USCENTCOM) forces returned to the United States from the Gulf War to major acclaim. During the most significant military parade the U.S. armed forces had conducted since World War II, known today as the National Victory Celebration, U.S. Army general H. Norman Schwarzkopf Jr. led 8,800 soldiers, sailors, airmen, and Marines down Pennsylvania Avenue in Washington, DC, for review by President George H. W. Bush on 8 June 1991.[1] It was an extraordinary moment and one that will likely not be repeated for some time to come.

Not all of the U.S. forces involved in Operation Desert Storm took part in the parade, however. Early in May 1991, the 5th Marine Expeditionary Brigade (5th MEB), except for some of its specialized combat forces, was on its way back to the United States when it was suddenly tasked with providing immediate humanitarian aid to the nation of Bangladesh. A massive tropical cyclone had struck the impoverished nation on 29 April 1991. The storm created a tidal wave that killed "an estimated 139,000 people and more than a million livestock." Muddy sea water inundated everything, causing catastrophic damage to infrastructure and lives. The subsequent famine and spread of disease that threatened Bangladesh could trigger "a humanitarian disaster of monumental pro-

---

[1] Thomas Ferraro, "Desert Storm Victory Celebration," UPI, 9 June 1991.

portions" if outsiders did not take "immediate action."[2] Consequently, Bangladeshi prime minister Khaleda Zia requested that the United States send humanitarian assistance to the region as soon as possible. The U.S. State Department immediately sent supplies through the U.S. Agency for International Development, then added both Navy and Marine Corps forces already in the vicinity to the effort. Initially, the 5th MEB remained in the Middle East region to assist with a possible evacuation of U.S. citizens from Ethiopia, where troubles were growing. By mid-May, the commander in chief of U.S. Pacific Command, U.S. Navy admiral Charles R. Larson, suggested to the U.S. defense attaché in Bangladesh that the 5th MEB would provide the best support for the mission. To better administer things, Larson appointed the III Marine Expeditionary Force (III MEF) commander, Major General Henry C. Stackpole III, to coordinate the activities of the entire amphibious relief effort, which was a fortuitous decision.[3] Stackpole's superb management and diplomatic touch throughout the entire operation, called Operation Sea Angel, was extraordinary. This action later became the model for how the U.S. military should conduct future humanitarian operations.

After meeting with Bangladeshi officials aboard USS *Tarawa* (LHA 1), Stackpole made the determination that distributing relief supplies was "the immediate issue," largely due to the massive amount of damaged infrastructure ashore. He believed that "helicopters, landing craft, small boats, and ground transportation assets were needed to move water, medicine, and relief personnel to the remote areas devastated by the cyclone." Stackpole intended that "the Marines and sailors would be the providers, while the Bangladeshi were the implementers."[4] This decision produced the additional benefit of reducing the U.S. footprint ashore in the majority Islamic nation while reinforcing the positive image of the government of Bangladesh toward its own citizens, which had been lacking previously. Operation Sea Angel ultimately required the around-

---

[2] Charles R. Smith, *Angels from the Sea: Relief Operations in Bangladesh, 1991* (Washington, DC: Marine Corps History and Museums Division, 1995), 1.
[3] Smith, *Angels from the Sea*, 25.
[4] Smith, *Angels from the Sea*, 47–48.

the-clock services of more than 7,500 military personnel for more than a month. During this time, the U.S. relief effort delivered thousands of tons of food, potable water, medicine, and, most importantly, helicopter air support to remote areas that the massive storm had isolated. The relief supplies reached an estimated 1.7 million people, thereby saving thousands of lives.[5]

As Operation Sea Angel indicated, the Marine Corps faced both challenges and opportunities in the new post–Cold War world after Operation Desert Storm. On 30 June 1991, General Alfred M. Gray Jr. retired from active duty after 41 years of service. General Carl E. Mundy Jr., a decorated veteran of the Vietnam War, succeeded him as the 30th Commandant of the Marine Corps. The choice of Mundy was an easy one. Before becoming Commandant, he had developed a sterling reputation while serving as the commanding general of the 4th MEB and leading Fleet Marine Force, Atlantic (FMFLANT) as a lieutenant general. Gray had also held both of these positions prior to his appointment as Commandant.[6]

Soon after Mundy's tenure as Commandant began, the Base Force issue, which the U.S. Department of Defense (DOD) and the Joint Chiefs of Staff had put on hold during the Gulf War, was quickly resurrected. Even before Operation Desert Storm's conclusion in 1991, the uniformed Services had reduced their personnel by 2.8 percent. That year, the Marine Corps alone was supposed to decrease its personnel from 196,652 to 194,040, a cut of 2,612 personnel from 1990. To support this reduction, the military Services tried to speed up voluntary retirements by changing the time-in-grade requirements from three years to two.[7] Despite this attempt, the Marine Corps had to form a selected early retirement board in 1991 that forced 46 colonels and 69 lieutenant colonels to retire.[8] One of the unfortunate victims of this board was the maneuver warfare acolyte Colonel Michael D. Wyly, who was forced into retirement

---

[5] Smith, *Angels from the Sea*, 85.

[6] George Frank, "New Marine Commandant Is 'Right Guy at Right Time'," *Los Angeles (CA) Times*, 3 June 1991.

[7] Bart Brasher, *Implosion: Downsizing the U.S. Military, 1987–2015* (Westport, CT: Greenwood Press, 2000), 63–64.

[8] Brasher, *Implosion*, 76.

only months shy of 30 years of active service. While DOD plans called for continued force reductions overall, the Marine Corps was allowed to maintain its three MEFs but were still set to lose "the equivalent of one active brigade and over 10 percent of their air assets."[9]

It was clear that while the need for military personnel was going down in the early 1990s, the percentage of minorities who comprised the Base Force was going up. During Operation Desert Shield and Operation Desert Storm, the United States sent approximately 570,000 troops to the Persian Gulf region. However, as defense analyst Bart Brasher noted, Black Americans made up a disproportionate percentage of U.S. forces. In all the Services, Black personnel "comprised 23.5 percent of the 570,000 military personnel deployed" and "made up 29 percent of enlisted personnel sent to the Persian Gulf" during the conflict. Among those personnel, 36 percent "served in combat specialties" in a conflict in which almost 43 percent of personnel were "in combat billets." Americans of African descent made up only 11 percent of the overall population of the United States at the time. With Black personnel comprising such a comparatively high percentage in the U.S. forces, including 18 percent of enlisted Marines, some U.S. military leaders had "concerns that casualties in a Gulf conflict would disproportionately affect" Black Americans.[10]

Nevertheless, military service continued to be an attractive option for many minority communities in America throughout the 1990s. In 1991, Commandant Mundy tasked Brigadier General Charles C. Krulak with the formation of a Force Structure Planning Group (FSPG). The FSPG was created to consider what the Marine Corps needed to do in the face of what appeared to be looming arbitrary force cuts called for by the bottom-up review. Krulak immediately went out and recruited, in his words, some of the "best minds" in the Marine Corps. Based out of loaned offices at Marine Corps Base Quantico, Virginia, the FSPG worked non-stop, seven days a week, for nine weeks straight. Krulak had been told that the Marine Corps would be reduced to approximately 159,100 Marines. However, he observed that as the Service got closer to that num-

---

[9] Brasher, *Implosion*, 71.
[10] Brasher, *Implosion*, 65–66.

ber, "it became obvious" that the Marine Corps would not "meet the national military strategy." Instead, the issue became "developing the case to add end strength" above the established 159,100 personnel. To accomplish this objective, Krulak stated, "We needed the rigor to convince DOD, the President, and Congress. We did the job right and we got the rigor and the number, which increased to approximately 177,000."[11] Basically, the FSPG proposed the most substantial changes to Marine Corps force structure since the Hogaboom Board recommendations in the 1950s.[12] More importantly, Krulak's experience with the FSPG gave him unparalleled knowledge about the interior workings of the Service. As he later noted, "it just gave me an unbelievable sense of what the Corps was all about," something he would lean on as he moved up the Marine Corps leadership.[13]

By 1992, the DOD's push for downsizing had heated up. By that year, the Marine Corps had dropped to its lowest number of personnel since 1961, having reduced its ranks to 184,529, a reduction of 9,511 personnel between 1987 and 1992.[14] To further assist the U.S. military Services with voluntary downsizing instead of a ordering a mandated reduction in force, which all the Services sincerely wished to avoid, the government instituted four major force reduction programs to reach the congressionally authorized ceilings. While the DOD considered but did not receive authorization for 15-year early retirement packages in 1992, it established the popular reduction incentives of the voluntary separation initiative and special separation benefit programs. The DOD hoped to get 60,000 voluntary separations from these in 1992, which it easily approached with 53,932 applicants in that year.[15]

---

[11] Gen Charles C. Krulak, USMC (Ret), Oral History Transcript, 29 November 2000 session, Dr. David B. Crist interviewer, Marine Corps History Division, Quantico, VA, 2003, 86, hereafter Krulak Oral History Transcript, 29 November 2000.

[12] The recommendations are summarized in Col John J. Grace, USMC (Ret), "The U.S. Marine Corps in 1991," U.S. Naval Institute *Proceedings* 118, no. 5 (May 1992). See *Report of the Fleet Marine Force Organization and Composition Board Report* (Washington, DC: Headquarters Marine Corps, 1956) for more on the Hogaboom Board report.

[13] Krulak Oral History Transcript, 29 November 2000, 89.

[14] Brasher, *Implosion*, 87–88. The reduction in force to 184,529 Marines was a 7.5-percent drop from the previous peak in 1987.

[15] Brasher, *Implosion*, 88.

While the DOD pushed for downsizing the Services, the Marine Corps faced growing responsibilities in contingency operations. Soon after the Gulf War ended, two immediate situations required the continued presence of Marines, among other Joint forces, in the Middle East. Operation Provide Comfort and Operation Southern Watch came about due to a vengeful Iraqi president Saddam Hussein lashing out against his longtime internal enemies—the Kurds of northern Iraq and the Shia Muslims of southern Iraq. Essentially, Hussein was conveying to both populations that despite his severe losses in the Gulf War, he still controlled Iraq. As award-winning correspondent Thomas L. Friedman of *The New York Times* believed, there was a "harsh, survivalist quality" that all Middle Eastern political leaders felt they must possess. Moreover, many Middle Eastern leaders and their internal followers were convinced that tribal allegiance took "precedence over allegiances to the wider national community or nation-state."[16] In this case, Hussein's "tribe" was clearly the Sunnis of western Iraq, thereby making the Shia Iraqis of the al-Basrah region and the Kurds in northern Iraq targets following his catastrophic battlefield defeat in 1991. Within months of the Gulf War's conclusion, the Iraqi Army launched a series of brutal attacks into northern Iraq that forced more than a million Kurds to flee their homes and congregate in squalid refugee camps in the mountains of northern Iraq and southeastern Turkey. Initially, there were at least 12 large camps that contained an estimated 45,000 people in each. Observers estimated that approximately 600 people died every day from exposure, malnutrition, and diseases.[17]

The first U.S. force on the scene for Operation Provide Comfort was the Army's 10th Special Forces Group, which received the "primary mission" of resupplying the refugees. They were later folded into what be-

---

[16] Thomas L. Friedman, *From Beirut to Jerusalem* (New York: Farrar, Straus, Giroux, 1989), 87–89.

[17] Col James L. Jones, "Operation Provide Comfort: Humanitarian and Security Assistance in Northern Iraq," in *U.S. Marines in the Persian Gulf, 1990–1991: Anthology and Annotated Bibliography*, comp. Maj Charles D. Melson, USMC (Ret), Evelyn A. Englander, and Capt David A. Dawson (Washington, DC: Marine Corps History and Museums Division, 1992), 191–92. According to Jones, approximately 1.3 million Kurds made their way to similar camps near the Iraq-Iran border, although "the fate of this group has yet to be determined." Jones, "Operation Provide Comfort," 191.

came known as Joint Task Force Alpha (JTF-A). Within days, the 24th Marine Expeditionary Unit (special operations capable) (24th MEU[SOC]), which had been in the middle of training exercises on the island of Sardinia, received orders to immediately cease its operation and transfer to northern Iraq as soon as possible to assist JTF Bravo (JTF-B), among other Joint forces coming to the region. Commanded by Colonel James L. Jones Jr., the future 32d Commandant of the Marine Corps, the 24th MEU(SOC)'s first element to appear was its aviation element, Marine Medium Helicopter Squadron 264 (HMM-264). As soon as the helicopters arrived, they began to "deliver over 1 million pounds of relief supplies and fly in excess of 1,000 hours without mishap."[18] Operation Provide Comfort, being "the first American-led humanitarian intervention to follow the Cold War," created "useful precedents for future operations." Particularly important, the operation illustrated "the utility of the joint/ multinational combat and support formations employed" and established "the nature of how they worked together."[19] In sum, this humanitarian operation foreshadowed the vast interagency affairs that such operations would become. While the DOD's abundant resources and capabilities still required the department to carry most of the burden, other government agencies were, by necessity, actively involved in the planning and implementation of these operations.

The 24th MEU(SOC) took on prominent roles as part of JTF-B. When U.S. Army major general Jay M. Garner arrived on scene to take command of JTF-B, he was temporarily without a staff, which required him to rely on the 24th MEU(SOC) headquarters staff until he could form his own. In fact, the 24th MEU(SOC) acted as "both a joint and multinational formation," as the unit integrated a British Royal Marine battalion and a Dutch Marine battalion as well as a U.S. Army airborne battalion. All these elements also relied on the MEU's "organic helicopters, trucks, artillery, and armored vehicles" for essential combat and service support. The Army also sent substantial engineering support and the en-

---

[18] Jones, "Operation Provide Comfort," 192.
[19] Gordon W. Rudd, *Humanitarian Intervention: Assisting the Iraqi Kurds in Operation Provide Comfort, 1991* (Washington, DC: Department of the Army, 2006), 231.

tire 18th Military Police Brigade to assist with handling refugees. Both the engineers and military police personnel proved invaluable throughout the entire operation.[20]

The JTF-B had numerous objectives at the outset of the operation. First, the 24th MEU(SOC) established a forward base to assist with sending supplies to the refugees in the mountains of northern Iraq and at Silopi, Türkiye, on 18 April 1991. Second, the MEU moved into Iraq to provide security in support of the overall humanitarian effort. The MEU's ground combat element, which included the British Royal Marine battalion and the Dutch Marine battalion, needed to establish this secure region without any incidents between the Iraqi Army and U.S. forces. This step was essential to convincing the refugees in Turkey to return to their villages before warm weather set in, which would dry up the mountain streams. Finally, JTF-B had to relieve the city of Zakho, Iraq, which the Iraqi forces had previously looted and then reoccupied with security forces posing as "police officers." Information had gotten back to JTF-B that the local Kurdish population of Zakho was being abused by the Iraqi police officers. The British provided the 45 Commando of the Royal Marines "a battalion-size light infantry formation fresh from a tour in Northern Ireland. . . . Specially trained for low-intensity warfare in urban settings, they provided [JTF-B] the ideal unit to take on Iraqi military forces in Zakhu [sic]." These forces were placed by Major General Garner under the operational authority of Colonel Jones's 24th MEU(SOC), and they were to conduct dismounted patrols on 21 April 1991. On 23 April, "four hundred Dutch marines under Lt.Col. Cees van Egmond joined JTF Bravo and were attached to Jones' 24th MEU." The Dutch Marines were "comparable" in size and training "to the British 45 Commando."[21]

Even though Coalition forces were building up outside of Zakho, "about 300 special police from Baghdad had slipped into [the city]" before Coalition security check points could be established. Gordon W. Rudd writes, "Essentially bullies, they intimidated the few civilians left in town." The 24th MEU(SOC) had to figure a way to get these troublemak-

---

[20] Rudd, *Humanitarian Intervention*, 232–33.
[21] Rudd, *Humanitarian Intervention*, 115–16.

ers to leave Zakho without triggering further violence. In fact, there were several altercations that took place, including one event in which several Iraqi soldiers challenged Colonel Jones and his sergeant major when they drove through the town. "Although they managed to escape," Rudd continues, "the sergeant major stated that it was the one time during the operation that he prepared to draw and use his pistol." Major General Garner was able to circumvent any further violence in Zakho when he informed the senior Iraqi security officer there that he needed "to get the police out of town or there would be trouble." The Iraqi officer agreed to remove most but not all the Iraqi police officers in Zakho. However, a grenade, "allegedly thrown by a Kurd, killed and wounded several Iraqis in their police station." This convinced the Iraqi commander to remove the remnant of the police, and it was not long before Garner set up his JTF-B headquarters "at a deserted Iraqi Army garrison in Zakhu [sic] and kept it there while he had forces in northern Iraq." Around this same time, JTF-B was reinforced by a "U.S. Army airborne battalion combat team (ABCT), organized around the 3d Battalion, 325th Infantry, from Vincenza, Italy and commanded by Lt.Col. John P. Abizaid."[22] One of the most important tasks that the Coalition forces completed was to get the airfield at Sersink, which had been damaged during Operation Desert Storm, back in order, as it was considered "the key supply point for JTF-B in northern Iraq."[23] Marines and sailors from the 3d Force Service Support Group (3d FSSG) under III MEF in Okinawa, Japan, came to the rescue with further logistical support. Operation Provide Comfort was truly a Joint and combined operation that came together just in time for the Kurds of Northern Iraq.

Nevertheless, many of the Kurdish refugees hesitated to leave the camps in southern Turkey. The Coalition eventually presented what was essentially an ultimatum to the Iraqis, creating a no-fly zone in northern Iraq that would be enforced by Coalition aircraft. Further, the Iraqi Army and secret police were prohibited from reentering the security zone. Soon, thanks to the efforts of the Battalion Landing Team 2d Bat-

---

[22] Rudd, *Humanitarian Intervention*, 116–18.
[23] Jones, "Operation Provide Comfort," 196.

talion, 8th Marines (BLT 2/8), the even larger city of Dahuk, Iraq, came under Coalition control. It was important to secure this city to convince the Kurdish refugees to return home, since so many of them had originated from Dahuk and its suburbs. Like the earlier situation in Zakho, the issue in Dahuk centered on convincing Iraqi security forces to leave town—something that Saddam Hussein had previously stated was unacceptable. This impasse was resolved when Coalition forces got the Iraqis to agree to the insertion of "noncombat" soldiers. In Major General Garner's mind, this meant an initial force of 80 military police personnel. The relative incident-free ease of securing Dahuk allowed Coalition forces "to accelerate the movement of refugees south into Northern Iraq, and it reduced and altered the support provided by the Civil Affairs and Combined Support Commands."[24] Shortly after, a veritable tidal wave of refugees was moving out of the mountains and through the Zakho Valley on their way to their former homes.[25] Fortunately, Colonel Jones had enough troops to ensure that the Coalition forces would secure the area for the returning Kurdish refugees.

Additional units eventually joined the 24th MEU(SOC) to assist in securing the Kurdish settlements in Iraq. British, Dutch, Spanish, and Italian organizations quickly sent in additional humanitarian support. The U.S. Department of State Disaster Assistance Relief Team, led by former Marine Fredrick C. Cuny, joined JTF-B to provide "capability of critical importance" by coordinating the humanitarian efforts of both governmental and nongovernmental organizations (NGOs). An Army airborne battalion, Army combat engineers, and a Navy Mobile Construction (Seabee) battalion eventually relieved some of the burden on the Marines and sailors of the 24th MEU(SOC), who had erected more than 1,100 tents for returning Kurdish refugees in 10 days.[26] The entire operation became a massive humanitarian effort for the Coalition powers on a scale they had never imagined.

---

[24] Rudd, *Humanitarian Intervention*, 181–82.
[25] Jones, "Operation Provide Comfort," 197.
[26] Jones, "Operation Provide Comfort," 194–95.

In sum, Operation Provide Comfort was a tremendous success. The Coalition developed contingency plans for further intervention if the Kurds began to be harassed by Saddam Hussein's security forces in the future. JTF-B had established a credible security zone in northern Iraq and even built "temporary transit centers" to further aid in Kurdish repatriation. Most importantly, Operation Provide Comfort paved the way for United Nations (UN) relief efforts and various philanthropic NGOs to effectively assist the Kurdish refugees after the Coalition security forces on the ground in northern Iraq were totally withdrawn.[27] By the time the last Coalition combat troops in BLT 2/8 and the Army's 3d Battalion, 325th Airborne Combat Team, left northern Iraq in July 1991, the forces had provided much-needed humanitarian assistance for hundreds of thousands of desperate and starving Kurds. As Jones noted, "People who had never dreamed of an operation of this magnitude were thrust together to make critical decisions. They overcame language, cultural, and ethnic barriers."[28] The Kurdish refugees had largely returned to their homes and were out of danger for the time being.

While the responsibility for Operation Provide Comfort fell under the cognizance of U.S. European Command, Operation Southern Watch, which began a full year after Coalition forces had withdrawn from northern Iraq, came under USCENTCOM's JTF Southwest Asia. Southern Watch began on 27 August 1992 ostensibly to force Iraq's compliance with United Nations Security Council Resolution 688 (UNSCR 688), which demanded that Saddam Hussein immediately end the repression of the Kurdish population in northern Iraq and Shia Arab population in southern Iraq and encouraged him to "open dialogue" that would "ensure that the human and political rights of all Iraqi citizens are respected."[29] Although UNSCR 688 never directly established a no-fly zone in southern Iraq, Hussein's predominate use of helicopter gunships against the Shia population caused the Coalition planners to conclude that they needed to rapidly establish and vigorously enforce a no-fly zone there. Due to

---

[27] Rudd, *Humanitarian Intervention*, 218–20.
[28] Jones, "Operation Provide Comfort," 198–99.
[29] United Nations Security Council Resolution 688, Iraq, S/RES/688, 5 April 1991.

military aviation being more significant to Southern Watch, the operation involved a substantial amount of Coalition aviation assets rather than the landing of ground forces, although elements of the U.S. Army's 24th Mechanized Division and I MEF were deployed to Kuwait in October 1994 after Hussein ordered Iraqi heavy armor units south of the 32d parallel in violation of the agreement. Starting in 1992, U.S. Air Force, Navy, and Marine aerial units took action to enforce the UN-mandated no-fly zone, including when an Air Force General Dynamics F-16 Fighting Falcon fighter shot down an Iraqi Mikoyan-Gurevich MiG-25 Foxbat aircraft that had violated zone on 27 December 1992. Although the Air Force was the most significant U.S. Service involved, flying more than 60 percent of all sorties, the operation "included extensive Navy and Marine aircraft sorties from the carriers *America* (CV 66), *Nimitz* (CVN 68), *George Washington* (CVN 73), *Carl Vinson* (CVN 70), *Enterprise* (CVN 65), *Kitty Hawk* (CV 63) and amphibious assault ship *Peleliu* (LHA 4)" starting in 1996. Throughout Southern Watch, which lasted until 2003 when the United States initiated Operation Iraqi Freedom, the U.S. forces did not lose a single pilot to enemy fire.[30]

Before Southern Watch, another significant USCENTCOM contingency took place in Mogadishu, Somalia, in January 1991. At that time, a civil war between armed militia factions under the command of various local warlords was ripping Somalia apart. The violence escalated to the point that the Somali government of Major General Mohamed Siad Barre collapsed in December 1990. This implosion caused the U.S. ambassador to Somalia, James Keough Bishop, to request that USCENTCOM, then preparing to conduct Operation Desert Storm, engage an immediate evacuation of the U.S. embassy and all U.S. nationals in the capital of Mogadishu. From 7 to 9 January 1991, Navy and Marine Corps forces carried out Operation Eastern Exit, a noncombatant evacuation operation "to rescue Americans and citizens of other nations" from the devastated capital city. "More accurately," as one historian writes, "it was an armed

---

[30] "Operation Southern Watch," GlobalSecurity, accessed 7 October 2019.

incursion, conducted without the permission of the local government and authorized to accomplish the mission by force of arms if necessary."[31]

At the time, most of the U.S. maritime forces afloat were conducting workups for an essential and widely publicized amphibious landing rehearsal called Sea Soldier IV. Ultimately, Sea Soldier IV became part of an elaborate ruse during Operation Desert Shield to trick the Iraqis into thinking that an amphibious assault was imminent (although the possibility of the Marines conducting an amphibious assault on the coast of Kuwait was still on the table at that time). Consequently, the commander of U.S. Naval Forces Central Command, Vice Admiral Stanley R. Arthur, ordered Rear Admiral John B. LaPlante, in charge of the command's amphibious forces, to prepare to conduct the evacuation operation. LaPlante's forces consisted of the 4th and 5th MEBs and the 13th MEU(SOC) spread out across 31 Navy amphibious vessels. Major General Harry W. Jenkins, the senior Marine afloat, was named overall commander of all embarked Marine Corps landing forces. Arthur did not envision the Marines storming ashore in Somalia since he needed most of them to remain on task in preparation for Sea Soldier IV. Consequently, he limited the Somalia amphibious task force to just two vessels.[32]

Jenkins placed Colonel James J. Doyle Jr., the commander of Brigade Service Support Group 4, in command of the forces for Operation Eastern Exit. Doyle was a highly decorated Marine from the Vietnam War and an extraordinary leader. He relocated from USS *Trenton* (LPD 14) to the larger USS *Guam* (LPH 9) and created a "special purpose command element—the 4th Marine Expeditionary Brigade, Detachment 1."[33] LaPlante, in turn, appointed one of his most trusted subordinates, Captain Alan B. Moser, to command the two-ship amphibious task force.

By 4 January 1991, the situation in Mogadishu had become so bad that Ambassador Bishop requested the immediate deployment of two platoons of Army paratroopers. Doyle thought that this was "a bad idea

---

[31] Gary J. Ohls, *Somalia. . . From the Sea* (Newport, RI: Naval War College Press, 2009), 2. To date, Ohls' superb book remains the definitive volume on the various operations that took place in Somalia, 1991–95.

[32] Ohls, *Somalia . . . From the Sea*, 25–28.

[33] Ohls, *Somalia . . . From the Sea*, 28.

because the space available for a landing zone was so small" and that the possibility of the paratroopers getting scattered outside the embassy and in Mogadishu was too risky. Further, *Guam*, steaming at 24 knots, was rapidly closing on the Somali coast and would likely arrive before any airborne assault could be organized and executed. Any debate about the paratrooper drop became moot as General Schwarzkopf "refused to authorize" the action.[34]

Once within long-distance helicopter range, Doyle ordered a detachment of 47 Marines from the 1st Battalion, 2d Marine Division, and a U.S. Navy Sea, Air, and Land (SEAL) team, both under the command of Lieutenant Colonel Willard D. Oates, to make a dawn insertion directly into the embassy compound. The 1st Battalion's commander, Lieutenant Colonel Robert P. McAleer, went along with the heliborne assault force despite Oates being the overall commander. Doyle remained on board *Guam* to directly coordinate the entire effort with Moser. Early on the morning of 5 January 1991, the Marines and the SEAL team loaded onto two Sikorsky CH-53E Super Stallion heavy-lift helicopters on *Guam* for the 400-nautical-mile flight into Mogadishu, which would include the helicopters conducting a problematic in-flight refueling operation for both the ingress and egress phases of the operation.[35]

Shortly after dawn, the two Super Stallions landed at the U.S. compound despite having some initial difficulty locating the embassy at low altitude. Prior to their arrival that morning, the embassy had already come under fire from about 150 Somali fighters, including some who had brought scaling ladders with them. The Marine Corps embassy guard valiantly held the perimeter against the significant Somali force, but the SEALs and McAleer's Marines arrived just in time. The command arrangement for the combined unit, with both Oates and McAleer on the ground, did not create any issues when they arrived because Oates went to work directly with the ambassador while McAleer commanded the security forces. The 1st Battalion provided enough reinforcements to drive off the Somalis while the SEALs secured the ambassador and his imme-

---

[34] Ohls, *Somalia . . . From the Sea*, 31.
[35] Ohls, *Somalia . . . From the Sea*, 32–33.

diate staff. Soon after, the Super Stallions lifted off with "61 evacuees, including all nonofficial Americans in the compound; the ambassadors of Nigeria, Turkey, and the United Arab Emirates; and the Omani charge d'affaires."[36] However, the danger had not ended with Super Stallions' departure.

Although Oates requested an additional 44 Marines to land in a second wave, Doyle believed that another helicopter insertion was too risky now that the Somalis knew that U.S. forces were in the region. With the embassy compound no longer under an imminent threat, he told Oates that he would have to accomplish the mission with the force already on the ground. Yet, the 1st Battalion and SEALs had the additional responsibility of protecting other consular officials in the city seeking refuge at the U.S. embassy, including the Soviet ambassador, who was eventually escorted to the compound by a combined force of Marines and SEALs. The additional evacuation of these individuals had been unplanned, but *Guam* now sat within the range of Boeing Vertol CH-46 Sea Knight medium-lift helicopters. After dark, to minimize the risk of Somali antiaircraft fire, Doyle and Moser dispatched 20 Sea Knights toward the U.S. compound. Although a local Somali police major attempted to prevent the refugees from boarding the helicopters to interfere with the return flights, quick thinking by the U.S. ambassador, who offered him cash and his choice of any of the now-abandoned embassy vehicles, allowed the Marines to continue the evacuation of 281 people, "including sixty-one Americans, thirty-nine Soviet citizens, seventeen British, twenty-six Germans, and various numbers from twenty-eight other nations." By 11 January, all the evacuees were offloaded at Muscat, Oman.[37]

The success of Operation Eastern Exit was overshadowed by the launching of Operation Desert Storm on 17 January 1991. Nevertheless, the U.S. military learned substantial lessons from this single two-day operation that once again proved that amphibious task forces were an extraordinarily flexible force choice and well-suited for the type of scenarios the United States would likely encounter in the new post-Cold

---

[36] Ohls, *Somalia . . . From the Sea*, 36.
[37] Ohls, *Somalia . . . From the Sea*, 38–39.

War era. Later, as the 31st Commandant of the Marine Corps, General Charles Krulak, drew on Eastern Exit as a "case study for understanding and implementing expeditionary concepts in the emerging new world order."[38] Indeed, although the opening of the massive Coalition air campaign against Iraq came just two weeks later, Eastern Exit must be considered one of the most highly successful short-notice contingency operations—one in which events moved at lightning speed—that the Navy-Marine Corps team conducted in the immediate post-Cold War era. Much of the credit for this accomplishment must go to LaPlante, Moser, and Doyle and their management of the entire affair.

Despite the successful evacuation at Mogadishu, Somali warlords, such as Mohamed Farrah Aidid, soon caused further trouble for U.S. interests in the region. Internecine warfare in Somalia created conditions that led to widespread destruction and famine throughout the country. Consequently, President George H. W. Bush ordered USCENTCOM, now commanded by Marine Corps general Joseph P. Hoar, to conduct Operation Provide Relief. Provide Relief was very similar to Provide Comfort in that its principal focus was to create a security environment in which relief services could more rapidly be given to the sick and starving population of Somalia. Marine Corps brigadier general Frank Libutti, yet another superb leader and Vietnam War veteran that USCENTCOM seemed to possess in abundance in those days, was placed in charge of the operation. He was initially sent to Somalia to observe and assess how to relieve the situation. Instead, while on the way over from Tampa, Florida, he was named commander of JTF Provide Relief. Its mission was to deliver and distribute tons of supplies, including food, medicine, and potable water, to the long-suffering Somali people who were dying at the rate of 1,000 people every day. Despite facing initial difficulties with the Kenyan government hosting him, Libutti completed his mission by forging relationships with NGOs already operating inside Somalia. These NGOs had the best picture of what was happening on the ground and provided Libutti with crucial information that he needed to complete his mission.

---

[38] Ohls, *Somalia . . . From the Sea*, 40.

Nevertheless, without bringing in a much larger security force, the issue of how to ultimately protect the supplies from marauding bands of heavily armed criminals proved challenging.[39]

In truth, Somalia in the early 1990s was one of the most lawless and dangerous places on the planet. With warlords and clan leaders rather than any truly effective internal governance dominating the country, random acts of violence against UN observers and aid workers grew. Those involved on the ground believed that the key to stability was getting food distributed to the starving populations, but, at least in the principal city of Mogadishu, more than "a thousand teenage gunmen, wired on khat (an herbal amphetamine) and galvanized by infamy" prevented its delivery.[40]

The khat issue should not be underestimated, as it had an insidious effect on nearly the entire Somali population. American journalist Jonathan Stevenson noted the significance of the khat trade during 1992. Abstaining from khat, Stevenson comments, would have "freed up at least six dollars per day per user." That amount of money could have purchased slightly more than 44 pounds of maize or rice, which could have fed "six people for a week" and could have boosted "an economy bled of currency by the khat traders themselves." No politicians in Somalia, however, "dared even suggest that the khat trade should be curtailed." Many benefited from the trade, and others believed that reducing it would increase the existing unrest. The drug had taken such strong hold that in September 1992, "a Somali aid employee of a French relief agency who was paid half in food instead of all in cash detained several expatriate doctors and nurses in an operating theater with an unpinned grenade until he got his way" solely because "food couldn't buy khat."[41]

Somali warlords, of course, dominated the khat industry. Khat needs volcanic soil to grow, which made neighboring Kenya an ideal location

---

[39] Eric F. Buer, *United Task Force Somalia (UNITAF) and United Nations Operations Somalia (UNOSOM II): A Comparative Analysis of Offensive Air Support* (Quantico, VA: Command and Staff College, Marine Corps University, 2001).

[40] Jonathan Stevenson, *Losing Mogadishu: Testing U.S. Policy in Somalia* (Annapolis, MD: Naval Institute Press, 1995), 9.

[41] Stevenson, *Losing Mogadishu*, 11.

for its legal production. However, khat loses its potency within 24 hours of being picked. As a result, warlords such as Osman Ali Atto funded the building of rudimentary airstrips in central Somalia to speed up its delivery. Atto and other wholesale khat dealers enjoyed an estimated 200-percent profit margin that "ends up in Rome bank accounts," growing the warlords' "war chest" but doing "Somalia's economy no good whatsoever."[42] To make matters worse, the warlords used khat to develop "genuinely lean-and-mean fighters." They gave the drug out "to their soldiers along with their daily food rations—which can be much smaller since khat suppresses the appetite—and nudg[ed] clan resentment toward outright hatred."[43] Stevenson noted that it was common for gunfights to break out between Somali factions even as khat shipments were being unloaded.

By December 1992, it was clear that a more significant effort was necessary to relieve the suffering of the Somali people. Despite his electoral loss to William J. "Bill" Clinton in November 1992, President Bush was determined to resolve the situation before he left office on 20 January 1993. Brigadier General Libutti warned officials that any larger operation in Somalia ran the risk of becoming a long-term nationbuilding effort. Due to this consideration, Bush put together an international coalition as part of a much larger action called Operation Restore Hope. He even agreed that the overall operation would fall under the cognizance of the UN Security Council with a Coalition force soon referred to as the United Nations-sanctioned Unified Task Force (UNITAF).

The UNITAF mission was not well thought out, and there were multiple reasons why sending forces into the Somali cauldron was not a good idea. Even U.S. State Department officials had grave misgivings. The U.S. ambassador to Kenya, Smith Hempstone, reportedly wrote in a confidential letter to the State Department, "If you liked Beirut, you'll love Mogadishu."[44] While Hempstone was happy that Restore Hope could

---

[42] Stevenson, *Losing Mogadishu*, 13. Atto was eventually arrested for the role he played in the June 1993 massacre of UN peacekeepers from Pakistan, which is discussed later in this chapter. See Stevenson, *Losing Mogadishu*, 14.
[43] Stevenson, *Losing Mogadishu*, 13.
[44] Ohls, *Somalia . . . From the Sea*, 76.

save thousands of lives in 1993, he believed that these same people might continue to face a grave risk the following year unless the United States committed to an extended stay.

Nevertheless, Bush went ahead with Restore Hope, wanting to have everything wrapped up by inauguration day, 20 January 1993. Lieutenant General Robert B. Johnston, the commander of I MEF and a veteran of the Vietnam War, Beirut, and Operation Desert Storm, was placed in overall command of the operation and was teamed with an extraordinary State Department official, Robert B. Oakley, who had been named roving ambassador by Bush. In addition to some earlier experience in Somalia, Oakley had also spent time in Vietnam and Beirut. Just days before the operation began, Libutti returned to Somalia to work directly with Oakley. Commandant Mundy offered Johnston the expert services of now-Brigadier General Anthony C. Zinni, who served as Johnston's operations chief throughout Operation Restore Hope. Fortunately for the entire effort, the 15th MEU/SOC, a special Marine air-ground task force (SPMAGTF) commanded by Colonel Gregory S. Newbold, was already afloat and steaming for the Somali coast. Although there was some acrimony between the Navy and Marines about having an understrength and equipped amphibious task force for Restore Hope, Newbold noted that "crises don't wait for adequate forces—you meet the crisis with what you have."[45] After securing its initial objectives of securing the port facility, the airfield, and the U.S. embassy compound, Newbold's SPMAGTF would then fold into follow-on forces from the 1st Marine Division, commanded by Major General Charles E. Wilhelm. The 1st Marine Division and BLT 2d Battalion, 9th Marines (BLT 2/9), had the mission of providing further security for the port and air facilities of Mogadishu and assisting I MEF forces built around the 7th Marine Regiment, commanded by Colonel Emil R. Bedard, that landed later.[46]

---

[45] Ohls, *Somalia . . . From the Sea*, 84.
[46] Col Dennis P. Mroczkowski, USMCR (Ret), *Restoring Hope: In Somalia with the Unified Task Force, 1992–1993* (Washington, DC: Marine Corps History Division, 2005), 15–16, 31–32, 37–38, 45.

Due to the sheer lawlessness in Mogadishu and its surrounding countryside, it quickly became clear to Johnston and Wilhelm that conducting a simple humanitarian relief operation was no longer in the cards. First Lieutenant Jason Q. Bohm, commanding the 81-millimeter mortar platoon in Weapons Company, 3d Battalion, 9th Marines, arrived in Somalia with his unit the second week of December, during which he found the intense smell of "drying camel carcasses hanging . . . outside butcher's huts" striking. Realizing that his company's mortars were not a likely force choice for teeming Mogadishu, Bohm suggested to his battalion commander, Lieutenant Colonel James P. Walsh, that his unit become the battalion's quick-reaction force. Walsh quickly agreed with Bohm's suggestion.[47]

This was a fortunate decision because Bohm's platoon was now required to respond to incidents taking place throughout the battalion's area of responsibility. Finding and removing weapons caches became one of its primary responsibilities. After conducting patrols in support of UNITAF in Mogadishu, Bohm and his platoon rejoined the rest of the battalion, along with other Coalition forces, in the interior city of Baidoa. This town served as an internal distribution point for humanitarian relief, making its security essential to the success of Operation Restore Hope. Despite its place as a hub for the humanitarian aid process, the U.S. forces gave it the nickname "City of Death" due to the "massive starvation in the area." Bohm noted that the space around Baidoa was randomly dotted with "small mounds of dirt that acted as graves for the thousands who had died." Because the graves seemingly had "no rhyme or reason to their placement," Bohm believed "it was almost as though people were buried right where they dropped."[48] Nevertheless, the scope of the entire relief and security operation in Somalia continued to expand.

Thanks to control over Mogadishu's port facilities and international airport, Johnston rapidly built up Coalition forces, which included the U.S. Army's 10th Mountain Division. Johnston divided the country into

---

[47] BGen Jason Q. Bohm, *From the Cold War to ISIL: One Marine's Journey* (Annapolis, MD: Naval Institute Press, 2019), 46–48.
[48] Bohm, *From the Cold War to ISIL*, 59.

Figure 62. Map of Somalia for Operation Restore Hope

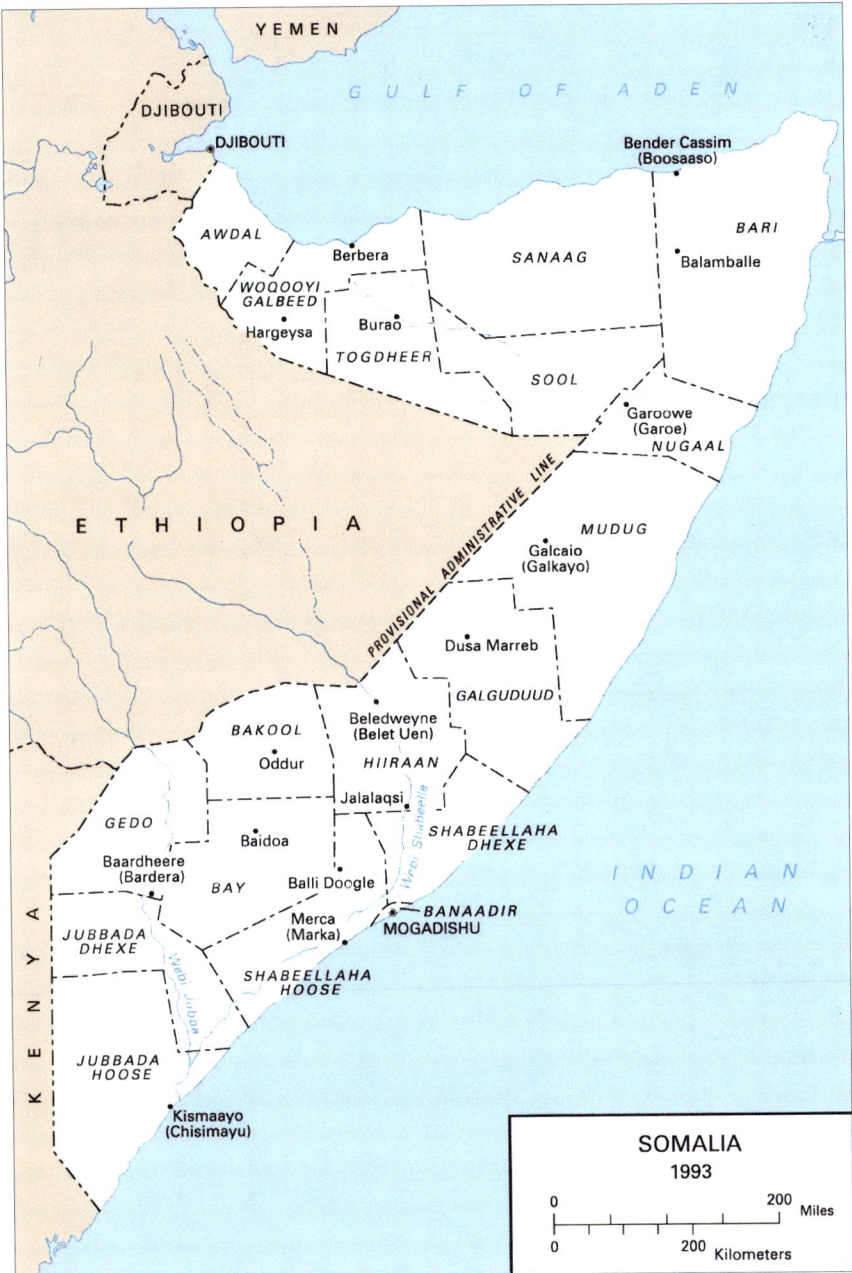

relief sectors but faced an unforeseen problem related to the colonial history of the region. The Somalis strongly objected to the presence of soldiers from former regional colonial powers such as France, Belgium, or Italy, which forced Johnston to adjust his plans. For example, UNITAF planners had initially tasked Italian troops to secure the city of Merka. Yet, many Somalis still resented Italy due to its attempted conquest of nearly the entire Horn of Africa in the 1930s under Italian prime minister and Fascist leader Benito Mussolini. As a result, "the local population protested strongly about the return of the Italians," which caused UNITAF leaders to give the assignment primarily to U.S. Army Forces Somalia. Somali warlords used the issue of colonialism as a "handy rallying call for the various factions when they organized protests against the presence or actions of UNITAF."[49]

One of the major issues that emerged during Operation Restore Hope was gun seizure missions. After years of lawlessness in Somalia, it seemed that nearly every person in the country, especially in densely populated Mogadishu, owned at least one fully automatic AK–47 rifle. Incidents of gun violence between local factions were frequent and usually deadly. The UNITAF forces believed that removing weapons caches and relieving some of the more violence-prone Somalis of their weapons would reduce this fratricide and provide better force protection. Further, Mogadishu teemed with vehicle-borne heavy machine gun teams referred to as "technicals." These vehicles, usually a pickup truck with a heavy gun mounted on its bed, became a constant concern for UNITAF.[50] Getting these highly lethal weapons systems out of the direct control of the warlords soon became a high priority. By early January 1993, Mohamed Aidid proclaimed that the Marines were unfairly disarming his men and warned of potential dire consequences if they did not desist. In fact, on 14 January, Somali gunmen shot and killed Private First Class Domingo

---

[49] Mroczkowski, *Restoring Hope*, 45. This volume is the best book covering the activities of all the UNITAF coalition forces in Somalia. Mroczkowski deployed with I MEF to cover the operation for the Marine Corps History Division.

[50] Jay E. Hines, Jason D. Mims, and Hans S. Pawlisch, *USCENTCOM in Somalia: Operations Provide Relief and Restore Hope* (MacDill AFB, FL: U.S. Central Command History Office, 1994), 25. Technicals gave those in the UN mission in Somalia the most concern.

Arroyo, a native of Puerto Rico, who had been on foot patrol near the Mogadishu airport. In the 10 days prior to Arroyo's death, the Marines had raided several warlord gun arsenals "and swarmed the Bakhara gun market in south Mogadishu, the largest in Somalia." Although disarming Somalis was not a primary mission for any portion of the Restore Hope Joint task force initially, it started to become one. To make matters worse, the UN secretary general, Boutros Boutros-Ghali, was a highly vocal advocate for total Somali disarmament, which led to "a global rumor" that the United States "had secretly agreed with Boutros-Ghali, in fact, to disarm Somalia."[51]

Early in Operation Restore Hope, Johnston let the Somali warlords know that no weapons were allowed in certain exclusion zones, specifically the areas around the port docks, the international airport, and the U.S. embassy compound. He also assisted Ambassador Oakley in convincing the most powerful warlords, such as Aidid, to place their technicals into cantonment areas, which were small, clearly defined areas designated for the encampment of personnel and equipment. While Boutros-Ghali considered disarmament a "'pre-requisite' to stability" in Somalia, no one seemed able to answer who would do the disarming.[52] Instead, Oakley and Johnston placed their hopes in reestablishing a credible police force or a national guard to fully restore order.

The diplomacy roles that Oakley, Johnston, and Libutti played cannot be underestimated. They all saw the proper coordination of diplomacy combined with the potential, credible threat of military force as paramount to the success of the overall operation. Johnston noted that throughout Operation Restore Hope, military commanders had to play more of a diplomatic role than a military one on numerous occasions. This kind of rare politico-military cooperation led to conferences on reconciliation and the establishment of protocols for weapons control and relief distribution. While Somalia was still a failed state in nearly every respect, the country began to return to a relative state of normalcy

---

[51] Stevenson, *Losing Mogadishu*, 64.

[52] Walter S. Poole, *The Effort to Save Somalia: August 1992–March 1994* (Washington, DC: Joint History Office, Office of the Chairman of the Joint Chiefs of Staff, 2005), 26–27.

by February 1993. Johnston noted that "every HRS [humanitarian relief sector] is different; commanders must be given broad missions. [They] will have to weave [their] way through a broad fabric of village elders and others. I'm pleased with what I see; commanders on the ground taking initiative and doing a splendid job."[53] For these kinds of situations, Johnston strongly recommended that flexible authority be given to unit commanders down to the lowest possible level. By March, UNITAF forces slowly disengaged from the region, turning over the entire effort to the United Nations Operations in Somalia II (UNOSOM II).

UNOSOM II ultimately became yet another UN-led peacekeeping disaster. After UNITAF forces departed Somalia in the spring of 1993, Aidid desired revenge for his alleged humiliation at the Coalition's hands, which included him sanctioning a June 1993 attack on Pakistani peacekeepers that lead to the deaths of 24 UNOSOM soldiers. Aidid's endorsement of the attack resulted in UNOSOM branding him a criminal—accurately so—and led to a U.S. Army Task Force Ranger incursion to bring him to justice that fall.[54]

The mission resulted in the capture of some high-value intelligence targets by the Rangers and the Army Delta Force, but it quickly descended into chaos, allowing Aidid and other high-value targets to escape arrest. In what became known as the Battle of Mogadishu (3–4 October 1993), the Rangers and associated Delta Force members had to fight their way out of the city against thousands of heavily armed—and now thoroughly enraged—Somali militia fighters. During the battle, two U.S. Army members of the elite Delta Force, Master Sergeant Gary I. Gordon and Sergeant First Class Randall D. Shughart, voluntarily fast-roped into the crash site of a downed Sikorsky UH-60 Black Hawk helicopter designated "Super-64" to defend its crew. Both Gordon and Shughart were killed and posthumously received the Medal of Honor for making the ultimate sacrifice on behalf of their fellow soldiers. In the two days of fighting, 18

---

[53] Mroczkowski, *Restoring Hope*, 111.

[54] The Task Force Ranger mission was later recounted in the superb book by Mark Bowden, *Black Hawk Down: A Story of Modern War* (New York: Atlantic Monthly Press, 1999), and later a major motion picture *Black Hawk Down*, directed by Ridley Scott (Culver City, CA: Columbia Pictures, 2001).

U.S. servicemembers lost their lives. Afterward, exultant Somalis at one point dragged the bodies of at least two U.S. soldiers through the streets in front of the international media.[55]

The outcome shocked President Clinton, who quickly ordered the Joint Chiefs of Staff to stop all efforts to arrest Aidid. The fallout was worse for Clinton's secretary of defense, Leslie "Les" Aspin Jr., who received significant criticism in the media over stories that he denied the Rangers armored forces they had requested before the "Black Hawk Down" mission, and he consequently resigned from office in December 1993. The U.S. experience in Somalia had been calamitous, but it proved a harbinger of even more extraordinarily complex post–Cold War contingencies to come. Alongside the deeply regrettable losses that Task Force Ranger incurred, the U.S. military learned important lessons about the difficulty of providing solutions to failed or failing states. Throughout its long institutional history, the Marine Corps had emphasized warfighting as its primary reason for being. Now, a more nuanced approach using both hard and soft power seemed appropriate for most contingency situations in the 1990s.[56]

Recurring humanitarian crises in Cuba and Haiti demonstrated similar issues. In September 1991, Jean-Bertrand Aristide, a former Catholic priest, became the president of Haiti. Soon after, a military junta under Haitian brigadier general Raoul Cédras overthrew Aristide's government. Aristide fled to the United States, where he promptly began a campaign to convince Clinton to take actions that would restore him to power. Rather than sending in troops, Clinton decided to levy strong economic sanctions against Haiti that seemingly left the nation's "tiny ruling class, which always seemed able to assure its own comfort, virtually unaffected."[57]

---

[55] Maj Timothy M. Karcher, USA, *Understanding the "Victory Disease": From the Little Bighorn, to Mogadishu, to the Future* (Fort Leavenworth, KS: U.S. Army Command and General Staff College, 2003).

[56] Karcher, *Understanding the "Victory Disease".*

[57] Col Nicholas E. Reynolds, USMCR, *A Skillful Show of Strength: U.S. Marines in the Caribbean, 1991–1996* (Washington, DC: Marine Corps History and Museums Division, 2003), 3.

Sensing a crisis in the making, the commander in chief of U.S. Atlantic Command (USLANTCOM), Navy admiral Leon A. Edney, crafted a plan to evacuate approximately 7,000 U.S. citizens from Haiti. To support the plan, Edney deployed 350 Marines to Guantánamo Bay, Cuba, from Camp Lejeune, North Carolina. However, Americans were not the people in immediate danger. Rather, refugees fleeing the oppressive Cédras regime in leaky, overcrowded boats were in the most peril. For example, the U.S. Coast Guard rescued one fragile 30-foot vessel that "held some 240 people, including women and children."[58]

The ultimate disposition of the refugees, once the Navy or Coast Guard rescued them at sea, became a central question. In the past, an Alien Migrant Interdiction Agreement between the United States and Haiti allowed the United States to simply return many of the migrants to Haiti. Executive Order 12807, signed on 23 May 1992, ended any screening process, making it even easier to send all migrants back to Haiti. Yet, a U.S. Supreme Court challenge prevented the United States from returning any migrants to Haiti "without at least reviewing their status in some way."[59] As a stopgap solution, the Navy and Coast Guard took the migrants to Guantánamo Bay, known to the Marines as "Gitmo." There, the United States provided the refugees temporary humanitarian assistance. Guantánamo was ideal as a temporary holding facility that afforded much needed humanitarian services while, at the same time, not being geographically connected to the United States.

Lieutenant General William M. Keys, then commanding FMFLANT, appointed Brigadier General George H. Walls Jr., commander of the 2d FSSG at Camp Lejeune, to lead JTF Guantánamo and assist the Navy with the reception of the displaced Haitians. Walls, who at that time was one of the few Black generals in the Marines Corps, was "a universally respected officer" who "possessed a gift for leading troops and trusting subordinates."[60] Walls soon named Colonel Peter R. Stenner as his operations chief. Most of the Marines who Walls placed in the task force

---

[58] Reynolds, *A Skillful Show of Strength*, 3–4.
[59] Reynolds, *A Skillful Show of Strength*, 5.
[60] Reynolds, *A Skillful Show of Strength*, 6.

were those capable of providing humanitarian assistance, such as medical personnel, engineers, and combat service support troops. By the time Walls and his task force arrived at Guantánamo Bay in late November 1991, more than 1,000 migrants were already there, with many more on the way. Providing basic shelter for these refugees was a massive undertaking, and the Marines typically "worked 18-hour days" to erect shelters for them.[61] Even a deployed Marine Corps artillery unit, Battery H, 3d Battalion, 10th Marines, that had recently returned from Operation Provide Comfort and arrived at Guantánamo Bay for a training exercise was dragooned into service assisting the task force and the base barracks Marines security force, commanded by Colonel Gary A. Blair. The other U.S. Services soon sent 400 additional personnel.[62]

Despite U.S. efforts, tensions grew between the refugees and military personnel. Growing impatient at the length of their confinement in the temporary camps at Guantánamo Bay, several hundred Haitians rioted at Camp McCalla II, located on the central portion of the naval base, on the evening of 14 December 1991. Other refugees could not understand why they were not being taken directly to the United States. When he arrived, General Walls was hit in the head by a rock but was otherwise not seriously harmed. Due to the continued unrest, Walls requested assistance from additional security forces from the United States as soon as possible. The Camp Lejeune base "air alert" unit, the 2d Battalion, 8th Marines, commanded by Lieutenant Colonel James C. Hardee, was on the ground two days later. Meanwhile, Walls created a plan for his now-reinforced mission to regain internal control over Camp McCalla II. He called the plan Operation Take Charge, which even included a phase in which the Joint force would fly in a battalion of Army military police personnel from Fort Lewis, Washington, to assist with the effort.[63]

At 0400 on 17 December 1991, Walls initiated Operation Take Charge. As he suspected, most of the malcontents were asleep, and the U.S. forces quickly regained total control over the camp. By 23 December, the Joint

---

[61] Reynolds, *A Skillful Show of Strength*, 8.
[62] Reynolds, *A Skillful Show of Strength*, 8–9.
[63] Reynolds, *A Skillful Show of Strength*, 12–15.

task force had gotten the situation settled to the point that the 2d Battalion, 8th Marines, returned to Camp Lejeune. By 1 January 1992, however, the number of migrants at Guantánamo Bay had swelled to nearly 11,000 people. A U.S. Supreme Court ruling that supported the Bush administration finally allowed the government to begin the forced repatriation of most of the migrants back to Haiti. The Marine Corps Fleet Anti-Terrorist Security Team company was sent to maintain order onboard the Coast Guard cutters taking the migrants to the Haitian capital of Port-au-Prince, especially important when many migrants became belligerent once they learned that they were returning to Haiti.[64]

In 1994, during yet another rising immigration tide from Haiti, Brigadier General Michael J. Williams, the commanding general of the 2d FSSG and a future Assistant Commandant of the Marine Corps, was ordered to form JTF 160 to deal with the growing crisis. To avoid another widely publicized riot ashore, this time, Clinton directed that the Services would screen all migrants for potential asylum while still at sea. This declaration proved problematic because it encouraged Haitians to flee on anything that would float, including inflated rubber inner tubes that were lashed together. Consequently, the hospital ship USNS *Comfort* (T-AH 20) was directed to anchor near Jamaica while the processing of those rescued took place. The 2d Battalion, 4th Marines (later renamed 2d Battalion, 6th Marines), under Lieutenant Colonel John R. Allen provided shipboard security. The Marine and Navy medical personnel encountered unanticipated problems due to the high number of migrants who arrived on board with infectious respiratory ailments such as tuberculosis. Because the ship's centralized ventilation system might spread these infections to the rest of the onboard population, including U.S. medical and security personnel, these individuals were placed in a ramshackle tent camp on the *Comfort*'s helicopter deck. This situation on deck reminded Allen of "the infamous Civil War prison camp at Andersonville." Nevertheless, the artificial deck camp served its purpose of

---

[64] Reynolds, *A Skillful Show of Strength*, 16–17.

isolating the infected persons, who accounted for approximately 12 percent of the ship's total migrant population.[65]

Additional issues emerged for the task force in other sectors. Due to the continuing influx of migrants, including 108 illegal Chinese migrants who the Coast Guard accidentally scooped up, the decision was made to create a land-based screening camp on Grand Turk Island in the Turks and Caicos Islands. This ultimately proved unworkable due to a series of events, such as the threat to the camp from tropical storms and its austere remoteness from just about everything in the region.[66] To further add to the task force's troubles, Jamaica balked at the early departure of *Comfort* because this meant that the nation would lose the revenue from the United States' payments for docking the *Comfort* there. While negotiations between Jamaica and the U.S. State Department dragged on, Allen's Marines had to put down at least one significant shipboard riot by the Haitian refugees. Allen's quick thinking, which resulted in the physical takedown of two of the most vociferous ringleaders, soon restored calm onboard. When the *Comfort* arrived at Guantánamo Bay, getting the migrants to leave proved nearly as complicated, but they were eventually persuaded to disembark. Some of the Haitians even gave Allen a handwritten "Certificate of Merit," written in French, that confirmed that "the Marines had accomplished their work."[67]

Once again, Guantánamo Bay became the central locus of the Haitian migration crisis. As had been the case two years earlier, the rapidly swelling tide of humanity in camps there created boiling tensions. This time, however, the camps held substantially higher numbers of migrants, averaging almost 20,000 resident migrants on any given day. To make matters worse, Cuban president Fidel Castro decided to no longer "detain would-be Cuban migrants determined to float to Florida," creating

[65] Reynolds, *A Skillful Show of Strength*, 27. In an impressive historical ceremony, John Allen and his regimental commander, Col Richard A. Huck, "led a route march from the pier [at Guantánamo] where the *Comfort* had docked to Cuzco Hills, where Sergeant [John] Quick had earned the Medal of Honor for signaling the fleet under fire in 1898. With Quick's sword at hand, Colonel Huck furled 2d Battalion, 4th Marines' battle colors and replaced it with 2d Battalion, 6th Marines colors." Reynolds, *A Skillful Show of Strength*, 31.
[66] Reynolds, *A Skillful Show of Strength*, 29.
[67] Reynolds, *A Skillful Show of Strength*, 31.

the potential for a second mass migration crisis. To circumvent another predicament, U.S. attorney general Janet Reno broadcast a radio message stating that Cuban migrants would "not be processed." Instead, she added ominously, "You are going to Guantanamo."[68]

The proclamation shocked Williams and his task force staff because of the ongoing situation with the Haitian refugees. Williams was told to prepare to receive several thousand Cuban migrants. By 27 August 1994, more than 14,000 Cuban migrants had already arrived, but USLANTCOM ordered Williams to prepare facilities to receive up to "45,000 Cuban migrants," which was eventually raised to 60,000—"a staggering new number." The flood of Cuban migrants created a new concern for the Marines. They believed that some of the migrants may be connected to Castro, who might be attempting to take Guantánamo Bay with a swell of alleged migrants who may or may not be seeking asylum in the United States. The new Guantánamo Bay Marine Barracks commander, Colonel John M. Himes, who formerly served as General Robert H. Barrow's military aide, "requested reinforcements to enable him to fight on two fronts: to hold the fence line and to protect the base's vital installation from rampaging [phony] migrants."[69]

By 12 September 1994, the U.S. forces had nearly lost control of Guantánamo Bay, with groups of Cuban migrants roaming parts of the base at will. A possible sabotage of the base's plumbing by the migrants caused a water shortage that exacerbated the problem for the Marines. Nonetheless, Allen and Lieutenant Colonel Douglas C. Redlich were determined to reassert control over the situation, launching an operation they called Clean Sweep. JTF 160 made critical upgrades for the Cuban migrants while also improving its internal security posture. Troublemakers among the migrants were separated from the larger population and returned to Cuban territory, whether they wished to be or not. Overall, JTF 160 learned valuable lessons about complex military and humanitarian operations in the new post–Cold War era. Allen, for one, believed that the future "key to success was to think in terms of low intensity

---

[68] Reynolds, *A Skillful Show of Strength*, 44–45.
[69] Reynolds, *A Skillful Show of Strength*, 47–48.

conflict and to continue to analyze the threat, remaining alert for changes and adapting to those changes."[70] This attitude later paid dividends for Allen when he commanded the North American Treaty Organization mission in Afghanistan during Operation Enduring Freedom.

Throughout the 1990s, Haiti proved an especially difficult conundrum for Joint planners. Since Cédras's military coup ousted the democratically elected regime in 1991, Aristide remained in exile in the United States, continually demanding that the U.S. government take action to reinstate his office in Haiti. Moreover, Cédras's rule was anything but benign. He regularly used his control of the Haitian military to crack down on dissidents. In three years under Cédras, "international observers estimated that more than 3,000 men, women, and children were murdered by or with the complicity of Haiti's then-coup regime."[71]

Although Cédras had signed an agreement with the United States to allow Aristide to return to power on 30 October 1993, he soon reneged on the deal. With the Somalia debacle still fresh, the United States faced another humiliating turn of events when an armed mob of Haitian civilians and military forces compelled the USS *Harlan County* (LST 1196), which carried 193 U.S. and 25 Canadian peacekeepers, to turn back from a peaceful landing at Port-au-Prince.[72] The entire fiasco was broadcast directly back to the United States.

Despite the problems that a defiant Cédras created, many people in the U.S. government also questioned Aristide's capacity to lead. In one instance, Senator Jesse A. Helms Jr. (R-NC) prodded the Central Intelligence Agency to declare that Aristide was "psychologically unstable, drug addicted, and prone to violence." In anticipation of possibly deploying forces to Haiti, USLANTCOM formed a Haiti intervention plan

---

[70] Reynolds, *A Skillful Show of Strength*, 60–61, 68.

[71] Kathleen Marie Whitney, "Sin, Fraph, and the CIA: U.S. Covert Action in Haiti," *Southwestern Journal of Law and Trade in the Americas* 3, no. 2 (1996): 321–22.

[72] Douglas Farah and Michael Tarr, "Haitians Block U.S. Troop Arrival," *Washington Post*, 12 October 1993; and Jared M. Tracy, "The USS *Harlan County* Incident: October 1993," U.S. Army Special Operations History Office of the Command Historian, accessed 27 July 2022.

Figure 63. Haitian president Jean-Bertrand Aristide

Source: official U.S. Air Force photo.

referred to as Jade Green.[73] While Jade Green needed some fleshing out, the Pentagon began developing, under tight security, Operations Plan 2370 (OPLAN 2370), "the forcible-entry option to return democracy to Haiti," in late October 1993.[74]

Nearly simultaneous to the development of OPLAN 2370, a "political-military plan" was formed for the USLANTCOM commander, U.S. Navy admiral Paul D. Miller. According to a plans officer with USLANTCOM at that time, U.S. Army lieutenant colonel Edward P. Donnelly Jr., the political-military plan for Haiti "was a first . . . because numerous government agencies and a unified command . . . participated in its creation." The National Security Council approved it, making it "authoritative to all." Additionally, it "further served to shape the Jade Green OPLAN that was rapidly coming to fruition."[75] In hope of a U.S. diplomatic breakthrough that would negate any potential use of force, planners also created an alternative OPLAN 2380 that called for an unopposed insertion of peacekeepers and humanitarian assistance on the island. Both operation plans contained widely divergent rules of engagement for any forces in Haiti. Eventually, the planners put together OPLAN 2375, a hybrid plan that bridged the more forceful OPLAN 2370 and the largely benign OPLAN 2380. OPLAN 2375 was supposed to cover the possibility that an incursion in Haiti would include both humanitarian and violent actions. As a result, this hybrid plan called for a modified force package with increased peacekeeping elements than the more warfighting-oriented OPLAN 2370. By this time in the early post-Cold War era, the Pentagon recognized the efficacy of Joint task forces. As had been the case with the successful Operation Just Cause in Panama, the chairman of the Joint Chiefs of Staff, Army general John M. D. Shalikashvili, ordered the formation of JTF 180, with the Army's XVIII Airborne Corps,

---

[73] Walter E. Kretchik, Robert F. Baumann, and John T. Fishel, *Invasion, Intervention, "Intervasion": A Concise History of the U.S. Army in Operation Uphold Democracy* (Fort Leavenworth, KS: U.S. Army Command and Staff College, 1998), 43.

[74] John R. Ballard, *Upholding Democracy: The United States Military Campaign in Haiti, 1994–1997* (Westport, CT: Praeger, 1998), 65. Ballard, a retired Marine Corps colonel who later became a defense expert on the employment of Joint forces during contingency operations, noted that MajGen John J. Sheehan was one of the key early planners for the Haiti intervention.

[75] Kretchik, Baumann, and Fishel, *Invasion, Intervention, "Intervasion,"* 44.

then commanded by Lieutenant General Henry H. Shelton, in the lead, while the senior unified combatant command, USLANTCOM, led by Admiral Miller, provided overarching guidance for the entire operation.[76]

The proposed plan for Haiti revolved around two central tasks. First, a large force would occupy the capital of Port-au-Prince, and then it would have the responsibility of establishing an enclave for returning Haitian immigrants.[77] This enclave was deemed necessary due to the huge numbers of Haitians being repatriated back to Haiti. To allow for a quick insertion, Miller used the deck of the nuclear-powered aircraft carrier USS *Dwight D. Eisenhower* (CVN 69), minus much of its air wing, to stage the Army ground forces. At the same time, Army Rangers were positioned aboard USS *America* (CV 66), again sans much of its air wing. This staging was a first for these aircraft carriers, essentially turning them into large and extremely expensive amphibious assault platforms. Miller created this concept, referring to it as adaptive Joint force packaging (AJFP). The idea of placing large numbers of Army ground forces onto the decks of a major naval combatant platform, such as a nuclear aircraft carrier, never sat well with the Navy's aviation component, nor did it with the Marine Corps, which saw it as a dangerous encroachment on its own mission. Other leaders, however, believed that the AJFP concept had the potential to assist the Services with adjusting to the "post-Cold War international security environment and smaller force structure." Particularly, the AJFP could secure "a greater degree of effectiveness to U.S. military operations at the lower end of the conflict spectrum." Nevertheless, AJFP supporters anticipated strong opposition to concept because opponents would argue that it would "create an additional layer of bureaucracy, [would] conflict with other, uncoordinated, force packaging initiatives, [would] limit force employment options to a fixed menu, and, finally, [would] never come to fruition." Consequently, once Miller

---

[76] Ballard, *Upholding Democracy*, 65–69.
[77] Ballard, *Upholding Democracy*, 65–66.

retired from active service on 31 October 1994, the Joint Chiefs of Staff quietly shelved the novel concept.[78]

Earlier in 1994, Les Aspin had resigned as the secretary of defense due to the fallout resulting from the Somalia debacle the previous year. The deputy secretary of defense, William J. Perry, who had previously served as the undersecretary of defense for research and engineering under President James E. "Jimmy" Carter Jr., replaced Aspin after being confirmed by a 97–0 Senate vote. The highly capable Perry led the DOD for the next three years, but he began his tenure having to deal with the Haiti crisis. In the summer of 1994, the Haitian military junta ordered human-rights monitors from both the UN and the Organization of American States out of the country. In response, the Pentagon ordered the USS *Inchon* (LPH 12) Amphibious Ready Group (ARG), which included 2,000 Marines embarked on *Inchon*, into the Caribbean on 6 July 1994 in case the United States needed to evacuate any U.S. civilians from Haiti. Additionally, USLANTCOM "activated a 'Haiti Response Cell' to deal with the situation 24 hours per day." At that time, it was clear that most of the U.S. military planners preferred the more complete but military-centric OPLAN 2370 over the less clear and diplomacy-minded OPLAN 2380.[79]

President Clinton, whose first term in office had been hampered by the vagaries of the Somalia operation, was determined to prevent the same thing from happening in Haiti. Consequently, the politically charged rules of engagement for the potential deployment of U.S. forces in Haiti soon took top priority. In a meeting at the Pentagon with the Joint Chiefs of Staff, Clinton expressed concern for the possibility of resistance to the installation of the rightfully elected Aristide. If the action required the use of force, Admiral Miller argued, the Joint Chiefs could adjust the rules of engagement as needed. Chairman Shalikashvili feared that elements of Cédras's military would initiate an insurgency

---

[78] Sean A. Bergesen, "Adaptive Joint Force Packaging (AJFP): A Critical Analysis" (master's thesis, Naval Postgraduate School, 1993), iii. The article, "Other Services Criticize AJFP Concept, but Air Force Approves," *Inside the Pentagon*, 21 April 1994, 13, provided specific reasons for why the Joint Staff soon rejected Miller's novel concept. In sum, it seemed it was only popular at Miller's USLANTCOM headquarters and with some elements of the Army.

[79] Ballard, *Upholding Democracy*, 85–86.

campaign similar to the one that occurred in Somalia once U.S. forces landed. Most of the Joint forces' leadership saw this possibility as remote due to the Haitian military having little to no support from the general population. Nevertheless, the U.S. forces scheduled for Operation Uphold Democracy, as the plan was now called, received three different rules of engagement cards that listed the parameters for the different situations they might face. Clinton was later heard lamenting, "If we were in the United States, we'd know what the difference was between a demonstration that turned into a riot and whether somebody robbed the corner 7-Eleven." The operation planners seemed to concur that operational speed could help diminish lawless activity once U.S. forces arrived. In sum, the Clinton administration was asking the military to take and occupy Haiti—by force if necessary—but preferred that ongoing diplomacy would allow the peaceful insertion of forces that could then maintain law and order throughout a country that no longer had a functioning military nor a stable police force.[80]

After all this planning, the Clinton administration suddenly put the operation on hold. At the last moment, a diplomatic delegation consisting of former President Carter, former chairman of the Joint Chiefs of Staff Colin L. Powell, and Senator Samuel A. Nunn Jr. reached an agreement with Cédras that would restore the Aristide presidency just as combat troops of the U.S. Army's 82d Airborne Division, 10th Mountain Division, and Army Rangers were en route to Port-au-Prince. The insertion, rather than the forced-entry combat operation that was initially planned, quickly morphed into what the generals of the 1990s called a "military operation other than war" (MOOTW), with the 10th Mountain Division now acting primarily as an occupation force. The combat forces were turned around midflight and quickly reconfigured, with the exception of much of the 82d Airborne Division, for the more benign insertion on 19 September 1994.[81]

---

[80] Bob Shacochis, *The Immaculate Invasion* (New York: Grove Press, 1999), 71. Shacochis is an award-winning journalist who spent 18 months covering Operation Uphold Democracy.
[81] Philippe R. Girard, "Peacekeeping, Politics, and the 1994 U.S. Intervention in Haiti," *Journal of Conflict Studies* 24, no. 1 (Summer 2004): 20–41; and "President Carter Leads Delegation to Negotiate Peace with Haiti," Carter Center, accessed 27 April 2023.

Prior to the diplomatic breakthrough with Cédras, the Marine Corps conducted two relatively large-scale show-of-force amphibious exercises in the Caribbean in the summer of 1994, which had been designed to also make an impression on Cédras. The first involved a mock amphibious landing in the Bahamas by the 24th MEU(SOC), commanded by Colonel Martin R. Berndt, in July. According to *The Washington Post*, Clinton administration officials apparently "called attention to the action" in order to "unnerve Haiti's military leaders and pressure them into leaving." In August, 1,000 Marines conducted the second exercise around the naval air station at Roosevelt Roads, Puerto Rico, to hypothetically "secure the airfield and prepare to evacuate civilians."[82]

Many of the Marines involved in both exercises had been deployed to Somalia previously and had been away from home for more than eight months. Eventually, the decision was made to replace the 24th MEU(SOC) with a SPMAGTF that was "built around the headquarters of the 2d Marines under Colonel Thomas S. Jones." The Camp Lejeune-based 2d Marines had been training for a Caribbean contingency for more than a year. The SPMAGTF contained nearly 2,000 Marines and sailors, including an artillery unit, Battery B, 1st Battalion, 10th Marines. Due to the need for artillery being highly unlikely in any contingency situation in Haiti, Jones ordered Battery B to form into a provisional rifle company. The 2d Marine regimental headquarters also had a more robust planning staff and significantly more logistical support than a deployed MEU(SOC). Jones and his staff believed that the SPMAGTF would most likely take on the mission of a noncombatant evacuation of U.S. civilians. Consequently, Jones conducted a series of amphibious exercises and landing rehearsals using the decks of the USS *Wasp* (LHD 1) ARG as staging grounds. Even so, Jones still did not know whether a potential Haitian incursion would be kinetic or peaceful. Further, the XVIII Airborne Corps headquarters had informed Jones to focus on Cap-Haïtien, Haiti's second largest city, as a potential objective because the Army had Port-au-Prince as its target.[83]

---

[82] Reynolds, *A Skillful Show of Strength*, 93–95.
[83] Reynolds, *A Skillful Show of Strength*, 95–97.

After the mission changed literally overnight, the Marines in the SPMAGTF no longer prepared to conduct a forced entry operation. While the Marines watched helicopters carrying the Army's 10th Mountain Division conduct an administrative landing from USS *Eisenhower* into Port-au-Prince on 19 September, the SPMAGTF was informed that it would conduct a nonhostile landing at Cap-Haïtien the following day. The rules of engagement, however, counted as a legitimate target any Haitian spotted carrying a weapon in an aggressive manner. Colonel Thomas C. Greenwood, the SPMAGTF's operations officer, noted that the Army had fired no shots during its incursion and told the Marines that "everyone we meet is an ally until they prove otherwise."[84]

To make landing operations as seamless as possible, Jones divided his Marines into two task forces: Irish and Hawg. He assigned Task Force Irish the mission of securing the port of Cap-Haïtien, while Task Force Hawg secured the nearby airfield. When the Marines arrived, thousands of Haitians poured out of their homes, many waving small American flags. Despite the largely pro-Cédras military and local police forces still being under arms, they did not overtly resist the Marines, primarily due to negotiations between Jones and the Haitian regional military commandant, Lieutenant Colonel Claudel Josephat, over the security to his forces and that of the local population. Nevertheless, Josephat was clearly not happy with the positive reception given to the Marines by the Haitian people. The first day of operations for the Marines went as smoothly as anyone could expect, but they remained on high alert that first night ashore.[85]

The Marines soon established checkpoints throughout Cap-Haïtien for the purpose of population control. After the landing, tensions grew between the local police, some members of the Haitian military, the largely pro-Aristide locals, and the Marines. In fact, between 21 and 24 September, the Haitian military and local police increased their attacks against their own citizens. On 24 September, Josephat informed Jones that a situation between Josephat's men and the Marines was becoming

---

[84] Reynolds, *A Skillful Show of Strength*, 103.
[85] Reynolds, *A Skillful Show of Strength*, 106–7.

Figure 64. U.S. military and political leadership meet with Col Thomas S. Jones after violence breaks out between U.S. forces and Haitian police during Operation Uphold Democracy

Source: National Archives and Records Administration.

increasingly troubling, but that he would do his best to control it. Soon, elements of a Marine patrol, Company E, 2d Battalion, 2d Marines, led by Captain Richard L. Diddams, found itself in a confrontation near a Haitian police station, where a growing crowd of Haitian civilians urged the Marines to "shoot the police." As the hostilities increased, two policemen aggressively pointed their weapons directly at Company E's platoon commander, 1st Lieutenant Virgil A. Palumbo, who immediately raised his own weapon and fired two shots into the policemen. Other members of Palumbo's platoon then opened fire on the Haitian police in and around the facility. The confrontation resulted in the deaths of 8 Haitian policemen and the wounding of at least 3 others, "two of whom died of their wounds before they could be evacuated." There was a single U.S. casualty, Navy boiler technician Jose Joseph, who had been acting as an interpreter and had been wounded in the leg. One Marine described the action, which lasted less than a minute, as being "like [the shootout] at the OK Corral."[86]

The following day, fallout from the police station firefight came swiftly. Although the rules of engagement clearly established that the Marines were justified in their use of deadly force in this situation, the commanders in Cap-Haïtien knew the incident would receive exceptional scrutiny from their political and military superiors. They were also concerned that the incident could initiate "a massive uprising" within the primarily peaceful Haitian military, which could create a "commensurate increase in the level of combat in Haiti."[87] Further, in what was becoming characteristic of many of these hybrid combat-humanitarian operations, situations on the ground often required military forces to make split-second life-and-death decisions at the local level, far from any senior controlling headquarters, which threatened to have far-reaching political and even military consequences. Most importantly, the incident clearly signaled to the Haitian military and police that the Americans would fight if confronted.

Despite the loss of life, the response of the Haitian people to the incident was exactly opposite of what might have been expected. Apparently, the Haitian police force was so unpopular that the day after the incident, 25 September, "mobs in Cap Haitien looted four police stations."[88] Additionally, it was not coincidental that the first airplanes carrying restive Haitian refugees from Guantànamo Bay landed in Port-au-Prince. The rapid repatriation of the Guantanamo refugees was also an implied mission of Operation Uphold Democracy. Senior U.S. leadership on the ground had concerns over how the Haitian government and general population would react to the apparent "forced repatriation" of Haitian citizens.[89] As a result, the Americans went out of their way to emphasize that all the migrants had willingly decided to return home.

---

[86] Reynolds, *A Skillful Show of Strength*, 111–12.
[87] Ballard, *Upholding Democracy*, 115.
[88] Kretchik, Baumann, and Fishel, *Invasion, Intervention, "Intervasion,"* 99.
[89] Ballard, *Upholding Democracy*, 116.

Operation Uphold Democracy seemed to expose multiple recurring themes that the U.S. military would face in MOOTWs, which were occurring with increasing frequency in the 1990s. In fact, some referred to the Haitian contingency event as an "intervasion," an operation that acts as a simultaneous humanitarian intervention and military invasion.[90] By the end of September 1994, 19,479 servicemembers had joined JTF 180, almost 5,000 more than U.S. planners had first estimated the operation would require after the completion of the assault phase. The use of OPLAN 2380 caused "a bunching effect in the numbers of the force that could not be avoided." Consequently, Admiral Miller directed that JTF 180 reduce its personnel numbers to 15,500 servicemembers as "intended for steady-state Uphold Democracy operations." With numerous moving parts across Haiti, the reduction in force could help the Joint task force combat potential "mission creep," something that had clearly plagued the Somalia operation. By this time, UN peacekeepers began arriving to assist the U.S. forces with basic security and humanitarian relief operations. The UN personnel, especially the international police monitors, greatly assisted the JTF commander with a "weapons buy-back program" while also starting to reform the Haitian police, which they considered "a critical requirement for the success of the operation."[91]

On 15 October 1994, President Aristide, accompanied by U.S. Secretary of State Warren M. Christopher, among others, triumphally entered the presidential palace in Port-au-Prince. Even so, sporadic riots continued to take place in the Haitian slums. Everyone involved in Haiti at the time believed that the nation needed nation-building and civil affairs assistance more than anything. General Shelton's more combat-centric JTF 180, however, was not prepared or mandated to support this mission. Consequently, the DOD ordered JTF 180 back to the United States, replacing most of the initial insertion force with soldiers from the Army's 10th Mountain Division, commanded by Major General David C. Meade, now the head of JTF 190. The force handoff went smoothly, which paid dividends for Meade later. Meade would successfully lead the tran-

---

[90] Kretchik, Baumann, and Fishel, *Invasion, Intervention, "Intervasion,"* 27–28.
[91] Ballard, *Upholding Democracy*, 117.

sition of the U.S.-led effort from combat operations to the longer-term nation-building and humanitarian relief mission and then, ultimately, to the UN Mission in Haiti, to which the United States continued to provide forces through 1996.[92]

During the height of the Haitian intervention, the Senate Committee on Armed Services held hearings on the situation there. Lieutenant General John J. Sheehan, the director of operations for the Joint Chiefs of Staff, and Deputy Secretary of Defense John M. Deutch, both of whom played a major role in formulating Operation Uphold Democracy, testified on 28 September 1994. During the committee's opening remarks that day, Senator J. Strom Thurmond Sr. (R-SC) warned both Sheehan and Deutch that he hoped they "had not forgotten the lessons of Somalia," nor the casualties there that "resulted from mission creep and military commanders trying to follow directions emanating from a confused foreign policy." Emphasizing his point, Thurmond declared, "We cannot allow that to happen again, in Haiti or anywhere."[93] Other members of the committee expressed similar concerns. These senators seemed to desire an established target date or window for the withdrawal of U.S. forces from Haiti to avoid mission creep like in Somalia. Deutch explained to the committee that the plan in Haiti was for a phased transfer of responsibility from solely U.S forces to the UN peacekeepers that would include some U.S. personnel for the nation-building effort, meaning that a set target date would make this transition more difficult. Despite Deutch's explanation, many of the committee members feared that the U.S. military commanders were slowly and inexorably being drawn into a mission creep situation in Haiti. Senator William S. Cohen (R-ME) specifically challenged Deutch when he stated that "back in the early 1980s there used to be a lot of people running around with T-shirts on and chanting that El Salvador was Spanish for Vietnam. If that was the case, then I would respectfully suggest to you that Haiti is patois for Beirut and Mog-

---

[92] Ballard, *Upholding Democracy*, 125–28.
[93] *Hearing on the Situation in Haiti, before the Senate Committee on Armed Services*, 103d Cong. (28 September 1994) (statement of Strom Thurmond, senator from SC), 2.

adishu."[94] Sheehan expressed little concern about the situation, noting that U.S. forces were already "feeding 1.2 to 1.3 million Haitians a day" while also ramping up other aspects of humanitarian assistance.[95] All in all, U.S. forces in Haiti seemed to have a solid hold over the state of affairs, especially after the issue of the police station gun fight subsided. Nevertheless, all of the committee members, regardless of political affiliation, retained concerns that Haiti could easily turn into another Somalia.

In 1995, after the bulk of U.S. forces had turned over operations to the UN, the Center for Naval Analyses put together a summary of material from a conference it held on military support to complex humanitarian emergencies, using Haiti as the central example. Sheehan participated in the conference, noting that "the success of a peacekeeping mission," based on the example that Uphold Democracy set as a successful operation, "is fundamentally a result of political decisions." To ensure these results, he argued, U.S. political leadership needs to "ask the right questions about the reasons for the mission and to make the right judgments with regard to that mission." Sheehan saw this effectively play out in Haiti. He pointed out that the political judgment to give the Cédras regime "a 30-day grace period to step down out of power" was the right one. Second, he believed that the "political-military plan reflected the right questions about the operation" and had been "well-integrated." The plan allowed for "excellent cooperation between various U.S. federal departments, agencies, and the military," while the "personalities and talents of the on-scene commanders were well suited" for peacekeeping and humanitarian operations.[96] Finally, he argued that President Clinton's executive decision to lift economic sanctions against Haiti greatly aided the overall success of the operation.

Nevertheless, Donald E. Schulz, then a professor of national security affairs at the U.S. Army War College, was not so sanguine, noting in 1996 that grave concerns remained in the United States. Despite signif-

---

94 *Hearing on the Situation in Haiti, before the Senate Committee on Armed Services* (testimony of John M. Deutch, deputy secretary of defense), 37.
95 *Hearing on the Situation in Haiti, before the Senate Committee on Armed Services* (testimony of LtGen John J. Sheehan, USMC, director of operations, Office of the Joint Chiefs of Staff), 13.
96 McGrady and Ivancovich, *Operation Uphold Democracy*, 35–37.

icant U.S. aid, including the intervention of military forces, he argued that congressional support for the entire operation "was extraordinarily thin and was conditioned on being able to move in, restore order and move out, while keeping U.S. casualties to an absolute minimum." These aims "structured serious limitations and irrationalities into the policy, which endangered the success of the entire operation."[97] Just a year later, however, Schulz wrote a postscript to his original study on Haiti, contending that the "political situation remained extremely fragile and volatile" and calling any possibilities for improved socioeconomic development "at best problematic." Primarily, Schulz worried that the United States deciding to suspend "over $130 million in foreign aid due to the failure of the Aristide administration to privatize state industries and rationalize the economy in accordance with the prescriptions of the U.S. Government and international financial institutions" stunted any existing or potential development.[98] Still, Schulz concluded that, at least in Haiti, "there are no easy answers." Believing that the United States relied heavily on "a tendency to view the world in simplistic terms," he contended instead that "in making judgments, the proper frame of reference is the country's past, rather than some idealistic—and invariably culture-bound—standard of what ought to be."[99]

Consequently, although Schulz admitted that Haiti remained in a fragile state, he argued that the country's improvement from its standards before the intervention to after the operation made Uphold Democracy "a qualified success." After Uphold Democracy, he noted:

> Aristide was restored and political power transferred from one duly elected government to another; a new police force has been created which, for all its limitations, is functioning reasonably well under extremely difficult conditions; though political violence and human rights violations persist, they are not the massive problem they have been in the past. Today, indeed,

---

[97] Donald E. Schulz, *Whither Haiti?* (Carlisle, PA: U.S. Army Strategic Studies Institute, 1996), vii–viii.
[98] Donald E. Schulz, *Haiti Update* (Carlisle, PA: U.S. Army Strategic Studies Institute, 1997), 1.
[99] Schulz, *Haiti Update*, 16.

Haiti has a government that not only has the broad support of the populace, but is clearly the most responsible and competent government the country has had in decades—and arguably ever.[100]

Africa continued to pose special challenges for U.S. forces throughout the 1990s. Prior to Operation Desert Storm, a growing civil war in Liberia required MEUs to conduct several major noncombatant evacuation operations for U.S. embassy personnel and civilians who found themselves trapped there. To facilitate the evacuations in late May 1990, Major Glen Sachtleben went ashore dressed in civilian clothes to conduct a semicovert assessment of potential landing beaches but did not find an ideal spot. Yet, the U.S. Department of State opposed any further beach surveys due to the concern that the Liberian government might confuse a beach evacuation for a full-scale amphibious invasion. Accordingly, the operation planners modified the evacuation plans to rely primarily on the MEU helicopter squadron HMM-261.[101]

To make matters worse, as the Marines of the 22d MEU lingered offshore in late July 1990, Liberia's president, the corrupt Samuel K. Doe, falsely accused the U.S. military attaché at the U.S. embassy of directly assisting rebel forces then fighting his own soldiers in the bloody civil war. Doe also began to foment anti-Americanism among the general population to garner further popular support for his now largely hated and increasingly brutal military forces. At this time, Doe's forces began randomly executing or mutilating their internal enemies, including attacking individuals in hospitals, schools, and churches. In early August, the U.S. embassy in Monrovia, in response to the violence, requested that the 22d MEU initiate a noncombatant evacuation operation of U.S. civilians from the embassy as well as U.S. citizens and foreign nationals who could get to the designated evacuation landing zones.[102] Most no-

---

[100] Schulz, *Haiti Update*, 21.
[101] Maj James G. Antal and Maj R. John Vanden Berghe, *On Mamba Station: U.S. Marines in West Africa, 1990–2003* (Washington, DC: Marine Corps History and Museums Division, 2004), 14–16.
[102] Antal and Vanden Berghe, *On Mamba Station*, 19–21.

tably, with the U.S. embassy remaining open, the MEU continued to additional security and further logistical support. The 22d MEU successful evacuated several thousand refugees largely without incident.

Being concerned that the contending forces in Monrovia might target the evacuees, the 22d MEU Commander, Colonel Granville R. Amos, wanted to complete the evacuation as rapidly as possible. Amos issued extensively detailed rules of engagement for his primary security element, BLT 2d Battalion, 4th Marines (BLT 2/4), "in the event that the host nation is unable or unwilling to provide the necessary protection to U.S. military forces." Amos's instructions, however, also provided the on-scene commanders great flexibility in the amount of force they deemed necessary, including the use of riot control agents. Nevertheless, the embassy directed that the Marines "allow vehicles to use the road if they showed no hostile intent and did not stop."[103] The following month, the 26th MEU(SOC) relieved the 22d MEU, and Amos's unit went back to Camp Lejeune for a long-deserved rest.[104]

Although on a smaller scale, the evacuation of citizens from Monrovia continued until February 1991, when a shaky ceasefire between the warring factions allowed the 26th MEU to depart that month. Nevertheless, full-fledged violence broke out once again in Monrovia a year later. Throughout the 1990s, West-African-based MEUs, often alongside other West African peacekeeping forces, were required to provide sporadic protection for U.S. citizens and other foreign nationals who found themselves entangled in seemingly never-ending factional fighting. Lieutenant Colonel Barry M. Ford, who participated in a noncombatant evacuation operation in Sierra Leone in 1997, noted that "the entire continent is really in desperate trouble" and that the U.S. forces would be "spending a lot of time in Africa."[105] Ford's prediction became significant throughout the late 1990s and early 2000s, as Marine units were deployed to Eritrea, Kenya, and Tanzania in East Africa and the Congo in Central Africa in 1998; Mozambique in Southeast Africa in 2000; and

---

[103] Antal and Vanden Berghe, *On Mamba Station*, 34.
[104] Antal and Vanden Berghe, *On Mamba Station*, 43–44.
[105] LtCol Barry M. Ford, quoted in Antal and Vanden Berghe, *On Mamba Station*, 103.

Liberia again in 2003.[106] The political and social complexity of nearly every post-Cold War intervention seemed to exponentially increase with each successive operation. Moreover, despite the Cold War being over, the global demand signal for amphibiously based U.S. Marines continued to rise rather than plummet.

As Marine units dealt with the turmoil at Guantánamo Bay and in Haiti and Liberia during the summer and fall of 1994, Commandant Mundy was dealing with another kind of insurgency at home. This revolution had to do with a DOD-generated national security study known as the "bottom-up review." The bottom-up review, led by then-Secretary of Defense Aspin, was an outgrowth of the Base Force concept that had been started under General Powell. Aspin was a longtime critic—previously as a former member of the House Committee on Armed Services—of the department he now led. During the 1992 presidential campaign, Aspin urged Clinton to consider additional military personnel cuts of 200,000 beyond even the austere Base Force plan. Instead, Clinton responded by calling for a "33 percent cut in military personnel to 1.4 million by 1997." The bottom-up review also changed the nation's longstanding national military strategy of being capable of simultaneously fighting two major regional contingencies, as the United States had done since World War II, to being able to do so "nearly simultaneously"—meaning U.S. forces would fight contingencies sequentially and within months of each other.[107]

While a reduction in force affected all the U.S. Services across the board, the percentage of women serving on active duty continued to rise. For example, by 1993, "15.5 percent of the Air Force enlisted component was female . . . [as were] nearly 13 percent of the Army and 11.5 percent of Navy enlisted personnel." Of all the Services, the more infantry-oriented and combat-centric Marine Corps had the lowest percentage of enlisted women at 5 percent of its total force. Partially due to the personnel cuts, Congress took steps to permit women serving in "even more assignments and specialties." That November, Congress "repealed the law

---

[106] Antal and Vanden Berghe, *On Mamba Station*, 103.
[107] Brasher, *Implosion*, 114.

preventing female sailors from serving aboard combat vessels," which created "136,000 additional posts to females."[108] Around this same time, Aspin dropped the restriction that prohibited women from flying combat aircraft.

The bottom-up review allowed the Marine Corps to keep its three-active-duty MEF and aircraft wing force structure, but Mundy had to fight hard to keep its overall strength from dropping below 173,000. In a letter to retired Marine generals, Mundy asked for their assistance in convincing the U.S. public and members of Congress that the Service represented a thin red line. Mundy used this phrase in a nod to Rudyard Kipling's famous "1890 poem about Tommy Adkins, the English version of G.I. Joe, who fought Britain's colonial wars."[109] Mundy believed that the force cuts coupled with ever-increasing deployments and contingency operations stretched the Marines to their financial breaking point, making him feel that the analogy to the Kipling-inspired and long-suffering Tommy Adkins was an apt one. Mundy argued that if he could convince Congress to establish a "hard funding floor" of $14 billion (USD) per year, what he called the "thin red bottom line," then the Marine Corps could continue to conduct multiple and often simultaneous contingency operations around the globe. To further strengthen his argument, Mundy referred to the Marine Corps as the "nickel force." At slightly less than 5 percent of the overall U.S. defense budget, the Marines provided far more combat power for the taxpayer dollar than any other Service.[110] By any measure, the Marine Corps represented a true defense bargain.

---

[108] Brasher, *Implosion*, 113. The full report can be found in Les Aspin, *Report on the Bottom-up Review* (Washington, DC: Department of Defense, 1993). See also the repeal of Title 10 U.S.C. Section 6015 on 30 November 1993.

[109] David Evans, "For a Thin Red Line of Marines, We Need Sam's Black Ink," *Chicago (IL) Tribune*, 17 September 1993; and Rudyard Kipling, "Tommy," Kipling Society, accessed 30 January 2023. Some analysts argued that Mundy's "nickel figure" did not account for the support that the U.S. Navy provided or "blue dollars in support of green," such as its service medical support and aviation acquisition. Others, however, pointed out that Marine Corps combat forces represented nearly 17 percent of all such forces the United States possessed in the 1990s. There was no doubt that the Marines were providing this force for far less than the other Services.

[110] "For a Thin Red Line of Marines, We Need Sam's Black Ink."

Nevertheless, by 1997, "the Corps had been cut by 26,383 Marines since 1987 (13.7 percent)."[111]

Even more disconcerting was DOD spending on procurement, which was slated to drop by 60 percent between 1990 and 1997. While the Marine Corps would finally get the Bell Boeing V-22 Osprey tiltrotor aircraft into its procurement budget, it was not going to get as many aircraft—only 360—that the Service thought was needed. By 1998, while all Marine Corps aviation platform levels had fallen by a small percentage (6.9 percent), the Marine Reserve had to reduce the number of squadrons it had operated since 1987 by half.[112] By 1997, according to defense expert Bart Brasher, more than one-half of all servicemembers had only been part of an "armed force that was constantly imploding."[113]

The immediate post-Cold War era also saw some significant doctrinal changes for the Navy-Marine Corps team. On 30 September 1992, Secretary of the Navy Sean C. O'Keefe, Chief of Naval Operations Admiral Frank B. Kelso II, and Commandant Mundy published a new maritime strategy document titled . . . *From the Sea: Preparing the Naval Service for the 21st Century*. This extraordinary and relatively concise paper was designed to replace the old Cold War–based *Maritime Strategy* of the 1980s. Compared to other strategy announcements, . . . *From the Sea* was revolutionary in scope, but it was not without controversy. For the first time in decades, the new strategy required the Navy to concentrate less on sea lines of communication and sea control and focus more on power projection. The document stated that "while the prospect of global war has receded," the United States was "entering a period of enormous uncertainty in regions critical to our national interests." Yet, the Services could "shape the future in ways favorable to our interests by underpinning our alliances, precluding threats, and helping to preserve the strategic position we won with the end of the Cold War." It further stressed that the Navy and Marine Corps now needed to provide naval expeditionary forces that were "shaped for joint operations," able to operate "forward

---

[111] Brasher, *Implosion*, 178.
[112] Brasher, *Implosion*, 185, 196.
[113] Brasher, *Implosion*, 178.

from the sea" and "tailored for national needs."[114] In sum, O'Keefe, Kelso, and Mundy established the future focus of national maritime strategy being on the world's sea littorals.

While the Marine Corps could not have been happier with ... *From the Sea* because it played directly into its core competency of amphibious warfare, components of the Navy—such its submarine service, which had thoroughly dominated the Service's senior ranks throughout much of the Cold War, and its carrier aviation—were not convinced that the era of defending critical sea lines of communication and sea control was entirely over. Consequently, two years later, Secretary of the Navy John H. Dalton, Chief of Naval Operations Admiral Jeremy M. Boorda, and Mundy signed off on *Forward ... From the Sea*. This document reversed much of the revolutionary fervor found in ... *From the Sea*. While the Navy and Marine Corps continued to improve its "readiness to project power in the littorals," the authors contended, the two Services still needed to "proceed cautiously so as not to jeopardize" their abilities to fulfill "the full spectrum of missions and functions for which we are responsible."[115] Dalton, Boorda, and Mundy believed that ... *From the Sea* provided "the initial step in demonstrating how the Navy and Marine Corps [would] respond to the challenges of a new security environment." Further, this modified strategy document ensured that sea control received its due and noted that "naval forces have five fundamental and enduring roles in support of the *National Security Strategy*: projection of power from sea to land, sea control and maritime supremacy, strategic deterrence, strategic sealift, and forward naval presence."[116] While the modified strategy still emphasized combat in the littorals, it reflected that the Navy's leadership worried about other issues of equal or even greater concern.

---

[114] Sean C. O'Keefe, Adm Frank B. Kelso II, USN, and Gen Carl E. Mundy Jr., ... *From the Sea: Preparing the Naval Service for the 21st Century* (Washington, DC: Department of the Navy, 1992), 1.

[115] John H. Dalton, Adm Jeremy M. Boorda, USN, and Gen Carl E. Mundy Jr., *Forward ... from the Sea* (Washington, DC: Department of the Navy, 1994), 8. It should be noted that the new document was pointing out that the Navy had more expansive national strategy concerns than indicated in ... *From the Sea*.

[116] Dalton, Boorda, and Mundy, *Forward ... from the Sea*, 10.

General Charles C. Krulak became the 31st Commandant of the Marine Corps on 1 July 1995. Krulak had a long family tradition connected with the Service. His father, Lieutenant General Victor H. Krulak, was a decorated World War II combat veteran and author of the acclaimed book *First to Fight: An Inside View of the U.S. Marine Corps*, and his godfather, Lieutenant General Holland M. Smith, had served as the senior amphibious force commander in the Pacific during World War II. Born at Marine Corps Base Quantico in 1942, the younger Krulak, following his father's route, attended the U.S. Naval Academy and was commissioned into the Marine Corps following his graduation in 1964. Soon after, Krulak served two tours of combat duty in Vietnam and was awarded the Silver Star and the Bronze Star with combat valor device and two Purple Hearts.[117]

As a brigadier general, Krulak led the 2d FSSG during Operation Desert Shield and Operation Desert Storm. During the height of preparations for Operation Desert Storm, Krulak discovered that he could better support the planned U.S.-led ground offensive by establishing a forward location near the Kuwaiti border. Consequently, he created a major forward logistics base at a deserted place in northern Saudi Arabia that he named al-Khanjar. As Krulak later stated, "we started building this miracle in the desert on 6 February and had it completed by 0100 on 20 February." This military hub indeed was an oasis. Once completed, al-Khanjar covered more than 11,000 acres. It became "the largest fuel point the Marine Corps had ever seen," storing 5 million gallons of fuel as well as 100,000 gallons of water. It also contained "the third-largest Navy hospital in the world in terms of operating rooms" within the compound. The importance of al-Khanjar convinced Krulak that "when historians and strategists and tacticians study the Gulf War, what they will study most carefully will be the logistics," declaring that it was "a war of logistics."[118] Undoubtedly, the creation of al-Khanjar greatly

---

[117] Author's personal conversation with Gen Charles C. Krulak, October 2017; and author's phone conversation with BGen Thomas V. Draude, August 2019. For a more detailed history of Krulak, see "Who's Who in Marine Corps History: General Charles C. Krulak, USMC (Retired)," Marine Corps History Division, accessed 8 January 2024.

[118] Gen Charles C. Krulak, quoted in Otto Kreisher, "Marines' Desert Victory," *Naval History* 30, no. 1 (February 2016).

assisted with the incredible success that the I MEF ground combat forces later achieved during their breach of the Iraqi minefields and the subsequent liberation of Kuwait City.

From his first day in office as Commandant, Krulak was a highly energetic leader. He wanted to build on the positive doctrinal changes that Commandant Alfred M. Gray Jr. had initiated. Krulak astutely recognized that the Marine Corps now needed to be mentally prepared for the complex operational challenges that it would face in the twenty-first century. He quickly set about preparing the Marines for the future. Most importantly, he centered his focus on the professional military education (PME) of enlisted Marines.

Krulak was a prolific writer throughout his time as a Marine officer and even into his retirement. One of the most important documents he authored was the *31st Commandant's Planning Guidance*, which focused on the twenty-first-century Marine Corps and noted that future crises for the United States would "place heavy demands on our nation's military services; demands that will require deep reservoirs of military skill, intellect, and innovation." Krulak believed that "this uncertain horizon" would require the retention of "a military force that can remain versatile yet act decisively in the face of such uncertainty." To ensure the existence of "a force that can quickly and surely anticipate change and adapt to a new reality," Krulak would maintain the Service's role as "that versatile, decisive force."[119]

Once he became Commandant, Krulak called on all the living past Commandants, especially Generals Louis H. Wilson Jr. and Mundy, actively seeking their advice. Krulak was amazed that Wilson remained actively engaged with emerging issues that impacted the Marine Corps, allowing him to provide timely and well-directed advice. Krulak recalled Wilson emphasizing that his position as Commandant gave him an advantage that "no other service chief has; the aura of being Commandant" and the "trappings that go with it." With Marine Barracks Washington, DC, and the Commandant's home being just 3 kilometers from the

---

[119] Gen Charles C. Krulak, "Commandant's Planning Guidance (CPG)," *Marine Corps Gazette* 79, no. 8 (August 1995): A-2.

Figure 62. Gen Charles C. Krulak

Source: official U.S. Marine Corps photo.

U.S. Capitol Building, he could use his proximity to "really win the grass roots of America" and Congress. Following Wilson's advice, Krulak would "host lunches and breakfasts. We took the parades and used them in different manners, all of it trying to develop a strength of support with the American people, and the Congress of the United States."[120]

Krulak emphasized the need for the Marine Corps to develop even more exceptional, innovative skills for the twenty-first century while also remaining a force in readiness for a wide variety of contingency operations. To get there, he wanted to fill the Service with "the world's finest military professionals—disciplined, motivated, dedicated warriors—stronger, smarter Marines, filled with the values that have served us well throughout our history, and infused with the agility of mind and body that will be required in future conflicts."[121] He also worked to ensure that Marines were "educated to act intelligently and independently, trained to seek responsibility, required to be accountable, and molded to act with boldness and individual initiative." He wanted the Service to be "a learning organization that creates individuals who not only can adapt to changing situations, but who can anticipate and even activate them."[122]

Krulak wished to get back to the concept of warfighting in an increasingly chaotic world. He noted that:

It goes back to the making of Marines to win battles. Our ethos is founded in our twin touchstones: the touchstone of valor and the touchstone of values. I wanted that touchstone of valor, that sense that we are warriors. We have a warrior ethos. Nothing is more important than our ability to fight our nation's battles and to win them. I just thought it was important. We had a very strong emphasis on war fighting under General Gray. Under General Mundy, there were different battles to fight, primarily battles involving end strength. What I was trying to do

---

[120] Krulak Oral History Transcript, 12 July 1999, Marine Corps History Division, Quantico, VA, 142.
[121] Krulak, "Commandant's Planning Guidance (CPG)," A-3.
[122] Krulak, "Commandant's Planning Guidance (CPG)," A-3.

was say, okay, we are always going to fight the end-strength battle, so let's get back to war fighting.[123]

Having observed the actions in Somalia, Haiti, Guantànamo Bay, and other locations such as Bosnia in the early to mid-1990s, Krulak published a pathbreaking document titled "The Strategic Corporal: Leadership in the Three-Block War" in the January 1999 issue of the *Marine Corps Gazette*. In the article, he laid out for his readers a fictitious contingency scenario, similar in scope to any number of confrontational situations that U.S. forces faced in the decade, that he called Operation Absolute Agility. Krulak observed that future young Marine leaders, even down to the junior noncommissioned officer level, would face tremendous pressure and responsibility to make appropriate and far-reaching decisions. He argued that "the rapid diffusion of technology, the growth of a multitude of transnational factors, and the consequences of increasing globalization and economic interdependence" blended together to "create national security challenges remarkable for their complexity." In slightly more than two decades, he predicted, "about 85 percent of the world's inhabitants will be crowded into coastal cities generally lacking the infrastructure required to support their burgeoning populations." These cramped conditions would lead to an explosion among "long simmering ethnic, nationalist, and economic tensions" that would "increase the potential of crises requiring U.S. intervention. Compounding the challenges posed by this growing global instability," he believed, "will be the emergence of an increasingly complex and lethal battlefield." An increased access to advanced weaponry would also "level the playing field," ending any form of technological superiority for the United States, and "adversaries, confounded by our 'conventional' superiority," will turn to asymmetrical actions to overcome any imbalances. Finally, he believed that the "ubiquitous media whose presence will mean

---

[123] Krulak Oral History Transcript, 12 July 1999, 149–50.

that all future conflicts will be acted out before an international audience" would create further complications for Marine Corps leadership.[124]

In sum, within a confined space on the future battlefield, Marines might be required to conduct humanitarian assistance operations, peacekeeping operations, traditional warfighting operations, or any combination of the three. These situations would become even more complicated because they would require junior noncommissioned officers to make critical decisions far away from senior leadership. In fact, the outcome of an entire operation might well rest on the "decisions made by small unit leaders, and by actions taken at the lowest level." Consequently, Krulak believed that the Marines needed to declare an end to the "zero defects" mentality of the past, in which junior leaders were not trusted to make critical decisions. The modern battlefield now virtually demanded that junior noncommissioned officers be better educated, mentored, and entrusted by senior leadership, while being held accountable for their actions.[125]

Pivoting off the white papers . . . *From the Sea* and *Forward . . . From the Sea*, Krulak, in cooperation with Chief of Naval Operations Admiral Jay L. Johnson, published another doctrinal concept related to the projection of naval power ashore titled *Operational Maneuver from the Sea* (OMFTS). Fully admitting that the document's initial assumptions would likely change, both Krulak and Johnson agreed that the littorals offered both a challenge and an opportunity.[126] The challenge was the need to develop a naval expeditionary force that could fight and win in usually confined littoral sea spaces. The opportunity was that any successful approach from the sea provided a future amphibious task force with tremendous flexibility related to where it could attack. Using the tenets of OMFTS, for instance, a fictitious amphibious task force emanating from Spain could make a successful attack anywhere along the entire eastern

---

[124] Gen Charles C. Krulak, "The Strategic Corporal: Leadership in the Three-Block War," *Marine Corps Gazette* 83, no. 1 (January 1999): 20.
[125] Krulak, "The Strategic Corporal," 20–22.
[126] Gen Charles C. Krulak, *Operational Maneuver from the Sea: A Concept for the Projection of Power Ashore* (Washington, DC: Headquarters Marine Corps, 1997), 2.

seaboard of North America, forcing a notional enemy "to be strong everywhere" and making things tremendously difficult for them.[127]

Krulak went on to establish a Warfighting Laboratory at Marine Corps Base Quantico. He was adamant that the Warfighting Laboratory act as an umbrella organization charged with the "development, field-testing, and implementation of future operational and functional concepts in potential doctrine, organization, training, education, and support solutions." From this lab, multiple cutting-edge conceptual experiments, such as Sea Dragon, emerged. Consequently, Sea Dragon—or Green Dragon, as it was first called—was really an output of the overarching Warfighting Laboratory effort that, in turn, led to further experimentation as needed. Krulak envisioned such experiments becoming the twenty-first-century version of the "Culebra" exercises that were so familiar during Major General John A. Lejeune's tenure as Commandant in the 1920s. For example, the Sea Dragon concept went on to require lesser follow-on experiments, such as Hunter Warrior and Urban Warrior. This latter operation focused on training for combat in urban spaces, which Krulak was convinced that Marines would face in greater frequency and ferocity in the early twenty-first century.[128] He was right.

One of Krulak's more exciting innovations was his creation of "the Crucible" exercise at both Marine Corps Recruit Depots at Parris Island, South Carolina, and San Diego, California. The Crucible had been established as an innovative capstone exercise for Marine recruits on the verge of graduating from one of the most rigorous military boot camps in the world. According to Krulak, he did not create the Crucible because he thought that existing Marine Corps recruit training was failing. Instead, he believed that changes in potential operating environments for these new Marines required a different type of training at the entry level. Krulak noted that "decentralized operations, high technology, increasing weapons lethality, asymmetric threats, the mixing of combatants and non-combatants, and urban combat will be the order of the day vice the exception in the 21st century." As a result, Krulak argued that Ma-

---

[127] Krulak, *Operational Maneuver from the Sea*, 8.
[128] Krulak Oral History Transcript, 12 July 1999, 169.

rine training should be of the "highest standard" and should instill them with excellent decision-making skills, self-confidence, and an "absolute faith in the members of their unit."[129] The Crucible gave new Marine Corps recruits all that and more.

By Krulak's time as Commandant, the past had become prologue. As the twentieth century—arguably the most violent and lethal century in the history of mankind—neared its end, Krulak believed that the Marine Corps had arrived at yet another strategic inflection point.

As the Marine Corps had accomplished with the concept of amphibious warfare prior to and during World War II, Krulak urged the modern-day Marine Corps, along with the other U.S. Services, to look ahead to the next strategic inflection point. He believed that potential threats in the early twenty-first century would not "be the 'son of Desert Storm'—it will be the 'stepchild of [the Chechen Wars]'," and that the "most dangerous enemy will not be doctrinaire or predictable." During his time as Commandant, Krulak forecasted the twenty-first century being as violent as the previous one but in a fundamentally different way. Instead of encountering force-on-force combat, the U.S. military would face more amorphous and confusing conflicts. Consequently, he noted that rapid global urbanization, global interdependence, competition for resources such as oil and potable water, the rise of China and India as comparable economic superpowers to the United States, and the intertwining of state sovereignty with technology and migrant flows would cause the new world order to become one of great disorder, making future security threats highly complex and difficult to manage.[130]

The high level of this complexity would require new Marines to embrace different skills that would turn them into what Krulak called the "Strategic Corporal." He believed that twenty-first-century conflicts would force Marines to "deal with a bewildering array of challenges and threats" that would "require unwavering maturity, judgment,

---

[129] Gen Charles C. Krulak, "The Crucible: Building Warriors for the 21st Century," *Marine Corps Gazette* 81, no. 7 (July 1997): 14.

[130] Gen Charles C. Krulak, "Operational Maneuver from the Sea," *Joint Forces Quarterly* 21, no. 1 (Spring 1999): 79.

Figure 66. Marine recruits at Parris Island, SC, conquer "the Crucible"

Source: official U.S. Marine Corps photo.

and strength of character" for them to succeed. "Most importantly," he argued, "these missions will require them to confidently make well-reasoned and independent decisions under extreme stress—decisions that will likely be subject to the harsh scrutiny of both the media and the court of public opinion." Under numerous potential circumstances, Krulak imagined that individual Marines would become the "most conspicuous symbol of American foreign policy," having a role in influencing immediate tactical situations as well as operational and strategic decisions. Consequently, the actions of these Strategic Corporals would "directly impact the outcome of the larger operation."[131]

This emerging doctrinal dilemma pushed Krulak to champion PME for all Marines. He especially supported increasing education among

---

[131] Krulak, "The Strategic Corporal: Leadership in the Three-Block War," *Leatherneck*, January 1999, 16.

the noncommissioned officer corps. Along with the Strategic Corporal, Krulak articulated an accompanying concept known as the "Three-Block War."[132] In this latter idea, he argued that every Marine must be prepared to assist refugees, to conduct combat operations in a mid-intensity combat environment, and to conduct peacekeeping operations "all on the same day" and "all within three city blocks."[133]

Despite its connection with the development of warfighting, the Three-Block War concept still had its critics. In 2005, for example, the Canadian Army, then under Lieutenant General Rick J. Hillier as chief of the Land Staff, thoroughly embraced the Three-Block War concept, with Hillier stating that it would "significantly alter how we structure, how we prepare, how we command, how we train, how we operate, and how we sustain ourselves."[134] In sum, the Canadian military fully embraced the Three-Block War, but it quickly proved difficult to fulfill. In 2006 and 2007, Canadian forces "applied this theory in Afghanistan's Kandahar province." During that time, Canadian soldiers "suffered a fatality rate" that was more than double that of the U.S. or U. forces there.[135] Two Canadian defense scholars, Walter Dorn and Michael Varey, argued that the concept tried to "simplify deeply complicated situations." They wrote that "the concept's one-size-fits-all approach risks losing sight of the special nature of many missions. By ignoring that some operations are primarily humanitarian, or peace support, or outright offensive combat, the concepts does not allow specificity of mission mandate, which is critical for mission clarity, both for Canadian forces personnel and local populations."[136]

---

[132] Krulak, "The Strategic Corporal," 14–17.

[133] Gen Charles C. Krulak, "The Three Block War: Fighting in Urban Areas," *Vital Speeches of the Day* 64, no. 5 (December 1997): 139.

[134] Walter Dorn and Michael Varey, "Fatally Flawed: The Rise and Demise of the 'Three-Block War' Concept," *International Journal* 63, no. 4 (December 2008): 968, https://doi.org/10.1177/002070200806300409.

[135] Franklin Annis, "Krulak Revisited: The Three-Block War, Strategic Corporals, and the Future Battlefield," Modern War Institute at West Point, 3 February 2020; and Dorn and Varey, "Fatally Flawed," 975.

[136] Dorn and Varey, "Fatally Flawed," 970–71.

In the United States, multiple defense leaders and analysts recognized some of the nuances within Krulak's Three-Block War. Marine Corps lieutenant general James N. Mattis and retired Marine Corps Reserve lieutenant colonel Frank G. Hoffman believed that the concept needed expansion to a fourth level, what they called the "Four-Block War" in 2005. They argued that infatuation with technology and alleged revolutions in military affairs in the United States reflected "an unrealistic desire to dictate the conduct of war on our own terms." By adding another dimension to Krulak's original concept, they addressed the "psychological or information operations aspects" of these new conflicts. "This fourth block," Mattis and Hoffman contended, "is the area where you may not be physically located but in which we are communicating or broadcasting our message."[137] Defense analyst Franklin C. Annis believed that "the nuances of Krulak's thinking have largely been abandoned outside the Marine Corps, if they were ever truly appreciated." Still, Annis argues, recognizing "how Krulak's ideas have been distorted" allows analysts to "chart the ways in which the strategic-corporal developmental philosophy is still relevant on the modern battlefield as a means of addressing war's increasing complexity." Sensationalized events that occurred during Operation Iraqi Freedom, such as the Abu Ghraib prisoner abuse incident, have given rise to a corrupted view of the Strategic Corporal, with soldiers being "forever associated with negative consequences" because they created a "fear of strategic 'mission failure' due to the actions of junior leaders" that also "caused vast degradation in the practice of mission command and eroded critical trust between leaders at various levels."[138]

U.S. Army colonel Thomas M. Feltey, while a student at the National Defense University's Joint Advance Warfare School, wrote an outstanding treatise on how the Strategic Corporal concept changed from its inception in 1997 to 2013. Fear of catastrophic mistakes made by junior leadership, Feltey argues, resulted in senior leadership establishing increasingly re-

---

[137] LtGen James N. Mattis and LtCol Frank G. Hoffman, "Future Warfare: The Rise of Hybrid Wars," U.S. Naval Institute *Proceedings* 131, no. 11 (November 2005): 18–19.
[138] Annis, "Krulak Revisited."

strictive management of the tactical environment, which in turn eroded the trust between the layers of command so important for success in an irregular warfare environment. Feltey analyzed four major—and, quite frankly, horrific—acts perpetuated by U.S. servicemembers on Vietnamese, Iraqi, and Afghan civilian populations. Based on the three examples from Iraq and Afghanistan, he came away believing that only the Abu Ghraib prisoner abuse incident had any long-term strategic significance for both of those war efforts.[139] Despite the horrors involved in the other instances, Feltey believed that Abu Ghraib was "so significant in the minds of combat commanders that their attitudes toward junior leaders and their perceptions of the strategic corporal changed," although not for good. Feltey concluded that the only way to reverse the situation was to restore an environment of trust between the two levels of leadership. However, Annis argued that "this relabeling of Krulak's concept [was] aimed at removing negative connotations inserted by others. Instead of repackaging the concept into a new term [empowering environment of trust], it would be far better to return to the positive notions of the strategic corporal by re-educating leaders about Krulak's original concept."[140]

Just like when commissioned and noncommissioned officers who emerged from the Vietnam War took proactive steps to reform the U.S. military, contemporary senior Marine leadership created programs in the post–Cold War world that promised to keep the Service on an appropriate learning and innovation track. Moreover, the Goldwater-Nichols Act of 1986 ensured that Joint warfare was here to stay, which is important to the future, as conflicts have foreshadowed a more chaotic and uncertain future. There is a clear need to take advantage of what all the U.S. Services can bring to the table to achieve operational success. Rather than anticipating another Operation Desert Storm, modern-day Marine Corps planners saw the operation as a unique anomaly. Instead, Three-Block War operations in Somalia and Haiti seemed to be the new nor-

---

[139] Col Thomas M. Feltey, USA, "Debunking the Myth of the Strategic Corporal" (master's thesis, National Defense University, 2015), 1–42.
[140] Annis, "Krulak Revisited."

mal. Even sustained major combat operations in Iraq and Afghanistan did not change this paradigm. Training Marines for hybrid-style warfare remains as relevant today as it was toward the end of the Cold War. Clearly, the Marine Corps needs more and better noncommissioned officers to ensure its success in the future. Education is the key, and Marine Corps schools at Marine Corps Base Quantico and other major Marine Corps installations must provide it.

CONCLUSION

The U.S. Marine Corps of the twenty-first century needs to conduct a serious review of its future role and mission within the U.S. national security enterprise or risk having the National Command Authority question its organizational existence and utility once again. Perhaps the current Marine Corps leadership could follow the example of General Carl E. Mundy Jr. when he faced the strong possibility of manpower and defense budget cuts after becoming the 30th Commandant of the Marine Corps following the Cold War. In response to the reductions, he gave Brigadier General Charles C. Krulak the task of forming a Force Structure Planning Group (FSPG). Another FSPG could help the Marine Corps prepare for the next 25 years as a rising and technologically advanced Chinese military, which is increasingly determined to keep the United States and its allies out of the South China Sea, likely becomes dominant.[1]

General David H. Berger, the 38th Commandant, seems to have initiated this debate and made some difficult choices about the Service's force structure. In a 2022 interview with *The Washington Post* reporter David Ignatius, Berger discussed the potential role of the Navy–Marine Corps team in Southeast Asia in the coming 8–10 years. With a rising China clearly on his mind, Berger predicted the possibility that the proliferation of precision-guided munitions (PGMs) will reach a point in which a

---

[1] *Marine Corps Force Structure Plan: The Final Report of the Force Structure Planning Group* (Washington, DC: Headquarters Marine Corps, 1991).

traditional Navy-Marine Corps amphibious operation will not be a likely option for any Joint task force commander. Berger argued that this situation "has led to a defensive posture on the part of some nations like China that believes they can deter, that they can control everything within the range of their long-range precision weapons." Berger went on to describe a picture of what has changed. The past trend for naval forces in such a PGM-dominated environment was to "stand off," or create some distance between their forces and these precision—and likely hypersonic—weapons. Instead, Berger stated, the Navy could employ another option of "standing in," meaning operating inside of the weapons engagement zone of an opponent. "If the U.S. needed to operate inside that regime," Berger argued, "it's not an either standoff or get inside. It's actually both." He contended that "the U.S. military has to be able to operate in great depth. And the Marine Corps' traditional, unique role is upfront and standing in." Berger saw Navy and Marine Corps forces being widely distributed while also having the ability to effectively operate either inside or outside the purported Chinese PGM umbrella. Yet, he had not given up entirely on amphibious warfare. "I think amphibious landings, amphibious assault, forcible entry, those things which Marines are known for, for 70 years, we'll continue to do, but we'll do them in a very different way," he stated. "And why? Because the character of war is changing. We need to change with it."[2]

The Cold War period in Marine Corps history is clearly replete with lessons for maritime strategists today. These lessons, especially after the conclusion of the Vietnam War, came fast and furious for Marine Corps senior leadership, starting with the 25th Commandant, General Robert E. Cushman Jr., and continuing to this day. The SS *Mayaguez* incident of May 1975 was one of the first, and worst, missions assigned to Marine Corps operational forces during the post-Vietnam War era. Several elements related to that operation stand out as fundamentally different

---

[2] Gen David H. Berger, interview with David Ignatius, "The Path Forward," *Washington Post Live*, transcript, 16 March 2022; and Maj William F. Dammin, "Force Design 2030 and the Marine Expeditionary Unit: Strategic Implications of Tactical Innovation" (master's thesis, Marine Corps University, 2022).

from other types of combat that Marines had experienced to that date. First, the operation was undertaken on incredibly short notice due to President Gerald R. Ford's direct demand for rapid action. This aspect illustrates the significance of the readiness of especially forward deployed operational forces. III Marine Amphibious Force (III MAF) on Okinawa, due to its proximity to Koh Tang Island, where U.S. forces thought the *Mayaguez* crew had been taken, was a logical choice for the operation. However, III MAF at the time had the reputation for being the least ready of all Marine Corps operational forces largely due to its traditional single-year deployment policy. The selection of a III MAF unit for this mission was a poor choice by the chairman of the Joint Chiefs of Staff, especially with the Marine Corps having a high-performing and complete combat brigade stationed in Hawaii that could have completed the mission. Finally, the Koh Tang Island operation required the close coordination of three separate Services—the Marine Corps, Air Force, and Navy—but suffered from poor management as "the White House and the Defense Department clashed over control of military operations."[3] The inter-Service cooperation of any of the Joint forces involved at Koh Tang was not much better.

Intelligence, Joint interoperability, and overall communications were significant issues as well. In fact, performance in all three of these critical areas throughout the *Mayaguez* crisis was extremely poor. One of the issues that emerged in most after-action reports on the affair noted that a major problem for Lieutenant Colonel Randall W. Austin's tactical operation was that he lacked the necessary intelligence for a successful operation. Disparate estimates of the Khmer Rouge forces facing the Marines on Koh Tang Island ranged from 20 poorly trained militia members to a heavily armed reinforced rifle battalion. The Joint cooperation between the three Services at Koh Tang was simply atrocious. The Marines, engaged in heavy combat ashore, could not initially talk directly to orbiting Air Force strike aircraft. The operation improved for the U.S. forces only after the arrival of a highly effective Air Force forward air controller

---

[3] David Vergun, "Lessons Learned from 1975 *Mayaguez* Incident," Defense.gov, 11 December 2018.

with a radio that could connect with Marine Corps forces on the ground. The White House's National Security Council staff reaching out directly to Air Force pilots flying bombing missions against suspected Khmer Rouge ships in and around the island in real time only exacerbated the situation.[4] The Marine Corps would face similar situations as those on Koh Tang during a longer period eight years later with its intervention in Beirut, Lebanon, in 1983.

Thanks to the revolution in worldwide telecommunications, from 1975 on, Washington, DC, bureaucrats developed a tendency to reach out and involve themselves directly in overseas operations, often to the detriment of the on-scene commanders. This situation improved once the military began using standing Joint task forces for contingency operations, with an on-scene commander, besides providing a unity of command, serving as an important filter for queries from Washington. The superior performance of the XVIII Airborne Corps during Operation Just Cause in Panama and Operation Uphold Democracy in Haiti provide two examples of how much progress had been made in Joint warfighting after 1975. The passage of the Goldwater-Nichols Act in 1986 provided additional motivations toward improving Joint warfare and virtually demanded that all the Services improve their cooperation levels or risk cuts to their funding and to promotion opportunities for their rising leaders.

Operation Urgent Fury in Grenada in 1983 indicated the limitations of conducting a Joint warfighting operation. The United States was exceedingly lucky to not have experienced yet another Koh Tang Island situation. Fortunately, the U.S. military applied overwhelming force, unlike when it used a single battalion at Koh Tang, against its opponents during Urgent Fury, and it also helped that the local population largely opposed the thuggish Marxist government of Bernard Coard and General Hudson Austin. Similarly, as had happened at Koh Tang Island, the U.S. forces held an extremely muddled intelligence picture of the situation in Grenada. The intelligence given to the U.S. units assigned to locate and evacuate U.S. citizens thought to primarily be located at a single med-

---

[4] Vergun, "Lessons Learned from 1975 *Mayaguez* Incident."

ical school campus turned out to be in error. More than half of all students in need of rescue were located elsewhere, including a completely unknown campus that was only discovered after combat operations were well underway. As at Koh Tang, inter-Service communication was exceptionally poor, and much of the activity ashore was coordinated on the fly or not at all. The only way the Joint task force commander could correct the situation was to physically divide the island up between a sector largely dedicated to Army forces and a second given to the Marines. When comparing Operation Urgent Fury with Operation Just Cause in Panama in 1989, the efficiency and effectiveness of Joint forces in each of these operations were unbelievably different.

One interesting strategic inflection point that emerged toward the end of the Cold War was a phenomenon that defense analysts labeled "hybrid warfare." In general, hybrid warfare consists of a fusion of conventional and unconventional elements of warfare that are "blended in a synchronized manner to exploit the vulnerabilities of an antagonist and achieve synergistic effects."[5] The intervention in Beirut acted as the first Cold War contingency that approached the hybrid warfare paradigm, but it faced additional issues that went beyond hybrid warfare. The War Powers Resolution, which Congress passed over President Richard M. Nixon's veto in 1973, created additional layers of complexity to the situation in Beirut. Although designed to avoid another seemingly unending war like that in Vietnam, the resolution unintentionally created confusion for the U.S. military as it avoided officially having troops engage in combat against an armed and usually unconventional opponent. Despite the casualties that the 24th Marine Amphibious Unit (24th MAU) incurred before the Marine barracks bombing on 23 October 1983, those Marines did not receive combat pay throughout most of the crisis because the Ronald W. Reagan administration purposely did not extend that privilege to them. At that time, the Reagan White House believed that congressional Democrats might invoke the War Powers Resolution, thereby taking control of U.S. policy in Lebanon, if the 24th MAU Marines received that special

---

[5] Arsalan Bilal, "Hybrid Warfare: New Threats, Complexity and 'Trust' as the Antidote," North Atlantic Treaty Organization, 30 November 2021.

pay they certainly deserved. Although the confusing local political situation in Lebanon had existed for centuries, U.S. policymakers who focused on Lebanon largely ignored this important consideration, which likely led to the disaster in Beirut in 1983. By February 1984, Reagan ordered all U.S. forces, with the exception of the reconstituted embassy guard, to leave Lebanon. The Syrians and the Islamic militias celebrated their departure, which reinforced the impression that the United States and its Europeans allies had no desire to engage in this new type of hybrid war.

Nevertheless, hybrid warfare has become the regular form of conflict in the modern era, something that Joint force commanders must recognize and address. Some defense analysts have argued that conflicts based in hybrid warfare have been the only ones that the United States has lost since its founding in 1776. Invariably, defeats in Vietnam, Lebanon, and Somalia did not necessarily come from the battlefield. Rather, they came from citizens of the United States losing their political will to carry on the fight. One analyst, Oxford University–educated retired Marine colonel Thomas X. Hammes, explained that the world has entered an era of *fourth-generation warfare* (4GW). This type of warfare "uses all available networks—political, economic, social, and military—to convince the enemy's political decisionmakers that their strategic goals are either unachievable or too costly for the perceived benefit."[6] However, since then, some now argue that the world has moved beyond 4GW to an even newer phase of warfare that incorporates a broader spectrum of conflict. Moreover, the definition of what constitutes *national security* is also changing. In considering 4GW, Max G. Manwaring, a defense analyst with the U.S. Army War College, provided additional context for the significance of this level of conflict. He noted that 4GW is "an evolved form of insurgency rooted in the fundamental precept that superior political will, when properly employed, can defeat greater military and economic power." By using political, economic, social, informational, and military networks, Manwaring argues, insurgent groups apply pressure to

---

[6] Max G. Manwaring, *The Strategic Logic of the Contemporary Security Dilemma* (Carlisle, PA: U.S. Army War College Press, 2011), 14–15; and Thomas X. Hammes, *The Sling and the Stone: On War in the 21st Century* (St. Paul, MN: Zenith Press, 2004), 1–15.

their foes to "convince enemy decisionmakers that their strategic goals are either unachievable or too costly for the perceived benefits," which leads the insurgents to directly assault "the minds of enemy decision- and policymakers to destroy their political will."[7]

The Russian military today exemplifies a concept known as *sixth-generation warfare*. This distinctly Russian view envisions a total battlefield incorporation of all forms of information, including targeting, intelligence, a wider use of unmanned weapons and systems, and, of course, a wide-scale proliferation of PGMs that have intercontinental range and can travel at extraordinary speed.[8] Although the Russian military is convinced that information dominance has become the central feature for future conflict, they do not hold to any one-size-fits-all theory. Instead, "in contemporary Russian military studies on the higher level of conflict understanding, including its political, informational, and social reflections is seen as battlespace, representing a holistic approach to the studies of war."[9] Basically, all aspects of society, including culture, politics, and information, matter as much as hypersonic cruise missiles during a conflict. Unless one fully understands this new information modality, even technologically superior forces still risk defeat at the hands of a far weaker opponent who is primarily armed with a stronger, more durable political will and a more successful media campaign.

The effort of the U.S.-backed Iraqi Army to retake the major city of Mosul, Iraq, which fighters from the Islamic State of Iraq and Syria (ISIS) had occupied and looted for nearly two years between 2016 and 2018, provides a significant case study on how information will play an ever-larger role in future conflict. During the height of the fighting, the Iraqis employed more than 100,000 troops, including a large armored division, to take down approximately 6,000 lightly armed but highly dedicated ISIS fighters. Consequently, ISIS's infamous propaganda bureau "went into information overdrive," which included the group celebrat-

---

[7] Manwaring, *The Strategic Logic of the Contemporary Security Dilemma*, 15.
[8] Vitaly Kabernik, "The Russian Military Perspective," in *Hybrid Conflicts and Information Warfare: New Labels, Old Politics* (Boulder, CO: Lynne Rienner Publishers, 2019), 55.
[9] Kabernik, "The Russian Military Perspective," 57–58.

ing each individual "suicide operation," creating " 'Breaking News' notices" for every enemy vehicle destroyed, and disseminating "front-line video clips" twice a day.[10] While ISIS was eventually driven from Mosul at great cost to the Iraqi Army, the "torrent of military misinformation," as well as the "twin streams of victimization and utopia-themed media," was spun so that the defeat did not affect how the outside world viewed ISIS. In essence, their defeat in Mosul no longer mattered since the propaganda messaging of a valiant defense was already established. Most notable during the Mosul media war was the near-total absence of ISIS execution videos that had long been a horrific staple of their media releases. This decision demonstrated that "the group's previous videographed ultraviolence was anything but mindless." Clearly, ISIS used these videos "selectively, adopted according to the specific situational exigencies of the time," but went against that practice "when those exigencies changed."[11]

As the United States reconfigures its forces for the twenty-first century, the nation now must institute a greater emphasis on the informational aspects of any future armed intervention or contingency operation. Even today, the United States remains shockingly slow to engage in information warfare. For example, one of the most watched news channels in the Middle East is the Arab language network *Al Jazeera*. Founded in 1996 by the Emir of Qatar, *Al Jazeera*'s news footage of political and military activity in the Middle East remains highly sought after even among other news outlets in Europe and Asia. *Al Jazeera* shifted to a 24-hour news broadcast in 1999. During the constant violent flare ups in the Middle East in the late 1990s, *Al Jazeera* was often the only international news network to have reporters and cameramen on the ground. As a result, it was usually their coverage that influenced most stories coming out of the Middle East. They even established a bureau in Kabul, Afghanistan, before the United States brought forces into the region in 2001. They controversially aired tapes provided by al-Qaeda leaders Osama Bin

---

[10] Charlie Winter, "The Battle for Mosul: An Analysis of Islamic State Propaganda," in *Hybrid Conflicts and Information Warfare*, 172.

[11] Winter, "The Battle for Mosul," 184–85.

Laden and Ayman al-Zawahiri. Amazingly, the United States had little to no response to these one-way messages except perhaps for the cruise missile U.S. forces expended on destroying the network's Kabul bureau office. While British prime minister Anthony C. L. "Tony" Blair appeared on the network on 14 November 2001 to explain the United Kingdom's reasons for joining the fight against al-Qaeda in Afghanistan, the most senior U.S. official to appear on the network was then-Secretary of State Colin L. Powell, and this did not take place until 2002.[12]

Due to the inherent conventional military advantage of the United States, especially after its 100-hour victory over the fourth-largest army in the world in Operation Desert Storm, many analysts saw opponents of the nation turning to irregular warfare as a means of leveling the playing field. This kind of irregular warfare would be more nuanced, however. Some irregular actions would be confined to legitimate forms of resistance, based on economic, political, cultural, and social issues. Other actions along those lines would be "more violent, corrosive, and ultimately, degenerative in effect." All these types of irregular warfare would have the common element of persistence and could have the ability to "erode American power, national will, and real influence over time through the imposition of increasing physical, psychological, and political costs."[13]

In February 1999, two former Chinese military officers turned strategic theorists, Qiao Liang and Wang Xiangsui, took things a step further. In their treatise *Unrestricted Warfare*, Qiao and Wang observed that "war has undergone changes of modern technology and the market system will be launched in even more atypical forms" in the future. In sum, they argued that "the new principles of war are no longer using armed force" alone to defeat an enemy. Instead, the new concepts of warfare would include "using all means," including both "military and non-military"

---

[12] "Al-Jazeera Offices Hit in U.S. Raid," BBC News, 13 November 2001; and James Risen and Patrick E. Tyler, "Interview with bin Laden Makes the Rounds," *New York Times*, 12 December 2001, B5.
[13] Nathan P. Frier, *Strategic Competition and Resistance in the 21st Century: Irregular, Catastrophic, Traditional, and Hybrid Challenges in Context* (Carlisle, PA: U.S. Army War College Press, 2007), 36.

as well as "lethal and non-lethal means" to achieve victory.[14] Nonmilitary operations allowed opponents of the United States to challenge U.S. power in such a way that the United States might hardly notice. Any opponents—China for instance—could resist through economic, social, or ecological means, or even focus on key marketplace vulnerabilities, such as semiconductor chips or energy, to get the United States to bend to their desires. All these forms of resistance would sit below the threshold of what has been categorized as armed conflict, potentially making the U.S. response limited to what would be perceived as a low-order threat. If undertaken across a broad spectrum of domains, however, such a threat might prove just as costly—and sometimes deadly—as any kinetic strike.

Most conventional powers understand the rules of warfare, at least those generally adhered to since 1648 after the Peace of Westphalia, fairly, but an emerging trend of cyberwarfare—denying or disrupting access to all forms of information systems across the electronic spectrum—has been occurring among these nations. Modern militaries have fully invested in informational connectivity for informing, using, and defending their forces as well as society at large. Information connectivity is central to modern society but also is its most significant vulnerability due to the fact that, as one analyst argues, "dependency on cyberspace has passed the point of no return."[15] The use of cyberwarfare also ties in nicely with hybrid-style wars. Now enemies of the United States, both of the state and nonstate actor variety, can initiate cyberattacks as part of a diverse broad-based warfighting strategy, and this can be done "either overtly, covertly, or clandestinely—all of which have been used in the immediate past." In fact, cyberwarfare is likely going to become the first weapon of choice for nonstate actors. The nonstate actor Wikileaks "orchestrated attacks on Master Card and Visa for cutting payment ser-

---

[14] Frier, *Strategic Competition and Resistance in the 21st Century*, 36–37. The translation of Qiao and Wang can be found in Liang Qiao, Al Santoli, and Xiangsui Wang, *Unrestricted Warfare: China's Master Plan to Destroy America* (Brattleboro, VT: Echo Point Books, 2015).

[15] Neno Malisevic, "Options for Tackling Current and Future Cyber Threats," in *Hybrid and Cyber War as Consequences of the Asymmetry: A Comprehensive Approach Answering Hybrid Actors and Activities in Cyberspace*, ed. Josef Schröfl, Bahram M. Rajaee, and Dieter Muhr (New York: Peter Lang Publishers, 2011), 187.

vices to their organization."[16] Additionally, the rules for this specialized type of warfare are frustratingly amorphous because they have not been written yet. For instance, there is no specific answer for whether state-sponsored cyberattacks against another state count as an act of war. If it did, the United States would have already been officially at war with the People's Republic of China and Russia by now. Yet, the international community has not established a commonly accepted definition for such behavior, allowing weaker states and groups to employ these "new dimensions of asymmetrical warfare" to their advantage.[17]

Cyberspace dependence is growing. Between December 2000 and August 2011, the number of internet users grew from 360 million to approximately 2 billion.[18] In fact, more than 5 billion people use the internet as of January 2023, with the total number of users globally growing by approximately 98 million in the past 12 months. The average global internet users spend almost 7 hours online each day.[19] The sheer amount of daily digital activity makes cyberspace an especially lucrative target. With the ease of social media manipulation, modern militaries face a tremendous information challenge that they are just now starting to address. The Arab Spring of 2011 exemplifies a revolutionary movement using cyberspace to the fullest. At the time, activists using cell phones and satellite communications started a massive uprising against the governments of Algeria, Tunisia, Libya, and Egypt. Although most of these protests were met with increasingly violent crackdowns, social media and digital technologies kept the movement going and acted as a workaround to the state-controlled media. Incredibly, the dissident's dominance of cyberspace resulted in the fall of the governments of Tunisia, Libya, and Egypt.[20]

---

[16] Max G. Manwaring, *The Complexity of Modern Asymmetric Warfare* (Norman: University of Oklahoma Press, 2012), xiii–xiv.

[17] Alexander Siedschlag, "Security Policy as an Analytical Approach," in *Hybrid and Cyber War as Consequences of the Asymmetry*, 12.

[18] Malisevic, "Options for Tackling Current and Future Cyber Threats," 187.

[19] "Digital around the World," DataReportal, accessed 8 February 2023.

[20] Bassant Hassib and James Shires, "Manipulating Uncertainty: Cybersecurity Politics in Egypt," *Journal of Cybersecurity* 7, no. 1 (2021), https://doi.org/10.1093/cybsec/tyaa026.

In fact, events throughout the immediate post-Cold War period and beyond blurred the lines between traditional concepts of war and peace. The persistent chaos in Haiti throughout the 1990s provides a classic example of this phenomenon. In 1994, it took the threat of the U.S. Army's 82d Airborne Division deploying to the country to force its military dictator, J. Raoul Cédras, to relinquish power. Even then, the danger remained that forces loyal to Cédras might start an insurgency against U.S. Joint forces that were providing emergency aid to the Haitian people. While the United Nations (UN) may seem more ideal for these humanitarian operations, it struggles to accomplish its goals in circumstances when the host country is unable to provide adequate security for aid workers and other international officials, as the case of Somalia has illustrated. In Haiti, the U.S. policy goal of regime change further complicated the humanitarian mission.

In 2006, Army general David H. Petraeus and Marine Corps general James N. Mattis combined forces to publish the *Counterinsurgency Field Manual*. Having been "much publicized," the book had reportedly been downloaded "more than a million times" in the following months as "friends and enemies across the world" wanted to "gain insight into how the United States understood irregular warfare," which could potentially be applied to counterinsurgency campaigns, especially those in Iraq at the time.[21] Petraeus believed that these forms of warfare forced U.S. personnel "to be prepared each day to be ready for a hand grenade or a handshake."[22] Mattis had an even deeper understanding of counterinsurgency missions, recognizing that "culture and history informed effective counter-insurgency campaigns." He argued that leaders and personnel who had "read *Angela's Ashes* and Desmond Tutu's writings," as well as studying "Northern Ireland and the efforts for rapprochement there, and in South Africa following their civil war" held just as strong an

---

[21] Russell Crandall, *America's Dirty Wars: Irregular Warfare from 1776 to the War on Terror* (New York: Cambridge University Press, 2014), 393, https://doi.org/10.1017/CBO9781139051606.
[22] Crandall, *America's Dirty Wars*, 13.

understanding of counterinsurgency as those who "read Sherman and, obviously, von Clausewitz."[23]

In conclusion, modern warfare has become in many ways as much about messaging as it is about advanced technologically driven military hardware. This trend emerged during the Vietnam War and its importance has reached a new level today. In 2012, the commander of the North Atlantic Treaty Organization's International Security Assistance Force (ISAF) in Afghanistan, Army general Stanley A. McChrystal, stated in an interview with *The New York Times* that the conflict was a "war of perceptions" rather than a "physical war in terms of how many people you kill or how much ground you capture, how many bridges you blow up." Instead, it took place "in the minds of the participants." With many channels providing 24-hour news coverage, "the term participants takes on a different meaning."[24]

McChrystal took things a step further. In his initial commander's assessment of the situation in Afghanistan, he noted that insurgent forces had "undermined the credibility of ISAF, the international community . . . and the Government of the Islamic Republic of Afghanistan" through "effective use of the information environment, albeit without a commensurate increase in their own credibility." Although he recognized this situation as "a critical problem for ISAF," he argued that "the consequences for [Afghan government] are even starker," making it necessary for both the Afghan government and the international community to "wrest the information initiative" from the insurgency.[25]

While modern commanders recognize the importance of information in modern warfare, many of these changes and trends in warfighting

---

[23] Dexter Filkins, "The Warrior Monk," *New Yorker*, 29 May 2017, 34–45. Mattis's references to *Angela's Ashes*, an award-winning autobiography by Irish expatriate Frank McCourt, and the writings of Bishop Desmond Tutu were to argue that soft power is often more influential than any military means that might be considered. McCourt's memoir documents his family's struggles in pre–World War II Limerick, Ireland. See Frank McCourt, *Angela's Ashes: A Memoir of Childhood* (London: Flamingo, 1997). Tutu became internationally recognized for his peaceful opposition to apartheid in South Africa.

[24] Greg Simons and Iulion Chifu, *The Changing Face of Warfare in the 21st Century* (New York: Routledge, 2018), 24.

[25] Simons and Chifu, *The Changing Face of Warfare in the 21st Century*, 29.

started during the Vietnam War. Army general William C. Westmoreland, the commander of U.S. Military Assistance Command, Vietnam from 1964 to 1968, had been promoted to the rank of brigadier general in 1952 at the age of 38, making him one of the youngest generals in the Army at the time. His prior experience and proclivities as a professional soldier, however, led him to try to fight a highly unconventional enemy—the National Liberation Front, or Viet Cong—using the conventional means he had at his disposal. Further, due to political considerations imposed on him, Westmoreland's methodology for fighting the war seemed to be a losing proposition from the start. In trying to show progress that did not exist, he lost the faith and trust of the U.S. domestic and even international media, ultimately losing the information war overall. Vietnam and the related information war forever changed the United States' warfighting paradigm.

After the Vietnam War, the Cold War was replete with strategic inflection points. Although the so-called "Vietnam Syndrome" affected the ways in which U.S. political and military leadership envisioned the use of its forces in overseas contingency operations after 1973, presidential administrations still used U.S. forces in a wide variety of situations. Before 1986, however, most contingencies operations that took place were fraught with missteps and coordination gaffes due to an inherent lack of Jointness. After the passage of the Goldwater–Nichols Act in 1986, the contingency operations and Joint warfighting skills greatly improved, a positive change that merits additional study. While the Goldwater–Nichols Act deserves credit for reforming the Armed Services, other areas of governmental soft power and even nongovernmental organizations also contributed to military operations overseas. In Somalia, for instance, various intergovernmental actors such as the UN and the African Union, nongovernmental organizations such as the International Red Cross and Doctors without Borders; and U.S. federal agencies such as the Department of State and the Department of Justice all ended up playing critical roles. All these elements require further study, especially as the emerging modalities of today's hybrid warfare trend practically demand it.

SELECT BIBLIOGRAPHY

Personal Papers, Letters, and Historical Reference Files

Barrow, Robert H. Historical Reference Branch Files, Marine Corps History Division, Quantico, VA.

Haynes, Frederick. Historical Reference Branch Files, Marine Corps History Division, Quantico, VA.

Marletto, Michael P., to Capt Charles P. Neimeyer, 24 October 1983. Author's collection, Quantico, VA.

Wilson, Louis H., Jr. Historical Reference Branch Files, Marine Corps History Division, Quantico, VA.

———. "New Directions Remarks." The District of Columbia Navy League 1975 Welcome Aboard Luncheon Minutes, 10 September 1975.

———. Personal Papers Collection, Marine Corps History Division, Quantico, VA.

Oral Histories

Oral History Collection, U.S. Marine Corps History Division, Quantico, VA.

Barrow, Robert H. 2015.

Cates, Clifton B. 1973.

Chapman, Leonard F., Jr. 1981.

Cushman, Robert D., Jr. 1984.

Geraghty, Timothy J. 1983.

Krulak, Charles C. 2003.

Twining, Merrill B. 1967.

Wilson, Louis H., Jr. 2008.

Author's Collection.

Ferraro, Peter J. 2019.

Funk, Robert C. 2019.

Smith, Norman H. 2019.

Government Publications

An Act to Amend the Military Selective Service Act of 1967; to Increase Military Pay; to Authorize Active Duty Strengths for Fiscal Year 1972; and for Other Purposes. Pub. L. No. 92-129 (1971).

Attrition in the Military—An Issue Needing Management Attention. Washington, DC: U.S. Government Accountability Office, 1980.

Background Information on the Use of U.S. Armed Forces in Foreign Countries. Washington, DC: Congressional Research Service, 1975.

Blanc Mont (Meuse-Argonne-Champagne). Washington, DC: War Department, 1922.

Dalton, John H., Adm J. Michael Boorda, USN, and Gen Carl E. Mundy Jr. *Forward ... From the Sea.* Washington, DC: Department of the Navy, 1994.

Exercise Northern Wedding/Bold Guard 78, Post Deployment Report, vol. 1. Camp Lejeune, NC: 2d Marine Division, Fleet Marine Force, Marine Corps, 1978.

Gray, Gen Alfred M. *Annual Report to the Congress, 1990.* Washington, DC: Department of the Navy, 1990.

Hearing before the Committee on Armed Services United States Senate, on Nominations of General Paul X. Kelley to Be Commandant of the Marine Corps, Richard L. Armitage to Be Assistant Secretary of Defense (International Security Affairs), Chapman B. Cox to Be Assistant Secretary of the Navy (Manpower and Reserve Affairs). 98th Cong., 1st Sess. 24 May 1983.

Hearing on the Situation in Haiti before the Senate Committee on Armed Services. 103d Cong., 2d Sess. 28 September 1994.

Hearings on Fiscal Year 1974 Authorization for Military Procurement, Research and Development, Construction Authorization for the Safeguard ABM, and Active Duty and Selected Reserve Strengths, Hearings before the Senate Committee on Armed Services. Part 2. 93d Cong., 1st Sess. 13 April 1973.

Hearings on Marine Corps' Recruit Training and Recruiting Program, before the House Armed Services Subcommittee on Military Personnel. 94th Cong., 2d Sess. 26 May 1976.

Hearings on the Unification of the Armed Forces, before the Senate Committee on Naval Affairs. 79th Cong., 2d Sess. 6 May 1946.

Krulak, Gen Charles C. *Operational Maneuver from the Sea: A Concept for the Projection of Power Ashore.* Washington, DC: Headquarters Marine Corps, 1997.

Laird, Melvin R. *Progress in Ending the Draft and Achieving the All-Volunteer Force: Report to the President and the Chairmen of the Armed Services Committees of the Senate and the House of Representatives.* Washington, DC: Government Printing Office, 1972.

Making Marines in the All-Volunteer Era: Recruiting Core Values, and the Perpetuation of Our Ethos. Quantico, VA: Lejeune Leadership Institute, Marine Corps University, 2018.

National Security Act of 1947. Pub. L. No. 117-103. 1947.

Navy-Marine Corps Instruction (NAVMC) 4544, Amphibious Operations—Employment of Helicopters (Tentative). Washington, DC: Department of the Navy, 1948.

O'Keefe, Sean C., Adm Frank B. Kelso II, USN, and Gen Carl E. Mundy, Jr. . . . *From the Sea: Preparing the Naval Service for the 21st Century*. Washington, DC: Department of the Navy, 1992.

Operation Plan 712: Advanced Operations in Micronesia. Washington, DC: Headquarters Marine Corps, 1921.

Osbourne, SSgt Richard E. *Sand and Steel: Lessons Learned on U.S. Marine Corps Armor and Anti-Armor Operations from the Gulf War*. Quantico, VA: Marine Corps Combat Development Command, 1993.

The President's Commission on an All-Volunteer Force. Washington, DC: White House, 1970.

Quinlan, David A. *The Role of the Marine Corps in Rapid Deployment Forces*. Washington, DC: National Defense University Press, 1983.

Report of the DOD Commission on Beirut International Airport Terrorist Act, October 23, 1983. Washington, DC: Department of Defense, 1983.

Report of the Fleet Marine Force Organization and Composition Board. Washington, DC: Department of the Navy, 1957.

Rescue Mission Report. Washington, DC: U.S. Joint Chiefs of Staff, 1980.

Rivlin, Alice M. *The Marine Corps in the 1980s: Prestocking Proposals, the Rapid Deployment Force, and Other Issues*. Washington, DC: Congressional Budget Office, 1980.

Small Wars Manual. Fleet Marine Force Reference Publication 12-15. Washington, DC: Headquarters Marine Corps, 1940.

Warfighting. Fleet Marine Force Manual 1. Washington, DC: Headquarters Marine Corps, 1989.

Warfighting. Marine Corps Doctrinal Publication 1. Washington, DC: Headquarters Marine Corps, 1997.

Wilson, Gen Louis H., Jr. *Statement of General Louis H. Wilson, Commandant of the Marine Corps before the Committee on Armed Services, U.S. House of Representatives on Marine Corps Posture, Plans, and Programs for FY 1979 through 1983*. Quantico, VA: Marine Corps University Library, 1978.

Monographs

Adkin, Mark. *Urgent Fury: The Battle for Grenada*. Lexington, MA: Lexington Books, 1989.

Alexander, Col Joseph H. *Across the Reef: The Marine Assault of Tarawa*. Washington, DC: Marine Corps History and Museums Division, 1993.

Allison, Fred H., and Kurtis P. Wheeler, eds. *Pathbreakers: U.S. Marine African American Officers in Their Own Words*. Quantico, VA: Marine Corps History Division, 2013.

Anderson, Martin, ed. *Registration and the Draft: Proceedings of the Hoover-Rochester Conference on the All-Volunteer Force.* Stanford, CA: Hoover Institution Press, 1982.

Andrade, Dale. *America's Last Vietnam Battle: Halting Hanoi's 1972 Easter Offensive.* Lawrence: University Press of Kansas, 2001.

———. *Trial by Fire: The 1972 Easter Offensive, America's Last Vietnam Battle.* New York: Hippocrene Books, 1995.

Antal, James G., and R. John Vanden Berghe. *On Mamba Station: U.S. Marines in West Africa, 1990–2003.* Washington, DC: Marine Corps History and Museums Division, 2004.

Atkinson, Rick. *The Long Gray Line.* New York: Henry Holt, 1989.

Aurthur, Robert A., and Kenneth Cohlmia. *The Third Marine Division.* Nashville, TN: Battery Press, 1988.

Axelrod, Alan. *Miracle at Belleau Wood: The Birth of the Modern Marine Corps.* Guilford, CT: Lyons Press, 2007.

Bailey, Beth. *America's Army: Making the All-Volunteer Force.* Cambridge, MA: Belknap Press of Harvard University Press, 2009.

Ballard, John R. *Upholding Democracy: The United States Military Campaign in Haiti, 1994–1997.* Westport, CT: Praeger, 1998.

Ballendorf, Dirk A., and Merrill L. Bartlett. *Pete Ellis: An Amphibious Warfare Prophet, 1880–1923.* Annapolis, MD: Naval Institute Press, 1997.

Bell, Douglas I. *Just Add Soldiers: Army Prepositioned Stocks and Agile Force Projection.* Carlisle, PA: U.S. Army Heritage and Education Center, U.S. Army War College, 2021.

Binkin, Martin. *America's Volunteer Army: Progress and Prospects.* Washington, DC: Brookings Institution, 1984.

Binkin, Martin, and Irene Kyriakopoulos. *Youth or Experience?: Manning the Modern Military.* Washington, DC: Brookings Institution, 1979.

Binkin, Martin, and Jeffrey Record. *Where Does the Marine Corps Go from Here?* Washington, DC: Brookings Institution, 1976.

Bitzinger, Richard A. *Assessing a Conventional Balance in Europe, 1945–1975.* Santa Monica, CA: Rand, 1989.

Bliven, Bruce, Jr. *Volunteers One and All.* New York: Reader's Digest Press, 1976.

Bohm, Jason Q. *From the Cold War to ISIL: One Marine's Journey.* Annapolis, MD: Naval Institute Press, 2019.

Bonk, David. *Chateau Thierry & Belleau Wood, 1918: America's Baptism of Fire on the Marne.* Oxford, UK: Osprey, 2007.

Bowden, Mark. *Black Hawk Down: A Story of Modern War.* New York: Grove Press, 1999.

Brasher, Bart. *Implosion: Downsizing the U.S. Military, 1987–2015.* Westport, CT: Greenwood Press, 2000.

Bristol, Douglas W., Jr., and Heather M. Stur, eds. *Integrating the Military: Race, Gender, and Sexual Orientation since World War II.* Baltimore, MD: Johns Hopkins University Press, 2017.

Brooke, Edward W., and Samuel Nunn. *An All-Volunteer Force for the United States?* Washington, DC: American Enterprise Institute for Public Policy Research, 1977.

Brown, Ian T. *A New Conception of War: John Boyd, the U.S. Marines, and Maneuver Warfare.* Quantico, VA: Marine Corps University Press, 2018. https://doi.org/10.56686/9780997317497.

Brown, Ronald J. *U.S. Marines in the Persian Gulf, 1990–1991: With Marine Forces Afloat in Desert Shield and Desert Storm.* Washington, DC: Marine Corps History and Museums Division, 1998.

Brunson, Col Richard G., USA. *The Inchon Landing: An Example of Brilliant Generalship.* Carlisle, PA: U.S. Army War College, 2003.

Brzezinski, Zbigniew. *Power and Principle: Memoirs of the National Security Adviser, 1977–1981.* New York: Farrar, Straus, and Giroux, 1983.

Burrowes, Reynold A. *Revolution and Rescue in Grenada: An Account of the U.S. Caribbean Invasion.* Westport, CT: New Greenwood Press, 1988.

Carely, Demetrios. *The Politics of Military Unification: A Study of Conflict and the Policy Process.* New York: Columbia University Press, 1966.

Chambers, John Whiteclay, II. *To Raise an Army: The Draft Comes to Modern America.* New York: Free Press, 1987.

Chandler, David P. *Brother Number One: A Political Biography of Pol Pot.* Revised edition. Boulder, CO: Westview Press, 1999.

Chun, Clayton K. S. *The Last Boarding Party: The USMC and the SS* Mayaguez, *1975.* Oxford, UK: Osprey, 2011.

Clancy, Tom, with Anthony Zinni and Tony Koltz. *Battle Ready.* New York: G. P. Putnam's Sons, 2004.

Clark, Robert D., Andrew M. Egeland Jr., and David B. Sanford. *The War Powers Resolution: Balance of War Powers in the Eighties.* Washington, DC: National War College University Press, 1985.

Clausewitz, Carl Von. *On War.* Edited and translated by Michael Howard and Peter Paret. Princeton, NJ: Princeton University Press, 1976.

Clifford, Kenneth J. *Progress and Purpose: A Developmental History of the United States Marine Corps, 1900–1970.* Washington, DC: Marine Corps History and Museums Division, 1973.

Cohen, Eliot A. *Citizens and Soldiers: The Dilemmas of Military Service.* Ithaca, NY: Cornell University Press, 1985.

Cole, Ronald H. *Operation Just Cause: The Planning and Execution of Joint Operations in Panama, February 1988–January 1990*. Washington, DC: Joint History Office, Office of the Joint Chiefs of Staff, 1995.

——. *Operation Urgent Fury: The Planning and Execution of Joint Operations in Grenada, 12 October–2 November 1983*. Washington, DC: Joint History Office, 1987.

Coletta, Paolo E. *The United States Navy and Defense Unification, 1947–1953*. Newark: University of Delaware Press, 1981.

Cooper, Richard V. L., and Bernard Rostker. *Military Manpower in a Changing Environment*. Santa Monica, CA: Rand, 1974.

Coram, Robert. *Boyd: The Fighter Pilot Who Changed the Art of War*. Boston, MA: Little, Brown, 2002.

Cosmas, Graham A., and Terrance P. Murray. *U.S. Marines in Vietnam: Vietnamization and Redeployment, 1970–1971*. Washington, DC: Marine Corps History and Museums Division, 1986.

Crandall, Russell. *America's Dirty Wars: Irregular Warfare from 1776 to the War on Terror*. New York: Cambridge University Press, 2014.

Cureton, Charles H. *U.S. Marines in the Persian Gulf, 1990–1991: With the 1st Marine Division in Desert Shield and Desert Storm*. Washington, DC: Marine Corps History and Museums Division, 1993.

Davis, Alphonse G. *Pride, Progress, and Prospects: The Marine Corps' Efforts to Increase the Presence of African-American Officers (1970–1995)*. Washington, DC: Marine Corps History and Museums Division, 2000.

Dawson, Capt David A. *The Impact of Project 100,000 on the Marine Corps*. Washington, DC: Marine Corps History and Museums Division, 1995.

Donnelly, Thomas, Margaret Roth, and Caleb Baker. *Operation Just Cause: The Storming of Panama*. New York: Lexington Books, 1991.

Dunham, George R., and David A. Quinlan. *The U.S. Marines in Vietnam: The Bitter End, 1973–1975*. Washington, DC: Marine Corps History and Museums Division, 1990.

Dunn, Peter J., and Bruce W. Watson, eds. *American Intervention in Grenada: The Implications of Operation Urgent Fury*. Boulder, CO: Westview Press, 1985.

Dunnigan, James F., and Austin Bay. *From Shield to Storm: High Tech Weapons, Military Strategy, and Coalition Warfare in the Persian Gulf*. New York: William Morrow and Company, 1992.

Dupuy, Trevor N., Grace P. Hayes, Gay M. Hammerman, Col John A. C. Andrews, USAF (Ret), Vivian Lyons, and James Bloom. *The Middle East War of October 1973 in Historical Perspective*. Falls Church, VA: NOVA Publications, 1976.

Eggleston, Michael A. *The 5th Marine Regiment Devil Dogs in World War I: A History and Roster*, Jefferson, NC: MacFarland, 2016.

Estes, Kenneth W. *Marines under Armor: The Marine Corps and the Armored Fighting Vehicle, 1916–2000.* Annapolis, MD: Naval Institute Press, 2000.

Fails, William R. *Marines and Helicopters: 1962–1973.* Washington, DC: Marine Corps History and Museums Division, 1978.

Fisher, Louis. *Presidential War Power.* 3d ed. Lawrence: University Press of Kansas, 2013.

Fisk, Robert. *Pity the Nation: The Abduction of Lebanon.* New York: Atheneum, 1990.

Flanagan, Edward M., Jr. *Battle for Panama: Inside Operation Just Cause.* Washington, DC: Brassey's, 1993.

Fleet Admirals, U.S. Navy. Washington, DC: Naval Historical Foundation, 1966.

Fleming, Keith. *The U.S. Marine Corps in Crisis: Ribbon Creek and Recruit Training.* Columbia: University of South Carolina Press, 1990.

Flynt, Maj William C., III, USA. *Broken Stiletto: Command and Control of the Joint Task Force during Operation Eagle Claw at Desert One.* Fort Leavenworth, KS: School of Advanced Military Studies, 1995.

Frank, Benis M. *U.S. Marines in Lebanon, 1982–1984.* Washington, DC: Marine Corps History and Museums Division, 1987.

Fredland, J. Eric, Curtis Gilroy, Roger D. Little, and W. S. Sellman, eds. *Professionals on the Front Line: Two Decades of the All-Volunteer Force.* Washington, DC: Brassey's, 1996.

Fridman, Ofir, Vitaly Kabernik, and James C. Pearce, eds. *Hybrid Conflicts and Information Warfare: New Labels, Old Politics.* Boulder, CO: Lynne Rienner, 2019.

Friedman, Brent A., ed. *21st Century Ellis: Operational Art and Strategic Prophecy for the Modern Era.* Annapolis, MD: Naval Institute Press, 2015.

Friedman, Thomas L. *From Beirut to Jerusalem.* New York: Farrar, Straus, Giroux, 1989.

Frier, Nathan. *Strategic Competition and Resistance in the 21st Century: Irregular, Catastrophic, Traditional, and Hybrid Challenges in Context.* Carlisle, PA: U.S. Army War College Press, 2007.

Fulkerson, Norman J. *An American Knight: The Life of Colonel John W. Ripley, USMC.* Spring Grove, PA: American Society for the Defense of Tradition, Family, and Property, 2009.

Gabriel, Richard A., and Paul L. Savage. *Crisis in Command: Mismanagement in the Army.* New York: Hill and Wang, 1979.

Geis, Mark B., and William A. D. Wallace. *Improving MPF Operational Effectiveness: Lessons Learned from Past Exercises and Operations.* Quantico, VA: U.S. Marine Corps, 1993.

Geraghty, Timothy J. *Peacekeepers at War: Beirut 1983—The Marine Commander Tells His Story.* Washington, DC: Potomac Books, 2009.

Guilmartin, John F., Jr. *A Very Short War: The* Mayaguez *and the Battle of Koh Tang.* College Station: Texas A&M University Press, 1995.

Greene, John Robert. *The Presidency of Gerald R. Ford.* Lawrence: University Press of Kansas, 1995.

Hallenbeck, Ralph A. *Military Force as an Instrument of U.S. Foreign Policy: Intervention in Lebanon, August 1982–February 1984.* New York: Praeger, 1991.

Hammell, Eric. *The Root: The Marines in Beirut, August 1982–February 1984.* Pacifica, CA: Pacifica Military History, 1985.

Hammes, Thomas X. *The Sling and the Stone: On War in the 21st Century.* St. Paul, MN: Zenith Press, 2006.

Hammond, Grant T. *The Mind of War: John Boyd and American Security.* Washington, DC: Smithsonian Institution Press, 2001.

Hammond, James W., Jr. *The Treaty Navy: The Story of the Naval Service between the World Wars.* Victoria, BC: Wesley Press, 2001.

Hart, Gary, and William S. Lind. *America Can Win: The Case for Military Reform.* Bethesda, MD: Adler and Adler Publishers, 1986.

Hays, Ronald E., II. *Combined Action: U.S. Marines Fighting a Different War, August 1965 to May 1971.* Quantico, VA: Marine Corps History Division, 2019.

Heck, Timothy, and B. A. Friedman, eds. *On Contested Shores: The Evolving Role of Amphibious Operations in the History of Warfare.* Quantico, VA: Marine Corps University Press, 2020. https://doi.org/10.56686/9781732003149.

Heinl, Robert D., Jr. *Soldiers of the Sea: The United States Marine Corps, 1775–1962.* Annapolis, MD: Naval Institute Press, 1962.

———. *Victory at High Tide: The Inchon-Seoul Campaign.* Baltimore, MD: Nautical and Aviation Publishing, 1979.

Hendricks, LtCol Douglas O. *Maritime Prepositioning Force in Theater Level Campaigning.* Fort Leavenworth, KS: U.S. Army School of Advanced Military Studies, 1991.

Herzog, Chaim. *The War of Atonement: October 1973.* Boston, MA: Little, Brown, 1975.

Hines, Jay E., Jason D. Mims, and Hans S. Pawlisch. *USCENTCOM in Somalia: Operations Provide Relief and Restore Hope.* MacDill Air Force Base, FL: U.S. Central Command History Office, 1994.

Hooker, Richard D., Jr., ed. *Maneuver Warfare: An Anthology.* Novato, CA: Presidio Press, 1993.

Huchthausen, Peter. *America's Splendid Little Wars: A Short History of U.S. Military Engagements, 1975–2000.* New York: Viking, 2003.

Isenberg, David. *The Rapid Deployment Force: The Few, the Futile, the Expendable.* Washington, DC: Cato Institute, 1984.

Jaffe, Lorna S. *The Development of the Base Force, 1989–1992.* Washington, DC: Joint History Office, Office of the Joint Chiefs of Staff, 1993.

Jeffers, H. Paul, and Dick Levitan. *See Parris and Die: Brutality in the Marine Corps.* New York: Hawthorn Books, 1971.

Jenkins, Harry W. *Challenges: Leadership in Two Wars, Washington DC, and Industry.* London: Fortis Publishing, 2020.

Johnson, Charles, Jr. *African Americans and ROTC: Military, Naval and Aerospace Programs at Historically Black Colleges, 1916–1973.* Jefferson, NC: MacFarland, 2002.

Jones, William K. *A Brief History of the 6th Marines.* Washington, DC: Marine Corps History and Museums Division, 1987.

Karcher, Maj Timothy M., USA. *Understanding the "Victory Disease": From the Little Bighorn, to Mogadishu, to the Future.* Fort Leavenworth, KS: U.S. Army Command and General Staff College, 2003.

Keiser, Gordon W. *The U.S. Marine Corps and Defense Unification, 1944–1947.* Baltimore, MD: Nautical and Aviation Publishing Company of America, 1996.

Kitfield, James. *Prodigal Soldiers: How the Generation of Officers Born of Vietnam Revolutionized the American Style of War.* New York: Simon and Schuster, 1995.

Klare, Michael T. *Resource Wars: The New Landscape of Global Conflict.* New York: Metropolitan Books, 2001.

Knappman, Edward W., ed. *South Vietnam: U.S.-Communist Confrontation in Southeast Asia, 1972–1973.* Vol. 7. New York: Facts on File, 1973.

Kretchik, Walter E., Robert F. Baumann, and John T. Fishel. *Invasion, Intervention, "Intervasion": A Concise History of the U.S. Army in Operation Uphold Democracy.* Fort Leavenworth, KS: U.S. Army Command and Staff College, 1998.

Krulak, Victor H. *First to Fight: An Inside View of the Marine Corps.* Annapolis, MD: Naval Institute Press, 1984.

Kuzmarov, Jeremy. *The Myth of the Addicted Army: Vietnam and the Modern War on Drugs.* Amherst: University of Massachusetts Press, 2009.

Laidig, Scott. *Al Gray, Marine: The Early Years, 1950–1967.* Arlington, VA: Potomac Institute Press, 2012.

Lamb, Christopher J. *The Mayaguez Crisis, Mission Command, and Civil-Military Relations.* Washington, DC: Joint History Office, Office of the Joint Chiefs of Staff, 2018.

Lehman, John F., Jr. *Command of the Seas.* New York: Charles Scribner's Sons, 1988.

Lehrack, Otto J. *America's Battalion: Marines in the First Gulf War.* Tuscaloosa: University of Alabama Press, 2005.

Lejeune, Gen John A. *The Reminiscences of a Marine.* Philadelphia, PA: Dorrance, 1930.

Lewis, Kevin N. *What If the Reagan Defense Buildup Is Over?* Santa Monica, CA: Rand, 1987.

Lind, William S. *The Maneuver Warfare Handbook.* Boulder, CO: Westview Press, 1985.

Livingston, Steven. *Clarifying the CNN Effect: An Examination of Media Effects According to Type of Military Intervention.* Cambridge, MA: Joan Shorenstein Center, Harvard University, 1997.

Lodge, Maj O. R. *The Recapture of Guam.* Washington, DC: Marine Corps Historical Branch, 1954.

MacGregor, Morris J., Jr. *Integration of the Armed Forces, 1940–1965.* Washington, DC: U.S. Army Center of Military History, 1981.

Mahoney, Robert J. *The* Mayaguez *Incident: Testing America's Resolve in the Post-Vietnam Era.* Lubbock: Texas Tech University Press, 2011.

Mansbach, Richard W., ed. *Dominican Crisis 1965.* New York: Facts on File, 1971.

Manwaring, Max G. *The Strategic Logic of the Contemporary Security Dilemma.* Carlisle, PA: U.S. Army War College Press, 2011.

———. *The Complexity of Modern Asymmetric Warfare.* Norman: University of Oklahoma Press, 2012.

McGrady, E. D., and John S. Ivanovich, eds. *Operation Uphold Democracy: Conflict and Cultures, A Summary of Material from CNA's 1995 Annual Conference.* Alexandria, VA: Center for Naval Analyses, 1997.

Melson, Charles D., and Curtis G. Arnold. *The U.S. Marines in Vietnam: The War that Would Not End, 1971–1973.* Washington, DC: Marine Corps History and Museums Division, 1991.

Melson, Charles D., and Wanda J. Renfrow. *Marine Advisors with the Vietnamese Marine Corps.* Quantico, VA: Marine Corps History Division, 2009.

Melson, Charles D., Evelyn A. Englander, and David A. Dawson, eds. *U.S. Marines in the Gulf War: Anthology and Annotated Bibliography.* Washington, DC: Marine Corps History and Museums Division, 1992.

Miller, John Grider. *The Bridge at Dong Ha.* Annapolis, MD: Naval Institute Press, 1989.

———. *The Co-Vans: U.S. Marine Advisors in Vietnam.* Annapolis, MD: Naval Institute Press, 2000.

Millett, Allan R. *Semper Fidelis: The History of the United States Marine Corps.* New York: MacMillan, 1980.

———. *Semper Fidelis: The History of the United States Marine Corps.* Revised ed. New York: Free Press, 1991.

Millett, Allan R., and Jack Shulimson, eds. *Commandants of the Marine Corps.* Annapolis, MD: Naval Institute Press, 2004.

Montross, Lynn, and Nicholas A. Canzona. *The Chosin Reservoir Campaign: U.S. Marine Operations in Korea, 1950–1953.* Washington, DC: Marine Corps Historical Branch, 1957.

Morris, David J. *Storm on the Horizon: Khafji—the Battle that Changed the Course of the Gulf War.* New York: Free Press, 2004.

Mroczkowski, Dennis P. *Restoring Hope: In Somalia with the Unified Task Force, 1992–1993.* Washington, DC: Marine Corps History Division, 2005.

———. *U.S. Marines in the Persian Gulf, 1990–1991: With the 2d Marine Division in Desert Shield and Desert Storm.* Washington, DC: Marine Corps History and Museums Division, 1993.

Nalty, Bernard C. *Strength for the Fight: A History of Black Americans in the Military.* New York: Free Press, 1986.

Neimeyer, Charles P., ed. *On the Corps: USMC Wisdom from the Pages of* Leatherneck, Marine Corps Gazette, *and* Proceedings. Annapolis, MD: Naval Institute Press, 2008.

Nelson, James Carl. *I Will Hold: The Story of USMC Legend Clifton B. Cates, from Belleau Wood to Victory in the Great War.* New York: Caliber, 2016.

Neustadt, Richard E., and Ernest R. May. *Thinking in Time: The Uses of History for Decision Makers.* New York: Free Press, 1986.

Ngo, Quang Truong. *The Easter Offensive of 1972.* Washington, DC: U.S. Army Center of Military History, 1979.

Nixon, Richard M. *No More Vietnams.* New York: Arbor House, 1985.

O'Brien, Cyril J. *Liberation: Marines in the Recapture of Guam.* Washington, DC: Marine Corps Historical Division, 1994.

O'Connell, Aaron B. *Underdogs: The Making of the Modern Marine Corps.* Cambridge, MA: Harvard University Press, 2012.

Ohls, Gary J. *Somalia . . . From the Sea.* Newport, RI: Naval War College Press, 2009.

Owen, Peter F., and John Swift. *A Hideous Price: The 4th Brigade at Blanc Mont, 2–10 October 1918.* Quantico, VA: Marine Corps History Division, 2019.

Palmer, Michael A. *Guardians of the Gulf: A History of America's Expanding Role in the Persian Gulf, 1833–1992.* New York: Free Press, 1992.

Paret, Peter, ed. *Makers of Modern: From Machiavelli to the Nuclear Age.* Princeton, NJ: Princeton University Press, 1986.

Perry, Ronald W. *Racial Discrimination and Military Justice.* New York: Praeger, 1977.

Petersen, Frank E. *Into the Tiger's Jaw: America's First Black Marine Aviator.* Annapolis, MD: Naval Institute Press, 1998.

Piedmont, Col John R., USMCR. *Det One: U.S. Marine Corps Special Operations Command Detachment, 2003–2006.* Washington, DC: Marine Corps History Division, 2010.

Pintak, Lawrence. *Beirut Outtakes: A TV Correspondent's Portrait of America's Encounter with Terror*. Lexington, MA: Lexington Books, 1988.

Piscitelli, Anthony J. *The Marine Corps Way of War*. El Dorado Hills, CA: Savas Beattie, 2017.

Pitt, Barrie. *1918: The Last Act*. New York: W. W. Norton, 1962.

Poole, Walter S. *The Effort to Save Somalia: August 1992–March 1994*. Washington, DC: Joint History Office, Office of the Joint Chiefs of Staff, 2005.

Puryear, Edgar F., Jr. *Marine Corps Generalship*. Washington, DC: National Defense University Press, 2009.

Randolph, Stephen P. *Powerful and Brutal Weapons: Nixon, Kissinger, and the Easter Offensive*. Cambridge, MA: Harvard University Press, 2007.

Rawlins, Eugene W. *Marines and Helicopters, 1946–1962*. Washington, DC: Marine Corps History and Museums Division, 1976.

Record, Jeffrey. *The Rapid Deployment Force and U.S. Military Intervention in the Persian Gulf*. Cambridge, MA: Institute for Foreign Policy Analysis, 1983.

Reynolds, Nicholas E. *A Skillful Show of Strength: U.S. Marines in the Caribbean, 1991–1996*. Washington, DC: Marine Corps History and Museums Division, 2003.

———. *Just Cause: Marine Operations in Panama, 1988–1990*. Washington, DC: Marine Corps History and Museums Division, 1996.

Rhea, Gordon C. *The Battles for Spotsylvania Court House and the Road to Yellow Tavern, May 7–12, 1864*. Baton Rouge: Louisiana State University Press, 1997.

Ricks, Thomas E. *Making the Corps*. New York: Touchstone, 1997.

Ringler, Maj Jack K., and Henry I. Shaw Jr. *U.S. Marine Corps Operations in the Dominican Republic, April–June 1965*. Washington, DC: Marine Corps History and Museums Division, 1970.

Rostker, Bernard. *I Want You!: The Evolution of the All-Volunteer Force*. Santa Monica, CA: Rand, 2006.

Rudd, Gordon W. *Humanitarian Intervention: Assisting the Iraqi Kurds in Operation Provide Comfort, 1991*. Washington, DC: Department of the Army, 2006.

Russ, Martin. *Breakout: The Chosin Reservoir Campaign, Korea, 1950*. New York: Penguin, 1999.

Ryan, Paul B. *The Iranian Rescue Mission: Why it Failed*. Annapolis, MD: Naval Institute Press, 1985.

Santelli, James S. *A Brief History of the 8th Marines*. Washington, DC: Marine Corps History and Museums Division, 1976.

Santoli, Al. *Leading the Way: How Vietnam Veterans Rebuilt the U.S. Military—An Oral History*. New York: Ballantine Books, 1993.

Schröfl, Josef, Bahram M. Rajaee, and Dieter Muhr, eds. *Hybrid and Cyber War as Consequences of the Asymmetry: A Comprehensive Approach Answering Hybrid Actors and Activities in Cyberspace.* New York: Peter Lang, 2011.

Schulz, Donald E. *Haiti Update.* Carlisle, PA: U.S. Army War College Press, 1997.

———. *Whither Haiti?* Carlisle, PA: U.S. Army War College Press, 1996.

Semper Fidelis: A Brief History of Onslow County, North Carolina, and Marine Corps Base, Camp Lejeune. Camp Lejeune, NC: U.S. Marine Corps, 2006.

Shacochis, Bob. *The Immaculate Invasion.* New York: Penguin, 1999.

Shaw, Henry I., Jr., and Ralph W. Donnelly. *Blacks in the Marine Corps.* Washington, DC: Marine Corps History and Museums Division, 1988.

Sherwood, John Darrell. *Black Sailor, White Navy: Racial Unrest in the Fleet during the Vietnam Era.* New York: New York University Press, 2007.

———. *Nixon's Trident: Naval Power in Southeast Asia, 1968–1972.* Washington, DC: Naval History and Heritage Command, 2009.

Shulimson, Jack. *Marines in Lebanon, 1958.* Washington, DC: Marine Corps Historical Branch, 1966.

Shulimson, Jack, and Maj Charles M. Johnson. *U.S. Marines in Vietnam: The Landing and Buildup, 1965.* Washington, DC: Marine Corps History and Museums Division, 1978.

Siegal, Adam, Karen Domabyl, and Barbara Lindberg. *Deployment of U.S. Navy Aircraft Carriers and Other Surface Ships, 1976–1988.* Alexandria, VA: Center for Naval Analyses, 1989.

Simmons, BGen Edwin H., USMC (Ret). *Over the Seawall: U.S. Marines at Inchon.* Washington, DC: Marine Corps History and Museums Division, 2000.

Simmons, Edwin H., and Joseph H. Alexander. *Through the Wheat: The U.S. Marines in World War I.* Annapolis, MD: Naval Institute Press, 2008.

Simons, Greg, and Iulian Chifu. *The Changing Face of Warfare in the 21st Century.* New York: Routledge, 2018.

Smith, Charles R. *Angels from the Sea: Relief Operations in Bangladesh, 1991.* Washington, DC: Marine Corps History and Museums Division, 1995.

Smith, Gen Holland M., USMC (Ret). *The Development of Amphibious Tactics in the U.S. Navy.* Washington, DC: Marine Corps History and Museums Division, 1992.

Sorley, Lewis. *Thunderbolt: From the Battle of the Bulge to Vietnam and Beyond.* New York: Simon and Schuster, 1992.

Sorley, Lewis, ed. *Vietnam Chronicles: The Abrams Tapes, 1968–1972.* Lubbock: Texas Tech University Press, 2004.

Spector, Ronald H. *U.S. Marines in Grenada, 1983.* Washington, DC: Marine Corps History and Museums Division, 1987.

Steele, Orlo K., and Michael I. Moffett. *U.S. Marine Corps Mountain Warfare Training Center, 1951–2001*. Quantico, VA: Marine Corps History Division, 2011.

Stevenson, Jonathan. *Losing Mogadishu: Testing U.S. Policy in Somalia*. Annapolis, MD: Naval Institute Press, 1995.

Suskind, Richard. *The Battle of Belleau Wood: The Marines Stand Fast*. Toronto: Macmillan, 1969.

Tuchman, Barbara W. *The March of Folly: From Troy to Vietnam*. New York: Random House, 1984.

Turley, Col Gerry H., USMCR (Ret). *The Easter Offensive: The Last American Advisors, Vietnam 1972*. Novato, CA: Presidio Press, 1985.

———. *The Journey of a Warrior: The Twenty-Ninth Commandant of the U.S. Marine Corps (1987–1991): General Alfred Mason Gray*. Bloomington, IN: iUniverse, 2010.

Twining, Merrill B. *No Bended Knee: The Battle for Guadalcanal*. Novato, CA: Presidio Press, 1996.

United States Marine Corps: Creating Stability in an Unstable World. Tampa, FL: Faircount, 2007.

Van Creveld, Martin. *Military Lessons of the Yom Kippur War: Historical Perspectives*. Washington, DC: Center for Strategic and International Studies, 1975.

Watson, Cynthia. *Combatant Commands: Origins, Structure, and Engagements*. Santa Barbara, CA: Praeger, 2011.

Weaver, Kendal. *Ten Stars: The African American Journey of Gary Cooper—Marine General, Diplomat, Businessman and Politician*. Montgomery, AL: NewSouth Books, 2016.

Westermeyer, Paul W., ed. *The Legacy of American Naval Power: Reinvigorating Maritime Strategic Thought*. Quantico, VA: Marine Corps History Division, 2019.

Westermeyer, Paul W. *Liberating Kuwait: U.S. Marines in the Gulf War, 1990–1991*. Quantico, VA: Marine Corps History Division, 2014.

———. *U.S. Marines in Battle: Al-Khafji, 28 January–1 February 1991*. Quantico, VA: Marine Corps History Division, 2008.

Westermeyer, Paul W., and Alexander N. Hinman. *Desert Voices: An Oral History Anthology of Marines in the Gulf War, 1990–1991*. Quantico, VA: Marine Corps History Division, 2016.

Westermeyer, Paul W., and Breanne Robertson, eds. *The Legacy of Belleau Wood: 100 Years of Making Marines and Winning Battles*. Quantico, VA: Marine Corps History Division, 2018.

Westheider, James E. *Fighting on Two Fronts: African Americans and the Vietnam War*. New York: New York University Press, 1997.

————. *The African American Experience in Vietnam: Brothers in Arms*. Lanham, MD: Rowman & Littlefield, 2008.

Wetterhahn, Ralph. *The Last Battle: The* Mayaguez *Incident and the End of the Vietnam War*. New York: Carroll and Graf, 2001.

Wise, James E., Jr., and Scott Baron. *The 14-Hour War: Valor on Koh Tang and the Recapture of the SS* Mayaguez. Annapolis, MD: Naval Institute Press, 2011.

Wright, James. *Those Who Have Borne the Battle: A History of America's Wars and Those Who Fought Them*. New York: Perseus, 2012.

Yates, Lawrence A. *The U.S. Military Intervention in Panama: Operation Just Cause, December 1989–January 1990*. Washington, DC: U.S. Army Center of Military History, 2013.

Articles

"A Conversation with General Barrow." *Sea Power* 21, no. 11 (November 1979).

Brodie, Bernard. "More about Limited War." *World Politics* 10, no. 1 (October 1957): 112–22. https://doi.org/10.2307/2009228.

Brewster, BGen Albert E., USMC (Ret). "The Commandant of the Marine Corps and the JCS." *Marine Corps Gazette* 92, no. 3 (March 2008): 58–60, 62–64, 66.

Chipman, Donald. "Admiral Gorshkov and the Soviet Navy." *Air University Review* 33, no. 5 (July–August 1982): 28–47.

Cushman, Gen Robert E., Jr. "Corps Operations Facing Austerity." *Marine Corps Gazette* 57, no. 8 (August 1973): 2–4.

Davidson, Roger H. "The Advent of the Modern Congress: The Legislative Reorganization Act of 1946." *Legislative Studies Quarterly* 15, no. 3 (August 1990): 357–73.

"Defense Issues 3: What's Next for the U.S. Marine Corps?" *Brookings Bulletin* 13, no. 1 (Winter 1976): 15–17.

Dodenhoff, G. H. "Why Not an Aviator Commandant?" *U.S. Naval Institute Proceedings* 125, no. 8 (August 1999).

Dorn, Walter, and Michael Varey. "Fatally Flawed: The Rise and Demise of the 'Three-Block-War' Concept in Canada." *International Journal* 63, no. 4 (Autumn 2008): 967–78.

Evans, David. "Schwarzkopf's 'Jedi Knights' Praised for Winning Strategy." *Chicago Tribune*, 1 May 1991.

Frank, George. "New Marine Commandant Is 'Right Guy at Right Time'." *Los Angeles Times*, 3 June 1991.

Galloway, George B. "The Operation of the Legislative Reorganization Act of 1946." *Political Science Review* 45, no. 1 (March 1951): 41–68.

Girard, Philippe R. "Peacekeeping, Politics, and the 1994 U.S. Intervention in Haiti." *Journal of Conflict Studies* 24, no. 1 (Summer 2004): 20–41.

Heinl, Robert D. "The Inchon Landing: A Case Study in Amphibious Planning." *Naval War College Review* 20, no. 5 (May 1967): 51–72.

Hittle, LtCol James D. "The Marine Corps and the National Security Act of 1947." *Marine Corps Gazette* 31, no. 10 (October 1947): 57–59.

Hopkins, Maj Gen John I. "This was No Drill." U.S. Naval Institute *Proceedings* 117, no. 11 (November 1991).

Jaffe, Lorna S. "The Base Force." *Air Force Magazine* 83, no. 12 (December 2000).

Kamps, Charles Tustin. "Operation Eagle Claw: The Iran Hostage Rescue Mission." *Air & Space Power Journal en Español* 18, no. 3 (2006).

Kelley, Paul X. "One Telephone Call Gets It All: Maritime Prepositioning for Crisis Response Enhancement." *Sea Power* 27, no. 11 (November 1984): 23–34.

Kopets, Capt Keith F. "The Combined Action Program: Vietnam." *Military Review* 82, no. 4 (July–August 2002): 78–81.

Kreisher, Otto. "Marines' Desert Victory." *Naval History Magazine* 30, no. 1 (February 2016).

———. Gen Charles C. "Commandant's Planning Guidance (CPG)." *Marine Corps Gazette* 79, no. 8 (August 1995): A-1–A-21.

———. "The Corps' Critics Are Wrong." *Washington Post*, 27 October 1985.

———. "The Crucible: Building Warriors for the 21st Century." *Marine Corps Gazette* 81, no. 7 (July 1997): 13–15.

———. "Through the Wheat to the Beaches Beyond: The Lasting Impact of the Battle of Belleau Wood." *Marine Corps Gazette* 82, no. 7 (July 1998): 12–17.

———. "The Strategic Corporal: Leadership in the Three Block War." *Marine Corps Gazette* 83, no. 1 (January 1999): 18–22.

———. "The Three Block War: Fighting in Urban Areas." *Vital Speeches of the Day* 64, no. 5 (December 1997): 139–41.

Mattis, James N., and Frank G. Hoffman. "Future Warfare: The Rise of Hybrid Wars," U.S. Naval Institute *Proceedings* 131, no. 11 (November 2005).

Mauskapf, Robert P., and Earl W. Powers. "LAVs in Action." *Marine Corps Gazette* 74, no. 9 (September 1990): 50–54, 57–59.

Millett, Allan R. "The U.S. Marine Corps: Adaptation in the Post-Vietnam Era." *Armed Forces & Society* 9, no. 3 (Spring 1983): 363–92, https://doi.org/10.1177/0095327X8300900301.

Owens, Harold M. "The All-Volunteer Force: Reaction or Alternative?" *Marine Corps Gazette* 57, no. 10 (October 1973): 24–30.

Packard, Nathan. "Congress and the Marine Corps: An Enduring Partnership." *MCU Journal* 8, No. 2 (Fall 2017): 9–37.

———. "Giving Teeth to the Carter Doctrine: The Marine Corps Makes the Case for its Strategic Relevance, 1977–1981." *International Journal of Naval History* 12, no. 2 (Summer 2015).

Polmar, Norman. "The U.S. Navy: Amphibious Lift." U.S. Naval Institute *Proceedings* 107, no. 11 (November 1981).

Prina, L. Edgar. "Wilson's Legacy, Barrow's Inheritance: A Combat-Ready Corps." *Sea Power* 21, no 6 (June 1979): 38–40.

Rems, Alan. "A Propaganda Machine Like Stalin's." *Naval History Magazine*, June 2019.

———. "Semper Fidelis: Defending the Marine Corps." *Naval History Magazine* 31, no. 3 (June 2017).

Smith, BGen Norman H. "Arctic Maneuvers 1984." *Marine Corps Gazette* 68, no. 12 (December 1984): 30–37.

Soper, LtCol James B. "Observations: Steel Pike and Silver Lance." U.S. Naval Institute *Proceedings* 91, no. 11 (November 1965).

Stevenson, Charles A. "Underlying Assumptions of the National Security Act of 1947." *Joint Force Quarterly* 48 (1st Quarter, 2008): 129–33.

Ulbrich, David J. "The Importance of Belleau Wood." *War on the Rocks*, 4 June 2018.

Weber, Capt Arthur S., Jr. "Unsolved Problem Areas." *Marine Corps Gazette* 59, no. 6 (June 1975): 41–42.

Whitney, Kathleen Marie. "Sin, Fraph, and the CIA: U.S. Covert Action in Haiti." *Southwestern Journal of Law and Trade in the Americas* 3, no. 2 (1996): 303–32.

Zabecki, Maj Gen David T., USA. "Review of *Generals of the Army: Marshall, MacArthur, Eisenhower, Arnold, Bradley*, edited by James H. Willbanks." *Parameters* 44, no. 2 (Summer 2014): 116–18.

Electronic Sources

"The 1991 Iraq War—The Battle at the UN." Association for Diplomatic Studies and Training, 20 October 2015.

"36 Years Ago the Military had Its Biggest Airborne Operation Blunder in History." PopularMilitary.com, 30 March 2018.

"Al-Jazeera Offices Hit in U.S. Raid." BBC, 13 November 2001.

Annis, Franklin C. "Krulak Revisited: The Three-Block War, Strategic Corporals, and the Future Battlefield." Modern War Institute at West Point, accessed 17 March 2022.

"Assistant Commandants of the Marine Corps." Marine Corps University, accessed 1 March 2023.

"Battle of Tarawa." History.com, 17 November 2009.

Beschloss, Michael. "LBJ and the Descent into War." HistoryNet, accessed 27 December 2021.

Bilal, Arsalan. "Hybrid Warfare: New Threats, Complexity and 'Trust' as the Antidote, 30 November 2021." *NATO Review*, 30 November 2021.

Boyd, John R. "Patterns of Conflict." Defense and the National Interest, accessed 30 August 2019.

Brewster, Albert E. "The Commandant of the Marine Corps and the JCS." Marines.mil, accessed 9 June 2019.

"Brief History of the United States Marine Corps." Marine Corps University, accessed 27 February 2023.

Carter, Phillip, and Owen West, "Dismissed!" *Slate*, 2 June 2005.

Crookston, Joseph A. "Marine Corps Roles and Missions: A Case for Specialization." GlobalSecurity, accessed 14 August 2019.

Davis, Matt. "Project 100,000: The Vietnam War's Cruel Experiment on American Soldiers." BigThink, 14 November 2018.

Fries, Carsten. "Battle of Iwo Jima, 19 February–26 March 1945." Naval History and Heritage Command, 16 March 2022.

———. "Operation Stalemate II: The Battle of Peleliu, 15 September–27 November 1944." Naval History and Heritage Command, 10 January 2020.

Givens, Adam. "Okinawa: The Costs of Victory in the Last Battle." National WWII Museum, 7 July 2022.

Hill, Melvin B., Jr. "Carl Vinson: A Legend in His Own Time." Carl Vinson Institute of Government, accessed 4 March 2023.

"The History of Marine Corps University." Marine Corps University, accessed 27 April 2023.

Huard, Paul Richard. "This Nuke Proved that Size Doesn't Matter." Medium, 14 February 2015.

"Iwo Jima and Okinawa: Death at Japan's Doorstep." National WWII Museum, accessed 8 December 2021.

Kalb, Martin. "It's Called the Vietnam Syndrome, and It's Back." Brookings, 22 January 2013.

"Military Reading Lists." U.S. Naval Institute, accessed 27 April 2023.

Morton, Col Matthew, USA. "We Were There: Reforger Exercises Designed to Counter Soviet Threat." Association of the United States Army, 24 March 2022.

Neller, Robert B. "ALMAR 09/18." Marines.mil, accessed 18 November 2021.

"Operation Southern Watch." GlobalSecurity, accessed 7 October 2019.

Patrick, Bethanne Kelly. "The Montford Point Marines." Military.com, accessed 7 June 2021.

"President Carter Leads Delegation to Negotiate Peace with Haiti." Carter Center, accessed 27 April 2023.

Reagan, Ronald W. "Remarks at the Recommissioning Ceremony of the U.S.S. *New Jersey* in Long Beach, California." Reagan Foundation, 28 December 1982.

Rees, Bradley L. "An Assessment of the Small Wars Manual as an Implementation Model for Strategic Influence in Contemporary and Future Warfare." *Small Wars Journal*, 29 April 2019.

Tracy, Jared M. "The USS *Harlan County* Incident: October 1993." U.S. Army Special Operations History Office of the Command Historian, accessed 27 July 2022.

Vergun, David. "Lessons Learned From the 1975 *Mayaguez* Incident." U.S. Department of Defense, 11 December 2018.

———. "Sports Heroes Who Served: Pro Football Player to Formidable Marine." U.S. Department of Defense, 15 September 2020.

Wolk, Herman S. "The 'New Look'." *Air Force Magazine*, 1 August 2003.

Newspapers and Magazines
Air Force Magazine
AP News
Army Times
Boston (MA) Globe
Chicago (IL) Tribune
Desert Sun (Palm Springs, CA)
Detroit (MI) News
Leatherneck
London Sunday Times
Los Angeles (CA) Times
Marine Corps Times
New York Times
New York Times Magazine
New Yorker
Newsweek
People
San Diego (CA) Union
Saturday Evening Post
Stars and Stripes
Time
United Press International (UPI Archives)
Washington (DC) Post
Washington (DC) Star

Theses and Dissertations

Bergesen, Sean A. "Adaptive Joint Force Packaging (AJFP): A Critical Analysis." Master's thesis, Naval Postgraduate School, 1993.

Brant, Bruce A. "Battlefield Air Interdiction in the 1973 Middle East War and Its Significance to NATO Air Operations." Master's thesis, U.S. Army Command and General Staff College, 1986.

Damian, Fideleon. "The Road to FMFM 1: The United States Marine Corps and Maneuver Warfare Doctrine, 1979–1989." Master's thesis, Kansas State University, 2008.

Dammin, Maj William F. "Force Design 2030 and the Marine Expeditionary Unit: Strategic Implications of Tactical Innovation." Master's thesis, Marine Corps University, 2022.

Feltey, Thomas M. "Debunking the Myth of the Strategic Corporal." Master's thesis, National Defense University, 2015.

Leahey, Michael J. "A History of Defense Reform Since 1970." Master's thesis, Naval Postgraduate School, 1989.

Packard, Nathan. " 'The Marine Corps' Long March': Modernizing the Nation's Expeditionary Forces in the Aftermath of Vietnam, 1970–1991." PhD dissertation, Georgetown University, 2014.

Vital, Mark D. "The Key West Agreement of 1948: A Milestone for Naval Aviation." Master's thesis, Florida Atlantic University, 1999.

INDEX

Abrams, Creighton W., Jr., 87–88, 93
Aidid, Mohamed Farrah, 382, 388–91
al-Khafji, Battle of, 353–58
all-volunteer force, 26–27, 156–59, 162–69, 178, 200
Allen, John R., 82, 394–97
Amos, Granville R., 274, 282, 412
Amos, James F., 82, 192, 218
Anderson, Earl E., 191–93
Aristide, Jean-Bertrand, 391, 397, 401–4, 407, 410
Armed Forces Qualification Test, 160–64, 168, 173
Armed Services Committee, U.S. House, 19, 37, 41, 140
Army of the Republic of Vietnam, 85–97, 107–9
Army Rangers, U.S., 275–76, 279–82, 340, 390–91, 400–2
Army War College, U.S., 9, 13, 409, 436
Aspin, Leslie, Jr., 203, 295, 391, 401, 413–14
Atlantic Command, U.S., 274, 277–79, 392, 396–401
Austin, Randall W., 120, 125–31, 433–34

Barrow, Robert H., 27, 56–61, 83, 175–78, 203, 213, 222–39, 244–46, 299, 311–12, 396
Base Force, 306–7, 369–70, 413
Belleau Wood, Battle of, 3–8, 15, 25–26, 29, 36, 182, 245
Binkin, Martin, 163, 201–3, 207–8, 226
Bishop, Maurice R., 271–72, 275
Bold Guard, Operations, 219–20, 294, 350
Bolden, Charles F., Jr., 56, 153
Boomer, Walter E., 349, 356–60, 365
boot camp, 57, 68, 176, 178, 224, 423

bottom-up review, 370, 413–14
Boyd, John R., 311, 315–23, 325
Bronze Star, 155, 245–46, 417
Bush, George H. W., 99, 155, 304, 307, 332, 334, 345, 348–49, 367, 382, 384–85, 394

Carter Doctrine, 211–12, 235
Carter, James E., Jr., 27, 168, 207, 210, 212, 214, 216, 226, 327, 332, 401–2
Cates, Clifton B., 43, 45, 63, 66
Cédras, J. Raoul, 391–92, 397, 401–4, 409, 442
Central Command, U.S., 143, 230, 297, 349, 367, 379
Central Intelligence Agency, U.S., 112, 120, 179–80, 263, 273, 397
Cheney, Richard B., 149, 307–8, 333, 345
Chosin Reservoir campaign, 51, 58, 61, 204–5, 238, 244
Clinton, William J., 156, 384, 391, 394, 401–3, 409, 413
close air support, 9, 129, 184, 187–88, 208, 342
Co-Van, 26, 85–86, 273, 283, 349, 351, 362
counterinsurgency, 11, 13, 84, 442–43
Crist, George B., 230, 272, 281, 297

Da Nang, 75, 77–78, 80, 107–9, 143
Department of Defense, U.S., 31, 100, 140, 207, 266, 296, 369
Desert Shield, Operation, 239, 310, 347, 350, 370, 379, 417
Desert Storm, Operation, 25, 28, 99, 149, 239, 310, 347, 354, 360, 364, 365, 367, 369–70, 375, 378, 381, 385, 411, 417, 424, 428, 439

Dong Ha, 88–91, 97
Doyle, James J., Jr., 379–82
drill instructor, 57, 172–75, 223
drug use in the U.S. military, 137–38, 169–71, 230
Druze, 250, 253–56, 258–59, 286

Eagle Pull, Operation, 107, 110, 114, 118, 286
Easter Offensive, 86–97, 136
Eastern Exit, Operation, 378–82
Edson, Merritt A., 36, 38, 40–41
Eisenhower, Dwight D., 20–21, 32, 41–42, 157
Elaborate Maze, Operation, 329
Ellis, Earl H., 9–10, 13
Erie, Carl R., 273, 280
European Command, U.S., 230, 239, 256, 266, 287, 377

Faulkner, James P., 273–74, 283, 287
Federal Bureau of Investigation, U.S., 240, 271
Feltey, Thomas M., 427–28
Fleet Marine Force, 13, 31, 33, 43, 63, 66–67, 82, 183, 202, 208–9, 299
Fleet Marine Force, Atlantic, 203, 224, 238, 290, 300, 302, 312, 369, 392
Fleet Marine Force, Pacific, 68, 78, 191
Force Structure Planning Group, 370, 431
Ford, Gerald R., Jr., 101, 114–20, 128, 131, 133
Forrestal, James V., 35–36, 43–45
fourth-generation warfare, 436
France, Morgan R., 254, 257, 265, 267
Frequent Wind, Operation, 108, 110, 114, 118, 286

Galvanic, Operation. See Tarawa, Battle of
Garner, Jay M., 373–76
Gemayel, Amine, 241, 256, 260–61
Geraghty, Timothy J., 247, 251–55, 257–69, 285, 287
Gerlach, Howard, 268, 270, 287

Goldwater–Nichols Department of Defense Reorganization Act of 1986, 27–28, 296–98, 307, 327, 330, 345, 428, 434, 444
Gray, Alfred M., Jr., 28, 68–70, 141, 204–5, 217, 219, 288–89, 300, 302–4, 307–9, 311–12, 314–15, 321–27, 351, 369, 418, 420
Guadalcanal campaign, 15, 35–36, 48, 50
Guam, Battle of, 46, 48–50, 61, 179
Guantánamo Bay, 392–96, 406, 413, 421,
Gulf War. See Desert Storm, Operation

Hart, Gary W., 311–13
Haynes, Fred E., Jr., 208–10
Headquarters Marine Corps, 10, 69, 163, 178, 180, 185, 297, 299–300,
historically Black colleges/universities, 56, 150, 153–54, 157
Hittle, James D., 36–37, 41–42
Hogaboom, Robert E., 66, 82, 184, 371
Holcomb, Thomas, 5, 7, 12–13
Holloway, James L., III, 95, 207, 215, 217–18
Hussein, Saddam, 271, 346–48, 352–53, 356, 358, 372, 376–78
hybrid warfare, 435–36, 444
hypersonic missiles, 28, 432, 437

I Corps tactical zone, 69, 77–78, 80, 82, 86
Iceberg, Operation. See Okinawa, Battle of
Inchon, Battle of, 19, 39, 51, 58, 70
Iranian Revolution, 27, 211
irregular warfare, 78, 439, 442
Islamic Amal, 255–56, 261, 263
Israel Defense Forces, 103–4, 239, 243, 247, 253, 260
Iwo Jima, Battle of, 17, 26, 33, 75, 179, 208, 210, 224

Jenkins, Harry W., Jr., 350, 359–60, 379

Johnson, John M., Jr., 107, 120, 127–28
Johnson, Louis A., 33, 45
Johnson, Lyndon B., 74, 114, 134
Johnston, Robert B., 385–86, 388–90
Joint Chiefs of Staff, 19, 32, 35, 39–41, 106, 117–18, 128, 168, 212, 217–18, 220–22, 230, 252, 255, 257, 262, 272, 278, 285, 295–97, 305–6, 327, 329–30, 345, 349, 351, 369, 391, 399, 401–2, 408, 433
Joint force air component commander, 297–98, 342, 357
Jones, David C., 222, 230, 296
Jones, James L., Jr., 218, 373–77
Joy, James R., 283, 285–87
Jumblatt, Walid K., 250, 253–55, 259
Just Cause, Operation, 28, 334–46, 365, 399, 434–35

Kelley, Paul X., 213–14, 232, 246–47, 252, 283–84, 288–89, 297–99, 325
Kelly, John F., 82, 326
Kennedy, John F., 19, 147
Keys, William M., 351, 356, 358, 364, 392
Khmer Rouge, 107, 100–13, 116, 120–24, 127–28, 130–33, 135, 433–34
Kissinger, Henry A., 96, 114, 116–17, 120
Koh Tang Island, 117–23, 126–34, 241, 433–35
Krulak, Charles C., 3, 7–8, 25, 28, 351, 359–60, 370–71, 382, 417–28, 431
Krulak, Victor H., 36, 38, 42, 45–46, 62–63, 68, 78, 80, 324

Lebanese Armed Forces, 247, 253–61, 270, 286–87
Lebanese Civil War, 240, 253, 255, 259–60
Lehman, John H., Jr., 226–28, 233–36, 252, 287, 295, 298
Lejeune, John A., 3, 7, 9–13, 28, 57, 329, 423
Libutti, Frank, 382, 384–85, 389
Lind, William S., 311–13, 319, 322–25

MacArthur, Douglas, 16–17, 32, 58
maneuver warfare, 28, 69, 100, 304, 311–26, 347, 365, 369
Marine air-ground task force, 105, 187, 218–19, 290, 294, 298, 330, 385, 403–4
Marine Corps Schools, 13, 36, 62, 68, 311, 320, 325, 429
Marine Corps University, 28, 146, 325
Martin, Edward H., 254, 257–58, 262–63
Mattis, James N., 82, 427, 442
Mayaguez incident, 113–34, 253, 432–33
McClure, Lynn E., death of, 171–76
McDonald, Wesley M., 273, 278–79, 283
McFarlane, Robert C., 250, 256–58, 260, 262, 295
Mead, James M., 239, 241–42, 251
Metcalf, Joseph, III, 274–75, 277–78, 280–81, 290
Military Assistance Command, Vietnam, 75, 82, 444
Miller, Paul D., 399–401, 407
Mountain Warfare Training Center, 204–6, 236, 294
Mundy, Carl E., Jr., 369–70, 385, 413–16, 418, 420, 431

National Command Authority, 24, 156, 218, 257–58, 346, 431
National Liberation Front, 74, 77–78, 80, 93, 97–98, 110, 444
National Security Act of 1947, 26, 31–32, 42–43, 221, 295
National Security Council, U.S., 114, 116–17, 119, 128, 133, 399, 434
Naval Academy, U.S., 9, 12, 56–57, 09, 153–54, 417
Naval Reserve Officers Training Corps, 150–52, 154
Neller, Robert B., 3, 343–44
Nixon, Richard M., 84–86, 94–98, 101, 114, 116, 134, 156–57, 159, 179–80, 193, 435
Noriega, Manuel Antonio, 306, 327–29, 332–34, 336, 338, 340–41, 344–45

North Atlantic Treaty Organization, 28, 68, 100–1, 106, 135, 156, 180, 182, 187, 203, 205, 208, 210, 212, 219–20, 225, 231–33, 247, 289–90, 292–95, 298, 304, 350, 443
Northern Wedding, Operations, 219–20, 294
Nunn, Samuel A., Jr., 163, 165–6, 168, 200, 295, 402

Oakley, Robert B., 385, 389
Officer Candidates School, 57–58, 151
O'Keefe, Sean C., 415–16
Okinawa, Battle of, 17–18
OODA loop theory, 317–19
Organization of the Petroleum Exporting Countries, 164, 210

Pacific Command, U.S., 230–31, 368
Palestine Liberation Organization, 239–40, 247, 256
Panamanian Defense Forces, 328–45
Pate, Randolph M., 45, 184, 208
Patterson, Robert P., 35–36, 42
People's Army of Vietnam, 23, 69, 77–78, 86–93, 97–98, 106–8, 110–11, 136, 300
Persian Gulf War. See Desert Storm, Operation
Petersen, Frank E., Jr., 52–56, 149, 152, 155
Pickel Meadow. See Mountain Warfare Training Center
Powell, Colin L., 305–7, 334–35, 345–46, 349, 402, 413, 439
precision-guided munitions, 28, 204, 366, 431–32, 437
prisoners of war, 96, 343, 348, 359, 362, 364
professional military education, 13, 25, 28, 418, 425
Provide Comfort, Operation, 372–73, 375, 377, 382, 393

Rapid Deployment Joint Task Force, 212–14, 230–35, 239, 246

Reagan, Ronald W., 27, 99–100, 198, 216, 226, 234, 239–41, 244, 247, 250, 261, 271, 282–86, 295–99, 304, 321, 327–28, 435–36
Record, Jeffrey, 163, 201–3, 207–8, 211, 226, 323–25
Redlich, Douglas C., 266, 396
Restore Hope, Operation, 310, 384–89
Ribbon Creek disaster, 172–73, 175
Richardson, Charles E., 335–44
Ripley, John W., 88–92
rocket-propelled grenade, 250, 264
Rolling Thunder, Operation, 74–75, 134
roll-on, roll-off discharge facility, 310
rotary-wing aircraft, 63, 65, 184
Rough Rider, Operation, 336
Royal Marines (British), 89, 205, 219, 246, 295, 373–74

Schlesinger, James R., 114, 117–18, 197, 200, 203, 296
Schwarzkopf, H. Norman, Jr., 278, 280–81, 352, 357–59, 365–67, 380
Scoon, Paul G., 271–72, 275, 280–81
Sea, Air, and Land team (U.S. Navy), 275, 279, 340–41, 344, 354, 380–81
Silver Lance, Operation, 70–72, 78
Smith, Norman H., 289–93
Smith, Oliver P., 58–59, 63, 204
Smith, Ray L., 273, 275, 285–86
South Vietnamese Marine Corps, 85–86, 88–89, 91–94, 106–7, 143
Southern Command, U.S., 329–31, 333, 345
Southern Watch, Operation, 372, 377–78
Stiner, Carl W., 257–58, 266, 333–40, 343, 346
Strategic Corporal, 421, 424–28
Sun Tzu, 300, 315, 317
Supreme Court, U.S., 99, 392, 394

Take Charge, Operation, 393
Tannous, Ibrahim, 256, 258, 266
Tarawa, Battle of, 14–17
Taylor, Maxwell D., 75

Teamwork 84, Operation, 289, 291–93
terrorism, 99, 218, 239–40, 242, 262–63, 266, 269, 284–87, 366
Tet Offensive, 93, 300
Thieu, Nguyen Van, 93–98, 111
Three-Block War, 421, 426–28
Thurman, Maxwell R., 168–69, 333–34, 345
Toohey, Patrick, 340–41
Truman, Harry S., 26, 31–32, 35, 39–45, 53, 149
Turley, Gerry H., 90–92, 219
Twining, Merrill B., 35–36, 42, 45, 204, 424

Uniform Code of Military Justice, 145–46, 169
United Nations, 58–59, 103, 310, 377–78, 383–84, 389–90, 401, 407–9, 442–44
United Nations Security Council, 346–48, 377, 384
Uphold Democracy, Operation, 402–10, 434
Urgent Fury, Operation, 27, 272–83, 330, 434–35

Vandegrift, Alexander A., 32–38, 41, 48
Vessey, John W., Jr., 252, 257, 285
Viet Cong. *See* National Liberation Front

War Powers Resolution, 98–101, 180, 193, 241–42, 435
Warner, John W., III, 139–40, 168
Warsaw Pact, 106, 233, 304, 313
Watkins, James D., 256, 298
Webb, James H., 299–300, 321
Weinberger, Caspar W., 234, 239, 284, 287, 299
Westmoreland, William C., 75–77, 80, 180, 444
Williams, Michael J., 394–96
Williams, Willie J., 56, 154–55
Wilson, Louis H., Jr., 27, 46–50, 56, 61, 68, 162, 166, 174–79, 190–210, 219–25, 229, 245–46, 299, 418–20
Woerner, Frederick F., Jr., 330, 333, 335
Wyly, Michael D., 312, 320–22, 326, 369

Zinni, Anthony C., 143–45, 314, 385
Zumwalt, Elmo R., Jr., 139, 180, 187–89

ABOUT THE AUTHOR

Following his graduation from the University of Maryland, Charles P. Neimeyer began his professional career as a military officer with the U.S. Marine Corps in 1976. Upon completion of The Basic School, he was assigned the military occupational specialty of field artillery officer. During his 20-year active-duty military career, he served in all three Marine Corps divisions, on the military staff at the White House for Presidents George H. W. Bush and William J. "Bill" Clinton, and as a military instructor at the Naval War College in Newport, Rhode Island. He retired from active service in 1996. He then returned to the Naval War College in 1997 as a professor of national security affairs and later served as the college's dean of academics from 1999 to 2002. In 2006, he became the director and chief of Marine Corps history in Quantico, Virginia. He remained in that capacity until his retirement from civilian federal service in December 2017. Upon completion of service, he received the Department of the Navy's Distinguished Civilian Service Award and the Marine Corps University Foundation Chapman Medallion. Since his civilian retirement, he has resumed teaching for the Naval War College's Fleet Support Program and specializes in strategy and warfare and theater security decision making. His books include *America Goes to War: A Social History of the Continental Army, 1775–1783* (1996), *The Revolutionary War* (2007), *On the Corps* (editor, 2008), and *War Comes to the Chesapeake: The British Campaigns to Control the Bay, 1813–1814* (2015). He currently resides with his family in Stafford County, Virginia.